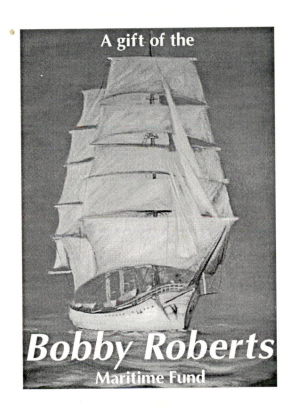

A gift of the

Bobby Roberts
Maritime Fund

TURTLE

TURTLE

David Bushnell's Revolutionary Vessel

Roy R. Manstan and Frederic J. Frese

WESTHOLME
Yardley

Frontis: The *Turtle* is launched into the Connecticut River, November 10, 2007. (*John Nilson*)

Westholme Publishing, LLC
904 Edgewood Road
Yardley, Pennsylvania 19067
Visit our Web site at www.westholmepublishing.com

First Printing May 2010
10 9 8 7 6 5 4 3 2 1

ISBN: 978-1-59416-105-6

Printed in United States of America

When Roy Manstan secured the *Turtle*'s hatch overhead while Fred Frese directed its launch, the spirit of two centuries of submariners joined this little vessel as it ventured out into the Connecticut River. It is to these countless adventurers from the past that the *Turtle* is dedicated.

For Roy, he also dedicates this book to his son Daniel and his daughter Sarah.

For Fred, he also dedicates this book to his grandmother, Anunziadine Cuomo, and his aunt, Laurinzine Carangelo.

One accident or another always intervened. I then thought, and still think, that it was an effort of genius, but that too many things were necessary to be combined to expect much from the issue against an enemy who are always on guard.
—George Washington

War is replete with mistakes because it is full of improvisations. In war we are always doing something for the first time. It would be a miracle if what we improvised under the stress of war should be perfect.—Vice Admiral Hyman Rickover

CONTENTS

Introduction xi

PART ONE: INTERPRETATION

1 *Turtle* Soup: Two Centuries of Interpretation   3

2 Science, Technology, and a One Room Schoolhouse   17

3 A Novel Idea Takes Shape   33

4 A Secret "Machine"   43

5 How to Feed a Two-ton *Turtle*   59

PART TWO: REPLICATION

6 The *Turtle* Project: Not Just Another Birdhouse   77

7 Propulsion: How to Move a Two-ton *Turtle*   85

8 The Hull: Was the *Turtle* More Than a Barrel?   97

9 Hatch and Life Support: How to Survive Inside a *Turtle*   109

10 Navigation: A Compass and a Depth Gauge Guided the *Turtle*   119

11 Firepower: The *Turtle* and Its Infernal Machine   129

PART THREE: INVESTIGATION

12 Walking in Bushnell's Shoes: Operational Testing of the *Turtle*   145

13 Propeller Thrust: Horsepower = Manpower   155

14 Hull Resistance: Moving a Two-ton *Turtle*   163

15 Transit Speed: A Tortoise or a Hare?   169

16 Maneuvering and Submerging: Engaging the Target Undetected   179

17 Pilot Endurance: A Mission Not for the Faint of Heart   193

18 Pilot Ergonomics: By Design or an Afterthought   205

PART FOUR: SPECULATION

19 The *Turtle* in the Connecticut River: A Convenient Location 215

20 The *Turtle* Did Not Operate Alone: Bushnell's Navy 225

21 Bushnell's Biggest Mistake: Too Few *Turtle* Pilots 233

22 September 6, 1776: The *Turtle* Sets Out on Its Mission 243

23 The Next Day: Why the *Turtle* Mission Failed 257

Epilogue 269

Appendix A David Bushnell to Thomas Jefferson 276

Appendix B Benjamin Gale to Benjamin Franklin 289

Appendix C Benjamin Gale to Silas Deane 292

Appendix D Ezra Lee to David Humphreys 298

Appendix E Ezra Lee to Charles Griswold 303

Appendix F George Washington and Thomas Jefferson 309

Glossary 311

Notes 331

Bibliography 347

Index 357

Acknowledgments 369

About the Authors 371

INTRODUCTION

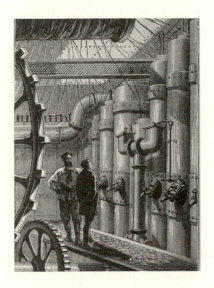

Once upon a time . . .

"A door opened, and I found myself in the compartment where Captain Nemo—certainly an engineer of very high order—had arranged his locomotive machinery.

'You see,' said the Captain, 'The electricity produced passes forward, where it works, by electro-magnets of great size, on a system of levers and cog-wheels that transmit movement to the axle of the screw. This one, the diameter of which is nineteen feet, and the thread twenty-three feet, performs about a hundred and twenty revolutions in a second.'

'And you get then?'

'A speed of fifty miles an hour.'"

—Jules Verne, *Twenty Thousand Leagues Under the Seas* (1876)

Such was the description that Captain Nemo provided scientist Pierre Aronnax onboard the *Nautilus*. Captain Nemo may have been the product of fiction, but there was nothing fictional or fanciful about David Bushnell (a true "engineer of very high order"), nor Ezra Bushnell, Ezra

Lee, and many others who created the historical legacy of the remarkable revolutionary submarine known as the *Turtle*.

Jules Verne had conducted extensive research on submarine technology before writing his novel set in 1866 and first published in 1869. By the time he was writing *Twenty Thousand Leagues Under the Sea*, the European press had been reporting news of the Confederate semi-submersible David-class torpedo boats and the sinking of the steam sloop *Housatonic* by the submarine *Hunley*. His research, however, was likely based on the French submarine *Plongeur*, launched in 1863, and an American submarine built and operated on Lake Michigan by Lodner Phillips in 1851. Although much smaller, the shape of Phillips's submarine bears a close resemblance to Verne's *Nautilus*.

The *Turtle* was created at a time in American history when circumstances engendered a burst of patriotic creativity. It was not electricity that drove the *Turtle*'s propeller, illuminated the interior, or pumped the ballast; it was pragmatic, dogged persistence and human sweat. David Bushnell created a real submarine vessel capable of launching an attack against a real man-of-war during a hot September at the beginning of a long war.

In dealing with historical events, authors extract and interpret information locked away in archives where they may find entire libraries of letters, documents, and published eyewitness accounts. Other researchers are limited to a few vague references to an event or even a single artifact uncovered from an archaeological context. The job of the authors of this book is to ensure that "Once upon a time . . ." becomes "On the night of September 6, 1776 . . . ," while avoiding "It was a dark and stormy night . . ." On the other hand, a little fantasy mixed with some adventure makes reading a book of facts a bit more interesting.

This book reviews the story of the *Turtle* in four parts: interpretation, replication, investigation, and speculation. The authors begin with a discussion of the many interpretations of David Bushnell's *Turtle* that have been put forward over the past two centuries. Yet the interpretation of historical events, including ours, can be skewed by personal opinion and first impressions, or influenced by the products of previous interpreters.

Our approach was to evaluate each of the few contemporary accounts sentence by sentence, sometimes word by word, from which we assigned a level of credibility and probable accuracy. To assist our interpretation, we included the social, political, and technological environment that may have influenced the evolution of Bushnell's ideas. We then endeavored to avoid the pitfalls of using a twenty-first-century mind-set to interpret eighteenth-century events. Based on these insights, we describe the production of a working *Turtle* replica and the operational tests that were performed, to investigate the capabilities of the original vessel. From the knowledge gained during the testing, we provide our own speculation about the *Turtle* story and what we feel was probable and what was unlikely.

DAVID BUSHNELL (1740–1826)

The British had the most powerful navy in the world, and intended to bring all of that power to bear against the rebellious colonies. The pragmatic Bushnell knew that his ideas would not win the war, but he was convinced that his concept of undersea warfare could influence its progress and outcome. He also knew that it wouldn't be a matter of merely carrying a set of drawings to Hayden's Shipyard on the Connecticut River and ordering such a vessel. To bring his ideas to reality, the job was on Bushnell's shoulders. The man's true genius was in his ability to understand all the complex elements that had to be accounted for in designing, building, and operating a vessel that had little precedent.

The only known image of David Bushnell is this woodcut produced for a rare edition of James Fenimore Cooper's *History of the United States Navy*, privately printed by Charles Bushnell in 1890. (*New-York Historical Society*)

David Bushnell was reserved but strong willed. Secure in the quality and efficacy of his designs, he was able to convince a small group of individuals to support his vision. Persistence, confidence, and sheer willpower led to the creation of a submarine with the capability to undertake a combat mission. That mission became a reality on September 6, 1776. With a relatively inexperienced pilot at the helm, the *Turtle* made its way to HMS *Eagle* anchored in New York Harbor. Several hours later, after daybreak, the *Turtle* returned to Manhattan unsuccessful in its attempt against the British flagship with Admiral Lord Richard Howe onboard. During the following four weeks, the *Turtle* made two attempts against British warships in the

Hudson River, both also unsuccessful and ending the career of this unique vessel. It was not a lack of courage or determination that foiled Bushnell, it was bad luck that stood in the way.

On May 6, 1779, David Bushnell was among a group captured by the British during a raid along the Connecticut coast. Unaware that this quiet civilian had been actively engaged in clandestine warfare against the Royal Navy, the British soon released him in a prisoner exchange. Had they known his true identity and activities, he may have suffered the same fate as his good friend and fellow Yale graduate, Nathan Hale.

Shortly after Bushnell was released, General George Washington offered him a commission as captain-lieutenant in a company within the newly formed Corps of Sappers and Miners, an element supporting Washington's Corps of Engineers. Bushnell was subsequently involved with the construction of siege works around Yorktown and witnessed the surrender of General Charles, Earl Cornwallis on October 17, 1781. After hostilities ceased, Bushnell spent the remainder of his service at West Point and nearby Constitution Island. He mustered out of the Corps in December 1783.[1]

Bushnell, however, is not remembered for his time with the Corps of Sappers and Miners, but for his invention of the world's first submarine, designed specifically for naval combat during the age of sail. It is the story of the *Turtle* that we seek to shed light on.

The source material

Throughout the more than two centuries since the *Turtle* maneuvered out from Manhattan "on the night of September 6, 1776," (it was not, by the way, "a dark and stormy night"), the story has been told and retold countless times and in several languages. One thing that they all have in common is a single primary source of information. In 1787, David Bushnell wrote a detailed description of the submarine and its mission in a letter to Thomas Jefferson. The letter was detailed as far as Bushnell was concerned, but only tantalizing to those of us who can't see the image of the *Turtle* that was on Bushnell's mind as he composed his letter. Most eyewitnesses considered the *Turtle* a passing curiosity and moved on to more pressing matters, leaving few related documents for future historians to ponder.

In 1798, Bushnell's letter was read before the American Philosophical Society under the heading "General Principals and Construction of a

Sub-marine Vessel, communicated by D. Bushnell of Connecticut, the inventor, in a letter of October, 1787, to Thomas Jefferson, then Minister Plenipotentiary of the United States at Paris." The contents of the letter were published in 1799 in the *Transactions of the American Philosophical Society*. The text of this published account is provided in facsimile in appendix A, and is the source of all Bushnell quotes used in this book.

According to a Bushnell genealogy, David Bushnell was "of slight build and nervous temperament . . . His reputation in Connecticut was that he was a man of very unassuming manners and exemplary character."[2] He valued the close friendships he made during his years at Yale, and in particular his mentor, Benjamin Gale of Killingworth, Connecticut. Gale wrote letters describing the *Turtle* to his friend and fellow scientist, Benjamin Franklin, and to Silas Deane, a Connecticut delegate to the Continental Congress from 1774 to 1776. Gale was certainly an informed observer, and his letters provide credible information that serve as corroborating evidence when interpreting Bushnell's description. More significant, however, are the hints in Gale's letters of a trusted relationship, whereby Bushnell could seek scientific and technological critique of his ideas. Gale's letter to Franklin is found in Clark (1964) and provided here as appendix B; his correspondence with Deane is in Trumbull (1870) and in appendix C.

The earliest and most credible published account reviewed during the construction of the replica *Turtle* was included in David Humphreys's *An Essay on the Life of the Honorable Major-General Israel Putnam* (1794, first edition published in 1788).[3] Humphreys (1752–1818) was a nineteen-year-old graduate when Bushnell arrived at Yale in 1771. They became friends and continued their relationship long after the war. We learn about Bushnell's association with Humphreys in a 1785 letter written by George Washington in response to Thomas Jefferson's inquiry about the submarine. Washington: "I cannot give you full information respecting Bushnell's projects for the destruction of ships . . ." Washington then proceeded to give his brief impressions of "Bushnell's projects," suggesting to Jefferson that "Humphreys, if I mistake not, being one of his converts, will be able to give you a more perfect account of it than I have done."[4] We believe it is this endorsement that led Humphreys, Washington's aide-de-camp, to intervene for Jefferson and persuade Bushnell to write the letter that eventually became the primary source used by all historians.

The other arguably credible source was Ezra Lee (1749–1821), the second of the three original *Turtle* pilots. In 1815, nearly forty years after the event, Lee wrote a letter to David Humphreys describing the *Turtle* and his experiences. The letter was eventually published in the *Magazine of American History*, vol. 29, January–June 1893. A copy of the manuscript letter is on display at the Thomas Lee House Museum in East Lyme, Connecticut. That letter has been transcribed by the authors and is included as appendix D. All of the Lee quotes in this book are from that transcription.

In 1820, Charles Griswold sent a letter to Prof. Benjamin Silliman recounting Ezra Lee's description. Later that year, Griswold's letter, apparently based in part on an interview with Lee, was published in an article titled "Description of a Machine, invented and constructed by David Bushnell" in *The American Journal of Science and Arts*, a transcription of which is included here as appendix E. The Griswold letter also appeared in Abbot's *Beginning of Modern Submarine Warfare Under Captain-Lieutenant David Bushnell*.

In 1823 James Thatcher, Continental army surgeon who served in the New York area in 1776, published *A Military Journal During the American Revolutionary War...* The lengthy title (see bibliography) ends with "from the Original Manuscript." Some authors have taken this to mean that his published account of Bushnell and the *Turtle* was taken directly from his journal and thus written contemporary with the event. The text, however, draws directly from Bushnell's letter to Jefferson and from Humphreys's book, which were both published long after the end of the war. Thatcher also uses the term "torpedo" when referring to Bushnell's submarine. It wasn't until 1807 when Robert Fulton coined the word torpedo in reference to submarine warfare technology.

The *Turtle* story found its way into other nineteenth-century books, including Barnes (1869) and Barber (1875), and by many more authors as submarine warfare began to mature at the turn of the twentieth century. Recent treatments of Bushnell and the *Turtle* can be found in Thomson (1942), Wagner (1963), Grant (1976), Roland (1978), Harris (2001), Diamant (1994 and 2003), and Lefkowitz (2006).

Robert Fulton, who experimented with his submarine *Nautilus* in France early in the nineteenth century, was certainly aware of the *Turtle*, but gives little credit to Bushnell for influencing his designs.[5] The French continued to experiment with submarine concepts throughout the nineteenth century as recounted in Hennebert (1888). Hennebert included descriptions of the *Turtle* and *Nautilus*, France's 1863 *Plongeur*, the Confederate David boats of the Civil War, the American *Intelligent Whale*, and John P. Holland's 1875 proposal for a one-man submarine. Later accounts of Fulton's *Nautilus* can be found in Sutcliffe (1909), Dickenson (1913), and Parsons (1922).

The voices of veterans from 1776 swapping sea stories in the comfort of Beers Tavern on the New Haven Green are long gone. Memories of a strange vessel that could maneuver under water were passed along to children and grandchildren, inevitably evolving and "improving" with time. It becomes difficult to extract the facts from what has become "once upon a time." Hints of these evolved accounts can be found in later books published on the topic, most notably *David Bushnell and his American Turtle* by Everett Tomlinson (1899).

Occasionally, some new tidbit of information is uncovered that provides more fodder for interpretation. Ezra Lee's request for his pension (submitted in 1820) includes a claim that his rheumatism was caused by being immersed in water up to his knees when operating the *Turtle*. Was this in fact the case? Or was it a bit of an exaggeration written over forty years after the event from frustration at not having received his promised pension?

GLOSSARY

Interpretation of historical evidence, whether from Bushnell, Gale, or Lee, (or even Tomlinson), depends on the level of credibility given to each author. After reading and rereading and reading again each account, a mental picture of the *Turtle* will emerge. A somewhat more subtle concern, however, is the interpretation of the words used by the writers. Sure, the descriptions were all written in English . . . I guess you would say "the King's English." While not always critical to building a credible interpretation of an object or event, some words carried different meanings or connotations in 1776 than they do today. The careful interpreter, therefore, must be mindful of that possibility when studying original source material. An interesting example that might confront a political scientist studying the eighteenth century is in the meanings of the words "statesman" and "stateswoman," which you can find in the glossary.

When compiling the glossary, we selected words we believed would help provide a better framework regarding Bushnell's world and illustrate how we arrived at our interpretation of the *Turtle* story. To gain a certain degree of confidence, albeit imperfect, in defining these words, we turned to dictionaries published during the last half of the eighteenth century.

The premier lexicographer in Bushnell's time was Samuel Johnson who, in 1755, published his *Dictionary of the English Language*, a massive two-volume folio edition. Johnson provided lengthy, if not entertaining, definitions for all of his 42,773 entries. He would illustrate the use of a word in a particular context by citing quotations from authors familiar to the readers of the day.

Johnson most frequently relied on quotations from Shakespeare. Yet other authors received a fair share of attention, including Joseph Addison, John Dryden, John Locke, John Milton, Alexander Pope, and Jonathan Swift; for scientific references, he provided quotations from Sir Francis Bacon, Robert Boyle, Nehemiah Grew, Isaac Newton, and others. When encountering somewhat obscure words, he would quote somewhat obscure authors; in defining the word "submarine," for example, he turned to the Reverend John Wilkins. He also occasionally quoted himself when lacking a better authority. The price of his folio edition, at 4£ 10s., limited its distribution and resulted in the publishers having Johnson produce a more affordable two-volume octavo edition. This popular version eliminated the lengthy quotations, but retained the names of authors Johnson used as sources.[6]

Our glossary definitions are primarily drawn from a copy of Johnson's 1767 octavo edition and a 1773 edition (the twentieth) of Nathan Bailey's *Universal Etymological Dictionary*. The first edition of Bailey's dictionary was published in 1721, followed by many "improved" editions throughout the last half of the eighteenth century and well into the nineteenth. This very popular dictionary was much more cut and dried than Johnson's, concentrating on the origins of the words rather than on literary sources.

When a word was not in these sources or required more explanation, other eighteenth and nineteenth-century references were used. When the Latin origin of words was of interest, Thomas Cooper (1578) and Adam Littleton (1684) were consulted. The source of each of the glossary entries is given at the end of the definition. In several cases a definition is provided from both Johnson and Bailey, each describing a word in slightly different terms. We have added commentary along with the verbatim definitions, ostensibly to add clarification, but in reality to keep the glossary from becoming so boring that the reader will simply not bother.

THE *TURTLE* PROJECT

With access to the same source material cited above, one of the current authors, Fred Frese, collaborated with journalist and historian Joseph Leary to produce a working replica of Bushnell's submarine for the American Bicentennial. Their replica was christened the *American Turtle* and launched with Leary at the helm on August 20, 1977. It is now on display at the Connecticut River Museum in Essex, Connecticut.

Twenty-five years later, in 2002, Old Saybrook [Connecticut] High School hired Frese. His task was to reinvent the school's technical arts program by providing an emphasis on mechanical skills. The intent was to enable students to engage in challenging projects, whether they were college bound or entering the trades. The design and construction of another working replica became the premier challenge and the *Turtle* Project was born.

Because of the place the *Turtle* holds in American history, and the Navy in particular, the Naval Undersea Warfare Center (NUWC) in Newport, Rhode Island, entered into an Education Partnership Agreement with the Old Saybrook School System. NUWC is a Division of the Naval Sea Systems Command and is responsible for Research, Development, Test, and Evaluation (RDT&E) of systems designed to support the Navy's undersea warfare mission. The staff consists of civilian employees of the Department of the Navy and active duty military who are specialists in particular submarine technologies. Author Roy Manstan, mechanical engineer and Command diving officer at NUWC prior to retirement in 2006, initiated the education partnership that facilitated access to members of the Warfare Center's technical staff.

Although conceived in 2001, the *Turtle* Project began in earnest during the fall semester of 2003. Four years later the completed replica was christened *Turtle* by Connecticut State Senator Eileen Daily and launched on November 10, 2007. In May 2008, the *Turtle* was subjected to a series of operational tests hosted by the Mystic Seaport Museum in Mystic, Connecticut. The tests were designed to investigate the operational capabilities of the replica and use the results to better understand what the original participants experienced.

Critical to the test and evaluation of our *Turtle* was how closely the replica approximated the original *Turtle*. Because there are no known contemporary drawings of Bushnell's "submarine vessel," the replication of the technologies he included relied almost entirely on our interpretation of the source material. In the chapters that follow, we include excerpts from the published accounts of Bushnell, Gale, and Lee. The referenced texts, provided in their entirety in the appendices, allow readers to see each excerpt in the context of the entire document and judge for themselves our interpretation of their descriptions.

With the exception of piloting the *Turtle* during its sea tests, the Old Saybrook High School students participated in all phases of the design and construction of their submarine. The educational experience was extraordinary. Some students were involved throughout their entire four-year high-school career.

This book retells the story of the *Turtle* one more time. The goal was to gain insight into the strengths and limitations of this unique vessel and explore the engineering genius of David Bushnell. Along the way we looked for what were likely myths, exaggerations, or simply products of great imaginations. Maybe we have improved those sea stories from 1776 that were passed from generation to generation. Maybe we have a few sea stories of our own.

PART ONE

INTERPRETATION

Connecticut governor Ella Grasso christens the Bicentennial replica *American Turtle*, August 1977. (*Connecticut River Museum*)

The intrigue and clandestine nature of submarine warfare has sparked the imaginations of inventors, entrepreneurs, adventurers, sailors, chroniclers, and historians. Sometimes with scant evidence, writers piece together a story that hopes to satisfy their curiosity. David Bushnell's "sub-marine vessel" is no exception, and may have resulted in the most diverse assortment of interpretations of any historical contrivance.

Yet some question whether Bushnell's *Turtle*, with all of the complexities in its design, could have ever successfully conducted a submarine mission, even suggesting that the *Turtle* was simply a myth. So, every new investigation into the *Turtle* story is important, providing another twist that may not solve the mystery but inserts one more piece into the puzzle. Fred Frese and Joseph Leary produced the first working replica of Bushnell's submarine, during the American Bicentennial. After its christening as the *American Turtle* in August 1977, their replica was operated under a variety of conditions, establishing that what Bushnell had described to Thomas Jefferson was, in fact, a vessel that could submerge and maneuver underwater: a submarine.

The following chapters begin with a sampling of *Turtle* interpretations that have proliferated throughout the technical and popular literature over the past 230 years. The origins of Bushnell's concepts of submarine warfare are then reviewed in light of the contemporary technology and social conditions that may have influenced his designs. A measure of Bushnell's true genius lies in his ability to convince both skeptics and supporters of the validity of his ideas and assemble a team willing to make the sacrifices that would eventually turn his vision into reality.

Chapter One

TURTLE SOUP

Two Centuries of Interpretation

Origin of the name turtle

The natural philosophers of the sixteenth and seventeenth centuries recognized three separate forms of tortoise. There were land tortoises (*Testudo terrestri*), freshwater tortoises (*Testudo aquatica dulci*), and sea tortoises (*Testudo marina*).[1] The word "turtle" had not yet become a popular term for this aquatic quadruped.

Eighteenth-century dictionaries noted that "turtle" evolved from the Latin word *turtur* describing a particular bird known as the Turtle Dove. The association of a bird with a reptile may have come from a group of sailors on a long voyage being convinced by their cook that the stew pot containing sea tortoise was palatable. "*Eat it*," he would have said, cajoling the crew, "*you'll love it. It tastes just like turtle* [dove]." Samuel Johnson corroborates this etymology. Imported from the West Indies, the main ingredient for turtle soup made this fare a delicacy among the English gentry, resulting in interesting recipes for mock turtle soup that could be enjoyed by the more common palate. (See the glossary.)

David Bushnell and Benjamin Gale never used the word "turtle." They described the shape of the vessel as resembling "two upper Tortoise shells joined together." When referring specifically to the submarine, Bushnell only used the term "sub-marine vessel" or simply "vessel." Gale, in his letter to Benjamin Franklin, described the vessel as "a new Machine for the Destruction of Enemy Ships," but typically used only the term "machine" in his correspondence.

It is in David Humphreys's biography of Major-General Israel Putnam, first published in 1788, that the names *Turtle* and *American Turtle* first appear. Humphreys began by describing "A Machine, altogether different from any thing hitherto devised by the art of Man, had been invented by Mr. David Bushnell, for submarine navigation," later referring to the machine as both "the Turtle" and "the American Turtle."2 Bushnell maintained a close relationship with Humphreys and it is likely that during conversations between these two (while possibly dining on turtle soup), one of them (probably Bushnell's young enthusiastic supporter) began referring to the vessel as the *"Turtle"* and, in a patriotic moment, the *"American Turtle."* Both names have become synonymous in reference to Bushnell's submarine.

By 1815, as Ezra Lee wrote in his letter to Humphreys, Bushnell's submarine had acquired a variety of references: "I will endeavor to give you some idea of the construction of this machine, turtle or torpedo, as it has been called . . . Its shape was most like a round clam, but longer, and set up on its square side." Clam? The Bicentennial replica built by Frese and Leary had been christened *American Turtle* in 1977. We therefore used Humphreys's *Turtle* (not Lee's "*Clam*") throughout the project, and when our vessel was christened and launched in 2007.

EDINBURGH ENCYCLOPAEDIA

When Bushnell's letter to Jefferson was published in 1799, the text began to circulate within the scientific and literary community. Thatcher's 1823 account of the *Turtle* draws heavily on Bushnell's letter as did the multitude of books on submarine warfare written after the Civil War. The first attempt to illustrate the *Turtle*, also based on Bushnell's "General Principals and Construction of a Sub-marine Vessel," is found in the American edition of the *Edinburgh Encyclopaedia* (Figure 1.1).3

Volume seven devotes over nineteen pages to diving and diving bell technologies, including a description of Cornelius Drebbel's submarine experiments where the author cites Reverend John Wilkins's chapter on submarine navigation (see Wilkins 1691). He then begins his discussion of Bushnell with: "The nearest approaches to realizing the bishop's [Wilkins] ingenious conceits, besides the experiment of Debrel [Drebbel], was made by Mr D. Bushnell of Connecticut, in America, who, in 1787, published a description of a submarine vessel of his invention, in which it was found very practicable to travel under water . . ."

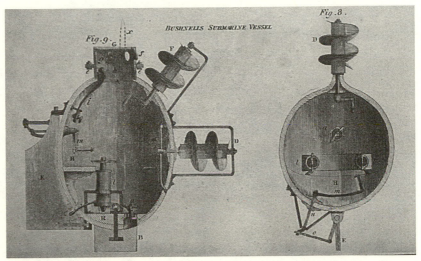

Figure 1.1. Engraving of the *Turtle* by Jacob Plocher (ca. 1820) is the earliest known depiction of Bushnell's "submarine vessel" appearing in the American edition of the Edinburgh Encyclopaedia. (*Brewster 1832*)

He is, of course, referring to Bushnell's 1787 letter to Thomas Jefferson that was not published until 1799. The author continues:

> The whole invention displays very great ingenuity and original-
> ity of idea. It is minutely explained in the publication alluded to;
> but as it is too complicated to be wholly understood from the ver-
> bal description, without greater attention than ordinary readers
> are disposed to give, our draughtsman has prepared two figures,
> viz. a vertical section, Plate CCXXXI [231] Fig. 9, and a hori-
> zontal section, Fig. 8, from the inventor's description, which,
> except for the letters of reference, is as follows . . . .

The author then provides a *nearly* verbatim recounting of Bushnell's letter. Included in the text are these "letters of reference" that the reader can use to compare the description of each part as written by Bushnell to the illustrations created by the "draughtsman." For example, "An oar D, formed upon the principal of the screw, was fixed in the fore part of the vessel." The letter "D" referred to in the figure above depicts the early and persistent misconception of a helical screw as the *Turtle's* mode of propulsion. Later in the article, the engraver and author also included their own version of Bushnell's steering mechanism.

A rudder E, hung to the hinder part of the vessel, commanded it with the greatest of ease. The rudder was made very elastic, and might be used for rowing forward. The tiller *m* was within the vessel at the operator's right hand, and passing behind him, was fixed at a right angle on an iron rod, or spindle *n*, Fig. 8 which passed through the side of the vessel. The rod had a crank on its outside end, which commanded the rudder by means of a rod *o*, extending from the end of the crank to a kind of tiller fixed upon the left hand of the rudder. Raising or depressing the first mentioned tiller, *m*, turned the rudder as the case required.

Take note of the sentence: "The tiller *m* was within the vessel at the operator's right hand, and passing behind him, was fixed at a right angle on an iron rod, or spindle *n*, Fig. 8 which passed through the side of the vessel." When transcribing Bushnell's original description, the Encyclopaedia author took the liberty of adding the phrase ". . . passing behind him . . ." to accommodate the concept expressed by the engraver. Compare the Encyclopaedia description of the tiller to that of Bushnell: "Its tiller was within the vessel, at the operator's right hand, fixed, at a right angle, on an iron rod, which passed through the side of the vessel . . ." This literary and artistic license produced an illustration with extraneous mechanical linkages that can confuse researchers and historians, who might use the Encyclopaedia concept as a credible *Turtle* model.

Regardless of its accuracy, as far as we have determined, the *Edinburgh Encyclopaedia* illustration is the first attempt to depict the *Turtle* using Bushnell's words, albeit somewhat misquoted. Because of this, it is important to attempt to determine when the engraver put his concept to a copper plate. Our estimate is between 1815 and 1820.

The eighteen-volume *Edinburgh Encyclopaedia* was delivered to subscribers sequentially between 1813 and 1832. Volume seven, containing the section on diving, was issued around 1820. The Encyclopaedia was profusely illustrated, with a set of plates accompanying each volume. Volume seven, for example, included twenty-nine plates numbered CCXII [212] through CCXL [240]. When the eighteenth and final volume was published and issued in 1832, the hundreds of plates were typically collected and bound into three separate volumes.

There is clear evidence that 1820 is the latest date that the *Turtle* illustration could have been crafted. The evidence comes from the person who engraved plates CCXXXI [231] with the *Turtle* and CCXXXII [232] with a variety of diving bells. At the bottom corner of each plate is the name of the engraver, or "draughtsman," Jacob J. Plocher. He was a well-known American engraver of scientific and medical illustrations during the early nineteenth century. Plocher engraved the plates for William Dunlap's *A History of the Progress of the Arts of Design in the United States,* and Charles Bell's *A Series of Engravings Explaining the Course of the Nerves*, both published in 1818. Jacob Plocher died December 27, 1820.

Lieutenant Francis M. Barber

Jacob Plocher's interpretation of Bushnell's submarine, sitting deep within the pages of the eighteen-volume *Edinburgh Encyclopaedia*, a set of books that occupied more than three feet of shelf space, did not receive wide distribution. It would be over fifty years before another attempt would be made to publish an illustration of the *Turtle*.

During the early 1870s, Lieutenant Francis M. Barber was one of several naval officers stationed at the U.S. Torpedo Station in Newport, Rhode Island, who presented a series of lectures on torpedo technology. Barber included his interpretation of the *Turtle* when he published his *Lecture on Submarine Boats and their Application to Torpedo Operations* in 1875 (figure 1.2). He also published his *Lecture on the Whitehead Torpedo* and the *Lecture on Drifting and Automatic Movable Torpedoes, Submarine Guns, and Rockets*.[4] Barber, a member of the Torpedo Corps, was stationed in Newport from 1872 to 1875. After he retired from the navy in 1895, he became an associate of John P. Holland, serving as a Holland Torpedo Boat Company agent selling submarines to Japan.[5]

Lieutenant Barber's version of the *Turtle* soon became the most well-known and copied illustration, finding its way into nearly every book published on the subject up to and including this one. But even Barber recognized that his interpretation of Bushnell's description resulted in only "a tolerably correct idea" of the *Turtle*.

The many faces of David Bushnell's sub-marine vessel

During the last quarter of the nineteenth century, the concept of submarine and torpedo warfare drew worldwide interest. France was a leading proponent, but England, Russia, Japan, Germany, Spain, Belgium, Austria, and Italy all experimented with various designs. The Greeks and

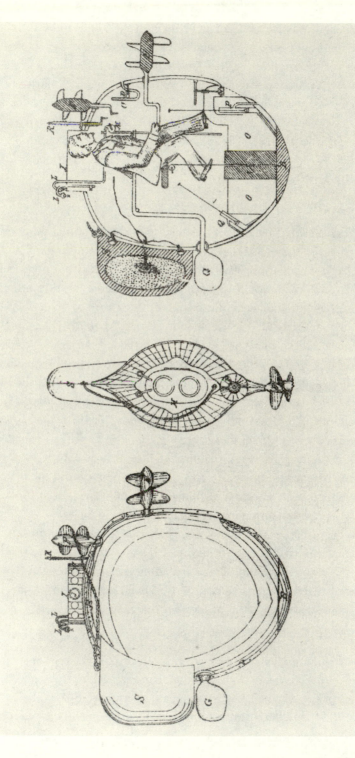

Figure 1.2. This illustration of the *Turtle* by Lieutenant Francis Barber is the most frequently reproduced interpretation of Bushnell's submarine, perpetuating the misconception of the use of a helical screw propeller. (*Barber 1875*)

Turks, longtime antagonists, also made brief attempts to incorporate submarines into their navies. In America, however, naval brass remained skeptical until John P. Holland's submarines and the Whitehead torpedo began to gain acceptance by these European countries.

With this technology came a renewed interest in the origins of submarine warfare, and the story of the *Turtle* was included in many technical and popular publications on the subject. Barber's *Turtle* had international appeal, appearing in several books published in France including Lieutenant Colonel Hennebert's 1888 *Les Torpilles*, and *La Navigation Sous-Marine* by G.–L. Pesce in 1906. The topic (and title) "La Navigation Sous-Marine" was also used by French authors Maurice Gaget and Maurice Delpeuch at the turn of the twentieth century.[6]

Bushnell's contribution to the evolution of submarine warfare was recognized throughout Europe. Belgian author L.-G. Daudenart published *La Guerre Sous-Marine et les Torpedos* in 1872. In 1880 C.L. Sleeman, former lieutenant in the British navy and later Commander of the Imperial Ottoman navy, published *Torpedoes and Torpedo Warfare*. The book *Submarine Boats* by Danish Lieutenant G.W. Hovgaard was published in England in 1887. Spain's Narciso Monturiol, designer of the submarine *El Ictineo,* mentions Bushnell in his 1891 book *Ensayo Sobre El Arte De Navegar Por Debajo Del Agua.*

The British took a particular interest in the *Turtle* and a few authors took a stab at their own interpretation. Alan Burgoyne published his two-volume *Submarine Navigation Past and Present* in 1903. Burgoyne was convinced that Barber's version of the *Turtle* was "nothing but the reflections of a vivid imagination." His reasoning was twofold: "Firstly, the presence of screw propellers, which we know were not invented till some thirty years later, and secondly, in one of the plans is a picture of a seated man dressed in a very stylish modern suit, sitting on the seat of the boat."[7] Figure 1.3 shows Burgoyne's rendition of Barber's illustration without the pilot.

Burgoyne was apparently unaware of the variety of American naval uniforms during the late nineteenth century. One style issued to non-commissioned officers included a double-breasted jacket and long tailored pants that covered the top of the shoes (figure 1.4).[8] Barber's *Turtle* pilot is essentially wearing the uniform of a petty officer of the time Barber was at the Torpedo Station. The engraver simply didn't bother to include any insignia on the sleeve, but there is a striking resemblance in the face.

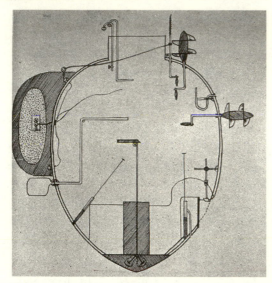 

Figure 1.3, left, Alan Burgoyne considered Francis Barber's interpretation to be the "reflections of a vivid imagination" and redrew the *Turtle* without its pilot (*Burgoyne 1903*). Barber's pilot was actually dressed in a regulation uniform of a late nineteenth-century petty officer, figure 1.4, right. (*Regulations Governing the Uniform of Commissioned Officers 1886*)

PROPULSION: THE BIG QUESTION

If how the pilot was dressed was a trivial issue, the question of what Bushnell used to propel his tiny submarine inspired argument among authors. Yet the authors were all depending on their own interpretation of what Bushnell and, in some cases, Ezra Lee had described. The beauty of Bushnell's invention was in the eyes of the beholder. As Alan Burgoyne put it: "The mode of propulsion employed has been the subject of some dispute." He had dismissed any possibility of a "screw propeller," whether a helix or bladed, on the *Turtle*. He was certain that "propulsion was obtained by the oars AA, fixed in the sides of the boat by watertight joints," a view shared by French author Maurice Gaget (1901), and included an illustration showing a pair of typical oars protruding from the sides of the *Turtle*.[9] British author Herbert C. Fyfe, in his *Submarine Warfare Past and Present* (first edition published in 1902), was noncommittal regarding the mode of propulsion, and included copies of both Barber's helical screw and the oar-driven *Turtle* (figure 1.5).

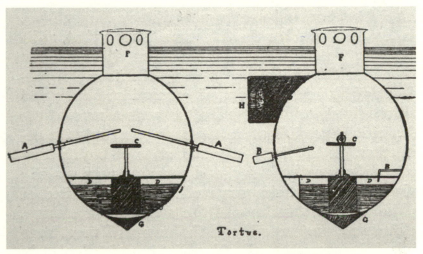

Figure 1.5. The *Turtle* shown with oars, a mode of propulsion proposed by some authors at the beginning of the twentieth century. (*Fyfe 1906*)

It may have been the authors of the book *Les Bateaux Sous-Marins Historique*, F. Forest and H. Noalhat (1900), who first proposed the rowboat theory. "The American worker David Bushnell invented the first submarine having actually navigated under serious conditions and given irrefutable results; he spent four years building this little ship that had the shape of a turtle which is the name that remained with it. If the shape to which it owes its name was unfavorable for speed, it did at least ensure great stability for this strange little craft." The authors then proposed the following: "Propulsion was achieved by means of two oars passing into two double tube casings of thick leather. These casings allowed the navigator to turn the oars either edgewise or flat, by drawing them backward or forward."[10]

The solution to the propulsion controversy was suggested in 1872 by the Belgian author L.-G. Daudenart in his book *La Guerre Sous-Marine et Les Torpedos*. G.-L. Pesce (1906) and his close associate Maurice Delpeuch (ca. 1902) included Daudenart's simple sketch of the *Turtle* with bladed propellers (figure 1.6) in their discussions of the propulsion question. Their objections to Daudenart's sketch were in the flattened shape of the hull and that the propellers were shown at the back of the hull and not at the bow.

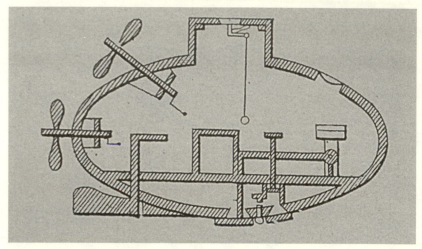

Figure 1.6. In 1872 Belgian author L. G. Daudenart produced a sketch of an oddly shaped *Turtle* with bladed propellers, although incorrectly placing them at the stern. (*Pesce 1906*)

Delpeuch: ". . . several French authors specializing in undersea navigation, and lately M. Alan Burgoyne, have recently claimed that the attribution of screw propellers to Bushnell's vessel was nothing but the product of an overactive imagination, and that the means for propulsion and submergence were none other than oars with which one sculled horizontally or vertically."

He then provided quotes from Bushnell and Ezra Lee:

Bushnell specifically states: '[An] oar formed upon the principal of the screw,' and E. Lee, even more clearly, states: 'An oar or paddle whose form was exactly that of windmill arms, which turned perpendicularly on an axis that projected out front.' After such quotations, is there any room for doubt? It is therefore propellers that served for motion and for submergence, and it even seems that Major Daudenart's representation of this is more exact than the one made by American officers, which represents the propellers in the form of Archimedian screws.

Delpeuch was emphatic, writing that:

BUSHNELL CAN LEGITIMATELY BE CONSIDERED AS THE FIRST INVENTOR TO APPLY THE SCREW PROPELLER IN NAVIGATION

Figure 1.7. One of the many early twentieth-century interpretations of *Turtle*. (*Field 1907*)

[capitalization is Delpeuch's]. The only difference with the current propeller [referring to Daudenart] is that his being placed at the front of the vessel vs. the current rear position pulled instead of pushed. Its utilization was thus inevitably defective: a minor detail that pales before the magnitude of the invention.[11]

G.-L. Pesce, also quoting Bushnell and Lee, thoroughly agreed "with the opinion of Delpeuch—contrary to that of Burgoyne—regarding the application of the screw propeller." Pesce also noted the errors in Daudenart's sketch, emphasizing that Bushnell's "screw propeller had been positioned forward, as it was later in the dirigible 'La France'."[12]

A contemporary of Alan Burgoyne, Brevet-Colonel Cyril Field of the Royal Marine Light Infantry, published *The Story of the Submarine* in 1907. Field created nearly one hundred illustrations for his book, including two of the *Turtle*, one of which is shown above. Note that in spite of evidence to the contrary, Field continued to use Barber's helical screw propeller. At least the rowboat theory had disappeared.

WORLD WAR I AND BEYOND

Submarine technology became increasingly important during the early years of the twentieth century, particularly with the deadly effectiveness

Figure 1.8. By World War I, some authors had begun to include bladed screw propellers with their illustrations of *Turtle*. (*Bishop 1916*)

of the German U-Boats of World War I, and authors continued to depict the *Turtle* as a way to introduce the topic. In the midst of the war, several editions of French author M. Laubeuf's *Sous-Marins et Submersibles* were published, and in England, Farnham Bishop's book *The Story of the Submarine* was published in 1916. Bishop included the Barber and Burgoyne drawings of the *Turtle*, but added an interpretation of his own that included a modern screw propeller (figure 1.8). Bishop provided his opinion of the controversy: "Before him was the crank of another propeller, or rather tractor, for it drew, not pushed, the vessel forward . . . the best known picture of the *Turtle* shows a bearded gentleman . . . boring his way through the water with two big gimlets. But Sergeant Ezra Lee of the Connecticut Line, who did the actual operating, described the forward propeller . . . as having two wooden blades or 'oars, of about 12 inches in length and 4 or 5 in width, shaped like the arms of a windmill.' Except in size, this device must have looked very much like the wooden-bladed tractor of a modern aeroplane."[13]

The popularity of submarine topics did not escape the authors of books for youth, particularly in America. The U.S. Navy purchased its first submarine from John P. Holland in 1900. In 1909, Victor G. Durham created The Submarine Boys Series. Titles such as *The Submarine Boys on Duty or Life on a Diving Torpedo Boat* and *The Submarine Boys' Lightning Cruise or The Young Kings of the Deep* enticed teenagers to enter the new submarine force. In 1917, Frederick and Virgil Collins published *The Boys Book of Submarines* with yet another interpretation of the *Turtle*, this one with a riveted steel hull (figure 1.9).

After World War I, there was renewed interest in the *Turtle* among the descendants of the original participants. A newspaper clipping (figure 1.10) shows a full-size model built in the 1930s by Judge Elmer Lynn

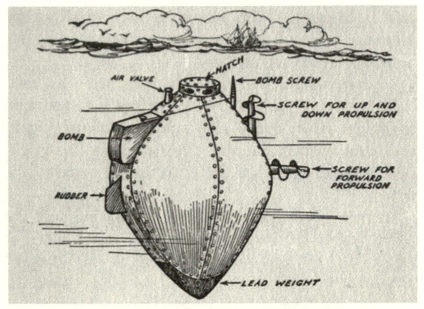

Figure 1.9. The hulls of many nineteenth-century submarines were built from riveted iron plates using steam boiler technology. Frederick and Virgil Collins decided that their *Turtle* would also include a riveted hull. (*Collins 1917*)

of Westbrook. In the photo with Judge Lynn are Mary Brockway Doane (L) and Abbie Pratt (R), great-great grandaughters of Phineas Pratt, who worked with Bushnell on the construction of the original *Turtle* and was one of his pilots. Note the similarity between Judge Lynn's replica and the miniature half model (figure 1.11) produced by Oscar Manstan at about the same time. Both of these models show the persistence of the concept of a helical rather than a bladed screw propeller.

The U.S. Navy had always understood the relevance of Bushnell's *Turtle* to the history of naval warfare. Beginning shortly after the Civil War with Lieutenant Commander J. S. Barnes (1869), followed by Lieutenant Francis Barber (1875), the *Turtle* story was well documented in 1881 by General Henry L. Abbot in his book *Beginning of Modern Submarine Warfare Under Captain Lieutenant David Bushnell*. During World War II, with submarines having become an essential component of naval warfare, a detailed history of Bushnell by David W. Thomson was published in the February 1942 issue of the *United States Naval Institute Proceedings*. In 1951, with the memory of German submarines

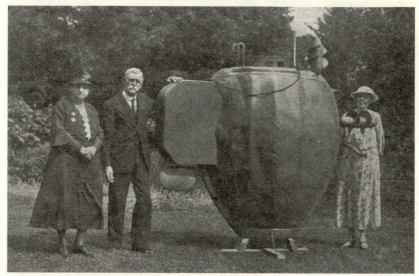

Figure 1.10, above, a newspaper clipping dated August 6, 1936, showing a full-sized *Turtle* model with its builder and two descendants of Phineas Pratt, one of the original *Turtle* pilots. (*Shoreline Times*) Figure 1.11, left, a miniature half model (3 1/2″) of the *Turtle* (ca. 1935) by Oscar "Ty" Manstan, grandfather of author Roy Manstan.

still fresh, an article about the *Turtle* (with yet another rendition of Barber's drawing) appeared in the *New Haven Register* with a headline that read in part: EZRA LEE OF LYME STAGED 1776 U-BOAT ATTACK ALONE.

Chapter Two

SCIENCE, TECHNOLOGY, AND A ONE ROOM SCHOOLHOUSE

WHAT WAS A "SUBMARINE"

Bushnell and his contemporaries used the phrase "sub-marine vessel" in reference to the *Turtle*. During the nineteenth century, the French also followed this convention using "bateau sous-marine" to describe this type of vessel. In this sense and in other applications, as in "submarine navigation," the word was considered an adjective. Terms such as "plunging boat" and "submarine torpedo boat" continued to be used throughout the nineteenth century. While some early twentieth-century authors, (e.g., Alan Burgoyne, Herbert Fyfe, Cyril Field, and others) began using "submarine" as a noun, it was still not recognized as such in William Dwight Whitney's 1903 edition of *The Century Dictionary, an Encyclopedic Lexicon of the English Language*. By 1909, however, *The Century Dictionary Supplement*, published by Benjamin J. Smith, uses the term as a stand-alone noun (see glossary).

The first suggestion for a military submarine occurred about 1578 by William Bourne, who published his concept in *Inuentions or Deuices. Very necessary for all Generalles and Captains, or Leaders of men, as wel by Sea as by Land*. More than four decades later, Cornelius Drebbel demonstrated his version of a submarine in the Thames in 1623, with hopes of selling the idea to King James I. Drebbel also designed a device for use as an underwater mine that he referred to as a "water petard."

Creative submarine concepts from the seventeenth century include those of Marin Mersenne in 1634 and 1644, the Rotterdam Boat by the French engineer De Son in 1653, and the oar-driven vessels of Giovanni

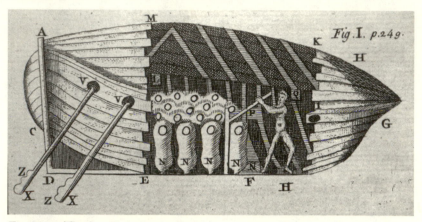

Figure 2.1. The first operational submarine was an oar-driven vessel built by Cornelius Drebbel, ca. 1620. While there are no contemporary illustrations of Drebbel's submarine, this view of Borelli's design from the 1670s is representative of these early concepts. (*Gentleman's Magazine, June 1749*)

Borelli in the 1670s and Denis Papin in 1695. The illustration of Borelli's submarine (figure 2.1) from the June 1749 issue of *Gentleman's Magazine* shows a wood-framed vessel that was probably leather covered and propelled by oars. Although the illustrator did not attribute the idea to its proper origin, it was taken directly from a book by Borelli, *De Motu Animalum*, published posthumously in 1680.

Although there are no contemporary drawings of Drebbel's submarine, Borelli's design is thought to be similar to what Drebbel had built where oars were the preferred mode of underwater propulsion. The article described their use: "Such a ship may be fitted with oars, VX, passing through holes V,V, in the sides of the vessel, and secured by goat-leather . . . the oars must have flexible broad ends Z, like the feet of geese and frogs, that they may expand when they take a stroke against the water, and furl themselves when they are drawn back."[1] Excellent sources of information on the history of submarines can be found in Harris (2001) and Roland (1978). We will consider Papin's submarine designs in relation to Bushnell in a later chapter.

THE REVEREND JOHN WILKINS

The first "interpretation" of submarine technology, other than literary references, was by the Reverend John Wilkins who, in 1648, devoted a chapter in his book *Mathematical Magic: or, the Wonders That may be*

Performed by Mechanical Geometry to the concept of creating "an Ark for submarine Navigation."[2] The novelty and mechanical complexity of Cornelius Drebbel's submarine must have intrigued Wilkins who, fifteen years later, would become one of the founders of the Royal Society of London. All of the following quotes are from chapter five of Wilkins's fourth edition published in 1691.[3]

He declared: ". . . that such a contrivance is feasible and may be effected, is beyond all question, because it has been already experimented here in *England* by *Cornelius Dreble*."

Wilkins continued on to elaborate the pros and cons of this submarine Ark. His list of the cons first lamented the "many noisom offensive things [i.e. human waste], which should be thrust out, and many other needful things [food and water], which should be received in." He then described the issues of propulsion and navigating underwater: "The second difficulty in such an Ark will be the *motion* or *fixing* of it according to occasion; The *directing* of it to several places, as the voyage shall be designed." Finally, he noted that "the greatest difficulty of all will be this, how the air may be supplied for respiration: How constant fires may be kept in it for light."

Wilkins assured his readers that there were solutions for these negative aspects of submarine travel. He further presented the case by listing five of the "many advantages and conveniences of such a contrivance."

1. 'Tis private; a man may thus go to any coast of the world invisibly, without being discovered or prevented in his journey.

2. 'Tis safe; from the uncertainty of Tides, and the violence of Tempests, which do never move the sea above five or six paces deep. From Pirates and Robbers which do so infest other voyages; from ice and great frosts, which do so much endanger the passages toward the poles.

3. It may be of great advantage against a Navy of enemies, who by this means may be undermined in the water, and blown up.

4. It may be of a special use for the relief of any place that is besieged by water, to convey unto them invisible supplies: and so likewise for the surprisal of any place that is accessible by water.

5. It may be of unspeakable benefit for submarine experiments and discoveries.

Wilkins also appealed to commercial shipping interests by noting the safety advantages of a submarine over a surface craft. It wasn't until the early twentieth century that a submarine (Germany's *Deutschland*) was built specifically for commercial transport, only to be converted for military use in 1916. Wilkins elaborated number 5 noting that the "submarine experiments and discoveries," while advancing knowledge of the submarine environment, also had significant commercial implications. "The discovery of submarine treasures is more especially considerable . . . [and] which may be much more easily fetched up by the help of this, than by any other usual way of the Urinators [see the glossary]." His emphasis, however, was to encourage this technology for clandestine military operations.

In most cases, submarine technology in the seventeenth century was intended as a means of delivering a weapon or a combat swimmer. Commercial interests, however, also encouraged resourceful individuals to enter the sea in a variety of contraptions. The first diving bells, referred to as "Compana Urinatoria," were proposed in the mid-sixteenth century. The "urinator" was expected to focus on the salvage of treasure and other cargo being brought back from the New World. Military interest in underwater salvage, however, was to retrieve the heavy armament that warships carried to the bottom. A British first-rate man-of-war may have had as many as one hundred cannons onboard.

When David Bushnell conceived his "sub-marine vessel" in 1771, there had been nearly two centuries of experimentation with underwater technologies, both for commercial and military applications. Bushnell was a voracious reader and undoubtedly had access to the personal libraries of his many mentors. When finally admitted to Yale in 1771, he was able to continue his scientific interests with the extensive library in New Haven. Once his ideas began to coalesce, Bushnell was able to draw inspiration from all of the scientific and popular literature that was available to him. His mentors and tutors continued to provide the sounding board he needed to confront each technological problem. Wilkins had summarized the ground rules: propulsion, navigation, respiration. Ignoring or underestimating the importance of these would be fatal to Bushnell's concept of submarine warfare.

THE SUBMARINE ENTERS THE LITERARY WORLD

> They write here one *Cornelius-Son*
> Hath made the *Hollanders* an invisible Eel
> To swim the Haven at *Dunkirk*, and sink all
> The shipping there.[4]

This excerpt from a play by Ben Jonson refers to the submarine that Cornelius Drebbel demonstrated to King James I on the Thames in 1623.[5] Jonson's play, *The Staple of News*, was performed in 1625 for James I, who would certainly have understood the reference to an "invisible Eel." The dialogue continues to stress the military application of this strange mechanical device, or "automa," although it appears that Jonson assumed that the design called for an auger to drill a hole through the hull.

> It is an *Automa*, runs under Water
> With a snug Nose, and has a nimble Tail
> Made like an A*uger*, with which Tail she wriggles
> Betwixt the Coasts of a Ship and sinks it straight.

Drebbel's foray into the world of submarine vessels continued to find references in the general and scientific literature. Wilkins's "Ark for submarine Navigation" referred specifically to Drebbel. Francis Bacon, writing his *New Atlantis* just prior to his death in 1626, described a world dedicated to the pursuit of knowledge and the advancement of mechanical arts, a world that included advanced weapons and submarines.

Bacon tells how the inhabitants of New Atlantis related their various capabilities to their English visitors.

> We have also Engine-houses, where are prepared Engines and Instruments for all sorts of motions. There we imitate and practice to make swifter motions than any you have; and to make them, and multiply them more easily, and with small force, by wheels and other means; and to make them stronger and more violent than yours are, exceeding your greatest Cannons and Basilisks. We represent also Ordnance and Instruments of War, and engines of all kinds; and likewise new mixtures and compositions of Gunpowder, Wildfires burning in Water and unquenchable; also Fireworks of all variety, both for pleasure and use. We imitate also flights of Birds; we have some degrees of fly-

ing in the Air; we have Ships and Boats for going under Water, and brooking of Seas.[6]

It seems that word of Drebbel's vessel had reached New Atlantis.

Bacon was convinced of the value of scientific thought and the expansion of knowledge through observation, validated by what became known as "experimental philosophy." His *Sylva Sylvarum, or Natural History in Ten Centuries* and *The Advancement and Proficiency of Learning: or the Partitions of Sciences* were published in multiple editions throughout the seventeenth and eighteenth centuries. His belief that a society will advance faster when knowledge is made available to the general public became a cornerstone of the philosophy of a society formed in 1663 by British scientists and intellectuals.

The Royal Society of London

Thomas Sprat first published *The History of the Royal-Society of London, for the Improving of Natural Knowledge* in 1667, only four years after the Society obtained its charter from King Charles II. To the seventeenth-century scientist, the term "philosophy" was synonymous with "knowledge." The Royal Society was about to shift the concept of knowledge as the exclusive property of the intellectual and political elite to encouraging accessibility by the society as a whole. To this end, the membership was encouraged to publish their work in the *Transactions of the Royal Society*, where they could describe their scientific experiments and discoveries in English rather than Latin.

Sprat, representing the Royal Society as a whole, understood that the study of natural philosophy was simply an extension of, and not in conflict with, moral and religious philosophy. "The Natural Philosopher is to begin, where the Moral ends."[7] Sprat also sought to reassure the movers and shakers of his day that they need not "suspect the *Change* which can be made by this *Institution* . . ."[8]

The Society, however, recognized that until the beginning of the seventeenth century, natural philosophy was essentially limited to observation. Without a system to validate what was observed, science would become dogma with little chance of advancement. The Royal Society proposed, "The *Third* sort of new Philosophers . . . who have not only disagreed from the *Ancients*, but have also propos'd to themselves the right course of slow, and sure *Experimenting.* . . . And of these, I shall only mention one great Man, who had the true Imagination of the whole extent of this Enterprize, as it is now set on foot; and that is, the *Lord Bacon*."9

The Royal Society was dedicated to the spread of the knowledge gained by their members, and in particular, the effect scientific experimentation had on the increase in productivity within the trades. Sprat asked his readers to "consider the purpose of the *Royal Society*, and the probable effects of *Experiments*, in respect to all the *Manual Trades . . .*" He explained "that it is not a vain or impossible Design, to endeavor the increase of *Mechanic contrivances . . .*" through the efforts of the Royal Society.10 He emphasized the Society's role in the productivity of artisans and tradesmen with: "*the surest increas remaining to be made in Manual Arts, is to be perform'd by the conduct of Experimental Philosophy*" [Sprat's italics].11

The Royal Society had set in motion an organization dedicated to promoting what is now referred to as the "scientific method." Members of the Royal Society, Edmund Halley in particular, were interested in devising new ways to explore the oceans (figure 2.2). Sprat provided a long list of "the Instruments they have invented" to carry on their new experimental philosophy on land and under the sea; the following are only a few.

A new Instrument for fetching up any Substance from the bottom of the Sea, whether Sand, Shells, Clay, Stones, Minerals, Metals.

A new Bucket for examining and fetching up whatever Water is to be found at the bottom of the Sea, or at any depth, and for bringing it up without mixing with the other Water of the Sea, through which it passes.

A Bell for diving under water to a great depth, wherein a man has continued at a considerable depth under water, for half an hour, without the least inconvenience.

Another Instrument for a Diver, wherein he may continue long under water, and may walk to and fro, and make use of his strength and limbs, almost as freely as in the Air.

A new sort of Spectacles, whereby a Diver may see any thing distinctly under Water.

A new way of conveighing the Air under Water, to any Depth, for the use of Divers.[12]

Sprat also expounded on the interest of the Royal Society in their investigations into the properties of fire, air, and water. For example, Sprat made specific note of the "*Experiments* about the Comparative Gravity of *Salt Water*, and fresh, and of several *Medicinal Springs* found in this Nation, and of the differing weight of the *Sea water*, in several Climats, and at several Seasons: of the weight of *Distill'd-water*, *Snow-water*, *May-dew*, [and] *Spring-water.*"[13]

From the beginning, Bushnell would need to convince his supporters that his vision of submarine warfare was credible. His ideas were certainly no less novel than those that Sprat described of the early members of the Royal Society, and it was through scientific argument that Bushnell was able to convince his most ardent champion, Benjamin Gale. In a letter to Benjamin Franklin, Gale wrote a lengthy description of Bushnell's "machine for the Destruction of Enemy Ships." Gale emphasized that "it is all Constructed with Great simplicity, and upon Principals of Natural Philosophy" adding that "his reasoning so Philosophically and Answering every Objection I ever made that In truth I have great relyance upon it."

Gale was a member of the American Philosophical Society and friend of Benjamin Franklin. Franklin and Gale's father-in-law, Jared Eliot, were members of the Royal Society and both had received awards in recognition of their scientific contributions. If Gale believed in the concept and was able to argue that Bushnell's vision of a submarine vessel was constructed using those "Principals of Natural Philosophy" pursued by the Royal Society, others with influence could be convinced of the legitimacy of ideas proposed by a Connecticut farmer.

Even Gale, however, hedged his bet as he ends his letter to Franklin with this apology: "I ask Ten Thousand Pardons for presuming to Trouble You with this Long Acct which I fear will Appear to You too Romantic to Obtain Belieff—but have Endeavored in the Strictest Sense to relate Facts Truly." Romanticism in the eighteenth century was not necessarily complimentary. Gaining credibility for his ideas was an uphill struggle for Bushnell.

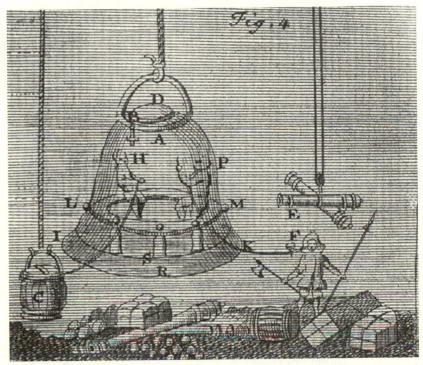

Figure 2.2. Most famously known for delving into outer space with his discovery of the comet that bears his name, Sir Edmund Halley also experimented with inner space technology. (*Martin 1747*)

Education, a valued commodity in colonial Connecticut

He who ne're learns his A.B.C.
Forever will a blockhead be....[14]

Throughout the colonial period, there was a general consensus that education should be a paramount responsibility of the growing society. In the 1715 compilation of the Connecticut laws, there is "An Act for Educating of Children" that begins

For as much as the Education of Children is of singular behoof and benefit to any People; and whereas many Parents and Masters are too indulgent, and negligent of their duty in that kind. It is therefore Enacted.... That the Select-men of every Town in this Colony in their several Precincts and Quarters, shall have a vigilant eye over their Brethren and Neighbors, and see that none of them suffer so much Barbarism, in any of their Families, as not to endeavor by themselves or others, to teach their Children and Apprentices, so much Learning, as may enable them perfectly to read the English Tongue . . .

If the "Parents and Masters" were found to be negligent in "the performing, what is by this Act required of them; every such Parent, Master, Select-men . . . shall pay a fine of *Twenty Shillings* . . ."[15] A fine of this size may have contributed to the popularity of *The New-England Primer* throughout the eighteenth century. This was a relatively inexpensive volume, and would have been purchased and shared among the families of an agricultural society (avoiding the twenty shillings fine). Tens of thousands of this small book were printed; very few surviving copies are known.

Laws also established that taxes collected by the "Treasurer of this Colony" would be used to support schools in every town. It was specified that towns "wherein there are Seventy House-Holders or Families . . . shall Keep, and Maintain One good, and sufficient School for the Teaching, and Instructing of Youth, and Children to Read, and Write."[16]

It wasn't just the laws that encouraged education. There was plenty of encouragement from the pulpit. "Let the people by all means encourage *schools* and *colleges*, and all the means of *learning* and *knowledge*, if they would guard against *slavery*. For a wise and *knowing* and a *learned* people, are the least likely to be enslaved." This was intended as a hint to the people about how to "*guard against tyranny in their rulers*" made by Samuel Webster, "Pastor of a Church in Salisbury" to the Massachusetts House of Representatives on May 28, 1777, in a passionate patriotic sermon.[17]

When Reverend James Dana's sermon before the Connecticut General Assembly was published, the colony was admonished that, "bad books, like bad company, corrupt good morals. Publications replete with levity, irreligion and obscenity have for many years had general circulation in Europe. A taste for such publications begins to prevail among the American youth, foreboding the entire loss of principals and manners unless seasonably checked." Dana was reminding his fellow citizens "to put the education of youth upon the best plan."[18]

Both David and Ezra Bushnell grew up in a typical Connecticut farming community and were members of the local Congregational Church. The Bushnell family lived in the Pachaug district of Saybrook where, according to law, there was a school that would provide the brothers with their early education. It would have become evident at a young age, however, that David and Ezra were destined for different educational paths. Ezra would take on the responsibilities of raising a family and running a successful farm, while David would become a scholar, inventor, engineer, doctor, and eventually a college professor. David Bushnell's access to the educational opportunities he would find at Yale provided inspiration for his vision of submarine warfare.

YALE COLLEGE

During the 1760s a series of tragedies struck David and Ezra Bushnell. First, their father died at the age of fifty-two, then their sister Dency, followed in 1769 by their sister Lydia. When their mother Sarah remarried a short time later, ownership of the farm fell to David and Ezra. It was then that David took the initiative to pursue his intellectual interests and sold his half of the farm to Ezra. At age twenty-nine, David was invited to live with the nearby Elias Tully family where, under the direction of the local Congregational minister, the Reverend John Devotion, he prepared for entry to Yale College in New Haven.[19] Figure 2.3 is a "Plan of New Haven and Harbor" drawn in 1775 by Ezra Stiles. The streets and many landmarks familiar to Bushnell, including Yale's Connecticut Hall, remain today.

For a young man raised on a farm, preparation for college meant constant reading and studying whatever his tutor and mentors suggested. During the two years he spent with the Tully family, David established relationships with Yale alumni that included his Saybrook tutor, Reverend Devotion, who had graduated in 1754. There were also Benjamin Gale (Yale, 1755) and Gale's father-in-law, Jared Eliot (Yale, 1706), both living in Killingworth, Connecticut, only a few miles from Bushnell's home. These individuals would all have readily loaned David books from their personal libraries.

Yale, established in 1701, was one of nine colleges in the colonies prior to the Revolution. The others were Harvard in 1636, William and Mary in 1693, the College of New Jersey in 1746 (now Princeton), Franklin's Academy or the Academy of Philadelphia in 1751 (now the University of Pennsylvania), King's College in 1754 (now Columbia), Rhode Island College in 1764 (now Brown University), Queen's College in 1766 (now Rutgers), and Dartmouth College in 1769. All of these institutions, well respected for the quality of education throughout the eighteenth century, have retained their reputation to the present day.

While a primary goal at Yale was to produce Congregational ministers, the liberal education offered there also prepared their graduates for careers that would ultimately benefit the colony, e.g., medicine, politics, and teaching, along with a supply of entrepreneurs. When the college opened its doors in 1701, it was located in the town of Saybrook, although the students typically studied under tutors located in other towns. In 1716 a decision was made by the college trustees to move the location to New Haven. The move created controversy, however, related in part to the library that formed much of the educational material. While certainly college property, the books and manuscripts constituted an exceptionally valuable resource to the inhabitants of Saybrook. They had been placed in the care of Lieutenant David Buckingham, who refused to allow them to be removed. Governor Gurdon Saltonstall found it necessary to order the county sheriff to take possession of the library materials and transport them to New Haven.[20] The governor's response, and the thought of losing the library, caused uproar among the residents. According to Edward Atwater, Buckingham refused the order, and when the sheriff met with significant resistance from the town, the governor ordered him to "impress men, carts, and oxen. Even then a mob collected in the night and took off the wheels from the carts, and broke down the bridges on the road to New Haven, so that before the library reached its destination, 250 volumes and many valuable papers were lost."[21]

The scientific literature at Yale

By the time David Bushnell entered Yale in 1771, the library had increased to about four thousand volumes. Although a listing of the library holdings at that time is not available, other lists published prior to and after the Revolution provide insight into the books and the topics of study available to students. There were, of course, books on religious and moral philosophy that Bushnell would have been required to study. His interests, however, were not so limited, and he would have concentrated his efforts on what was then referred to in the scientific community as natural and experimental philosophy.

Thomas Clap, president of Yale College at the beginning of the Revolution, noted that "we have a good library . . . well furnished with ancient authors; such as the Fathers, Historians, and Classicks, many

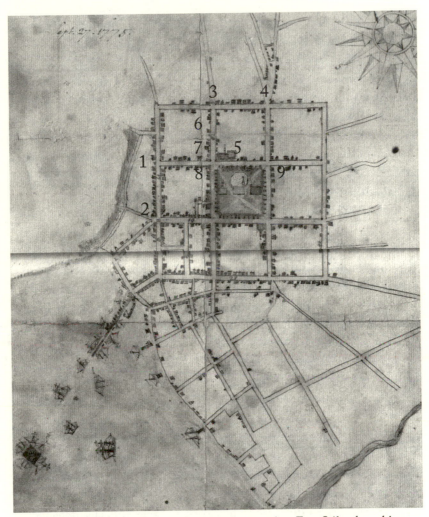

Figure 2.3. New Haven was a popular college town when Ezra Stiles drew this map dated September 27, 1775, the year David Bushnell graduated from Yale College. Bushnell knew the town well: 1. College Street; 2. Church Street; 3. Chapel Street; 4. Elm Street; 5. Yale College; 6. Isaac Doolittle "Clockmaker;" 7. Yale President Thomas Clap; 8. Beers Tavern; 9. Tory Tavern. (*The New Haven Museum and Historical Society*)

valuable books of Divinity, History, Philosophy, and Mathematicks . . ." Concern over the future of this well-furnished library caused nearly three-quarters of the volumes to be removed in 1777 when New Haven was anticipating an invasion by the British. The library holdings were not returned until 1783, unfortunately missing about 750 volumes.[22]

From Yale's *Catalog of the Library* ... (Clap 1743), students had access to the works of authors such as Francis Bacon, John Locke, Alexander Pope, Jonathan Swift, and Joseph Addison, along with the popular publications *The Tatler*, *The Guardian*, and *The Spectator*. Even though books of religious doctrine and catechism were paramount, Yale students could also read a wide variety of topics including, surprisingly, Joseph Glanville's controversial *Saducismus Triamphatus, or, Full and Plain Evidence Concerning Witches and Apparitions Treating of their Possibility* and Francis Hutchinson's answer in *An Historical Essay Concerning Witchcraft, With Observations Upon Matters of Fact.*

The 1743 catalog listed the library holdings according to various topics. Those of interest to Bushnell's designs included: "Arithmetic," "Algebra," "Geometry," "Trigonometry," "Optics," "Conic-Sections," "Mixture of all sorts of Mathematics," and "Natural Philosophy." From over a hundred books under these headings were authors such as Watt (*Philosophical Essays*), Derham (*Physicotheology*), Galileo (*Systema Mundi*), Boyle (*Philosophical Works* and *New Experiments Physico-Mechanical*), Hooke (*Posthumous Works*), and De Sagulier (*A Course of Experimental Philosophy*). There were, of course, many works by and about Isaac Newton including his *Principia Mathematica.*

When Nehemiah Strong (Yale, 1755) was hired to teach mathematics and natural philosophy in 1770, the college had made a commitment to provide students with the ability to concentrate their studies on subjects leading to careers other than the ministry. It was good timing for Bushnell whose interests were definitely secular. There is no record of how much Bushnell confided in Strong regarding his interest in submarine warfare. According to Louis Tucker, "Yale officials detected in Strong a lack of enthusiasm for the Revolutionary cause and, conversely, an apparent sympathy for the British position."[23] While not an overt Tory, Strong understood that the students and faculty were nearly universal in their support for American independence. From Bushnell's standpoint, Strong was the only game in town, and was responsible for guiding his students in their studies of natural philosophy.

Bushnell's credibility would be based on his becoming as knowledgeable as possible in science and mathematics. Of the utmost importance to Bushnell and other inventive minds were the *Transactions of the Royal Society of London*. Abridgements of the individually printed *Transactions* were published in book form in various editions throughout the eighteenth century. The Yale library included all of the volumes, each of which described in detail the results of experiments and observations made by the most prominent scholars and scientists. These books would have drawn Bushnell, and other curious students, to spend hours at the library where he would have found inspiration and scientific validity to his ideas.

Among the books that Nehemiah Strong would have recommended to his students, one in particular may have influenced Bushnell's thoughts. In 1747 Benjamin Martin published his *Philosophia Britannica: or a New and Comprehensive System of the Newtonian Philosophy, Astronomy and Geography, in a Course of Twelve Lectures . . .* Known also as "Martin's Philosophy," this was as close to a standard college textbook as could be had in the eighteenth century. Published after the 1743 catalog of library holdings, several editions of the *Philosophia Britannica,* including the two-volume first edition, are listed in the 1791 catalog suggesting that at least one edition was available when Bushnell attended Yale. Martin covered several topics directly related to the design of components that Bushnell included in his submarine.

Throughout the *Turtle* Project, we referred to the scientific and popular literature available to Bushnell and his collaborators. In this way, we attempted to replicate the various mechanical systems with only the information and level of scientific knowledge that Bushnell had access to in the eighteenth century. But what were the conditions that inspired such creativity from these American patriots?

Figure 3.1. Located on the New Haven Green at the corner of College and Chapel streets, Beers Tavern was a gathering place for Yale students. (*Terry 1937*)

Chapter Three

A NOVEL IDEA
TAKES SHAPE

A staple in the diet of colonial Americans, New England rum became an important stimulant to the expansion and encouragement of a patriotic work ethic. On February 19, 1777, the Connecticut Council of Safety voted "to draw on the Pay-Table for the sum of £108 13*s* 4*p*, for 326 gallons of rum bought of Zabdiel Rogers and company, for use of the cannon foundery at Salisbury."[1] It must have been a cold winter.

On the New Haven Green within sight of Yale College, the chapel, and the home of the college president was Beers Tavern (figure 3.1). This was an establishment known for its patriotic sympathies, and a popular meeting place. New Haven town meetings were also held at Beers Tavern where, on the morning of April 22, 1775, Benedict Arnold, Captain of the Governor's Guards, confronted the selectmen. It was three days after the skirmish at Lexington. Arnold and his troops intended to join the American army assembling in the countryside surrounding Boston. He demanded, and ultimately received, ammunition from the town's arsenal. Three months later George Washington, having been voted Commander-in-Chief by the Second Continental Congress, spent the night at Beers Tavern en route to his troops surrounding Boston.[2]

Beers Tavern was also a place where, for the cost of a draught or two, Yale students could assemble, complain about their professors, the food at the commissary, and that learning Latin had little relevance to modern thought. It was also where serious discussions were held concerning politics and the rising sentiment against the Crown. It would have been familiar to Bushnell's many young protagonists, who became active in the American military during the war. In addition to David Humphreys (Yale, 1771, aged 19) and Nathan Hale (1773, aged 18), there were Abraham Baldwin (1772, aged 17) and Joel Barlow (1778, aged 24), both chaplains in the Continental army, and Richard Sill (1775, aged 20), a lieutenant in Colonel Parsons's Brigade, when Ezra Lee volunteered to replace Ezra Bushnell as pilot. Another famous graduate that Bushnell no doubt knew was Benjamin Tallmadge (1773, aged 19), who was a major in the Continental Dragoons.

Little is known of Bushnell's personality other than he was dedicated to the pursuit of knowledge. According to Henry Howe in his book *Memoirs of the Most Eminent American Mechanics*, "Bushnell . . . is only remembered as being a very modest, retiring young man, shunning society and bound down to his books."[3] At Nathan Hale's recommendation, Bushnell had become a member of Yale's prestigious but secretive Linonian Society. It was a place where members could discuss their literary and intellectual interest as well as, no doubt, politics. Hale had compiled a list of the books in the Society's small library.[4] Most of the titles consisted of works of poetry and literature by Shakespeare, Milton, Addison, and others, plus British and Roman histories.

The Linonian Society was a good match for an intellectually curious and "retiring young man," yet we don't know if Bushnell occasionally joined the more boisterous students who frequented Beers Tavern. Sympathetic listeners would have been available at either location, and with an ample supply of rum on hand, everyone's sense of adventure and creativity would have increased as the evening wore on. One can only speculate as to how Bushnell's ideas evolved and with whom he may have confided among the students and faculty. Nonetheless, public discussions had to remain vague. On Elm Street directly across the New Haven Green from Beers Tavern was Tory Tavern (see figure 2.3), an establishment obviously known as a hangout for British sympathizers, now the home of Yale's Elihu Society.

THE POPULAR LITERATURE

While Bushnell's credibility among the scientists, or "natural philosophers," who mentored his ideas would have been based to a large extent on his familiarity with the scientific literature and the works of individuals such as Robert Boyle and Robert Hooke, he may also have turned to the popular literature.

With an increase in literacy came a desire for variety in reading material. Printing in the colonies was an expensive proposition, and few families could afford books. Much of the literature available to the general population was religious in nature. As towns grew in population and diversity, interest in life beyond the family farm increased. By the first half of the eighteenth century, locally published almanacs and newspapers helped fill the need for information that transcended moral and religious doctrine.

Those who could afford the luxury of a personal library may have subscribed to one of the many monthly magazines from England. Three of the more popular were *The Gentleman's Magazine and Historical Chronicle*, *The London Magazine or Gentleman's Monthly Intelligencer*, and *The Universal Magazine of Knowledge and Pleasure*. Some carried exceptional titles: *The Universal Museum and Complete Magazine of Knowledge and Pleasure Containing the Greatest Variety of Original Pieces, on the Most Curious & Useful Subjects in every Branch of Polite Literature, Trade and Commerce; and various Parts of Science and Philosophy*. If you wanted something with a much shorter title and not published in London there was *The Scots Magazine* from Edinburgh, and for those interested in literary topics, there was *The Monthly Review*.

These publications provided a wide assortment of articles covering politics, history, literature, mathematics, science, travel, and topics of general interest. Most were profusely illustrated with woodcuts and engraved plates of birds and beasts, castles and kings, and detailed foldout maps for those with a worldview. For the mechanically inclined, which brings us to Bushnell, there were always depictions of inventions and unique "machines" capable of all sorts of activities that could stimulate the imagination.

THE HYDRASPIS AND OTHER CONTRIVANCES

The illustration, figure 3.2, is from the August 1747 issue in volume 17 of *The Gentleman's Magazine*. It accompanies an article titled: "An Account of the Hydraspis; or, Water-Shield: a Machine, by the help of which a Person may walk on the Water without fear of sinking." The article notes that "the inventor of this machine, a German of considerable rank and erudition, owns himself indebted . . . to his observations on the swimming of geese and ducks . . ."[5] It appears that any contrivance or topic, whether obscure or revolutionary, was provided with equal

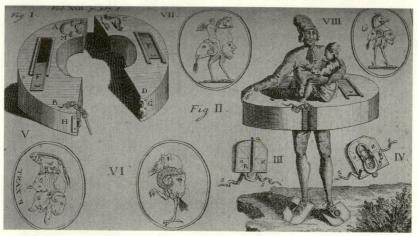

Figure 3.2. The "hydraspis" is an example of the creativity and ingenuity of eighteenth-century inventors that found its way onto the pages of popular magazines available to David Bushnell and his contemporaries. (*Gentleman's Magazine, December 1747*)

space. Among the wide variety of topics that readers enjoyed, volume 17 included illustrated articles about a rhinoceros, Mount Vesuvius, and a unique "diving ship."

Eighteenth-century magazines fed the curiosity of their readers with occasional articles about strange "diving machines" and "diving engines." An example was the one-atmosphere suit invented by John Lethbridge in 1715 and used by him and his associate, Thomas Rowe, for over forty years with great success. Lethbridge's device is described in other sections of this book (see chapters 5 and 9).

If a Bushnell skeptic, however, wanted to point to an example of the effects of a poorly designed diving machine, he might pull down from the shelf a copy of *The Annual Register for the Year 1774.* In this publication, the reader would find "The Authentic Account of a late unfortunate Transaction, with respect to a Diving Machine . . ." What follows is a three-page account of a Mr. J. Day who devised a vessel in which he planned to remain submerged at a depth of one hundred feet for twelve hours. "The ship went gradually down in 22 fathom water, at 2 o'clock on Tuesday June 28, in the afternoon, being to return at 2 the next morning." Day's vessel never returned and after repeated attempts to locate him the search was given up. The article concludes: "Thus ended this unfortunate affair. The poor man . . . put his life to the hazard upon his own mistaken notions. Many and various have been the opinions on this strange, useless, and fatal experiment, though the more reasonable and intelligent part of mankind seem to give it up as wholly impracticable."[6]

DENIS PAPIN

During the nearly two hundred years that had passed since William Bourne first published his concept of a submarine, a curious variety of "diving engines" appeared on the scene—inspired by scientists, mechanics, artisans, entrepreneurs, adventurers, and "visionaries" who had proposed, built, marketed, and operated them, and in which some had taken their last breath. Caution and skepticism must have been always on the minds of Bushnell's companions. With his brother's life at stake, Bushnell would never have proceeded without having thoroughly thought through each design and answering each skeptic's questions. He may have pointed to the article in *The Gentleman's Magazine* about the "diving ship" designed by Denis Papin.

Denis Papin (1647–1712) became a member of the Royal Society of London in 1663, at the age of sixteen. His mathematical genius was recognized by many in the scientific community, providing him with opportunities to work with Christian Huygens, Robert Hooke, and Robert Boyle. Much of his work (with Boyle in particular) was associated with the study of the properties of air, known then as the field of "pneumatics." Papin conducted his own experiments studying the effect of air on the properties of gunpowder. After twenty-five years of scientific research, much of it published in the *Philosophical Transactions of the Royal Society*, he became a professor of mathematics at Marburg in Germany.

While carrying an impressive scientific résumé, Papin was also an inventor. Threats of war among and against the many principalities that comprised seventeenth-century Germany induced the ruling class to search for unique weapons. Papin's creative abilities were recognized by Charles, Landgrave of Hesse-Cassel. "The prince, being told of the extraordinary conveniences of the famous diving ship, constructed by *Drebel*, commanded one of the like kind to be attempted [by Papin]."

THE SUBMARINES OF PAPIN AND BUSHNELL

In 1695, Papin published *Recueil de diverses Pieces touchant quelques nouvelles Machines . . .* , a treatise describing his many inventions, including two versions of the submarine he was commissioned to produce for the

Landgrave of Hesse-Cassel. It is odd that this particular prince would be interested in submarine warfare considering that Hesse-Cassel is 180 miles (as the eagle flies) from the coast. While Papin's *Recueil de diverse Pieces . . .* may not have crossed Bushnell's desk, a popularized account was likely available in the private library of one of his mentors or possibly on the Yale library shelves. The December 1747 issue of *Gentleman's Magazine* cited above contained a detailed description of Papin's design and included an illustration (figure 3.3) taken directly from his 1695 book.

Papin and Bushnell both employed design features that included the shape and strength of the hull, its ventilation and air supply, the hatch, the mode of attack, a depth gauge, how the operator attained neutral buoyancy and submerged, the method of propulsion, and the bilge pumps for controlling water ballast. The quotes below comprise the entire Papin article followed by a brief comparison to Bushnell's design. The letter designations refer to the illustration. Note that the *Gentleman's Magazine* contributor was translating the French text from Papin's original treatise and likely added much of his own interpretation.

"AA is a wooden tub of an elliptic or oval figure, 6 feet in height, as many the greater diameter, and 3 feet in the lesser;" Bushnell's hull was also elliptical and of nearly the same dimensions. According to Benjamin Gale, the *Turtle* did not exceed six feet in height and was no longer than seven and a half feet, including the length of the rudder.

"BB is the Hessian rotary sucker and forcer, which attracts the external air through the pipe CC." Bushnell also used a ventilation system that drew fresh air through a pipe located in the hatch. The introduction of fresh air forced the respired air to exit the hull through a second pipe.

"Again DD is a great hole, which serves as a door;" A hatch at the top of both vessels implies that entry was intended to occur with the submarine afloat rather than prior to launch. This may seem an obvious feature, but the occupants of diving bells, a technology much more common than submarines, entered prior to being lowered into the water. The method of making their respective hatches watertight was never discussed.

"EE a great cylindrical copper vessel 6 feet long, and a foot and a quarter in diameter, whose aperture is within the tub AA, and is closed, as also the hole DD, in the most exact manner, with plates fitted for the purpose, and screw'd very tight. PP is a prop to support the cylinder arm. FF is a pump, by which the men shut up in the tub AA may introduce

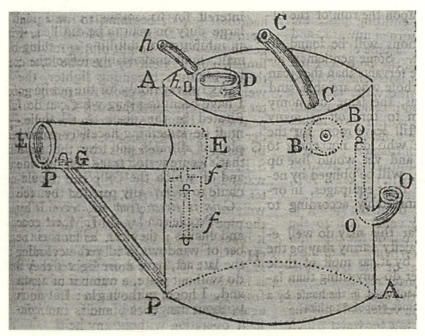

Figure 3.3. Many of the features included in Denis Papin's submarine can also be found in Bushnell's *Turtle*. Papin and Bushnell had to contend with the same issues of mobility, structural strength, buoyancy, air supply, and depth control. (*Gentleman's Magazine, December 1747*)

air into the cylinder EE, to repel water from the hole G, through which the man inclined in the cylinder will be able to destroy the enemy's ships." The writer doesn't mention the weapon his submariner was intending to use to destroy a ship, nor how it would be deployed. It may have been some form of "water petard" that Drebbel had devised for King Charles I. What is not left to the reader's imagination is that the individual lies inside a fifteen-inch diameter cylinder with a hatch secured behind him, much like a modern torpedo tube. Bushnell's method of deploying a weapon was much more operator friendly.

"The pressure on the sides is prevented by the round form of the tub, and the weights put in the bottom to sink the tub, and resist the water on that part; and as for the upper part, the pressure of the incumbent water is not very considerable, because it is not necessary that the tub descend very deep." Papin doesn't describe the depths that his vessel was designed to withstand, probably not planning to submerge much below the surface. He may simply have intended that the weapon be attached just below the targeted ship's waterline. There is some ambiguity here in that Papin also included a depth gauge and may, in fact, have considered submerging to the keel of a warship. He depended, as did Bushnell, on the inherent strength of a curved hull design to withstand the external pressure. The heavy ballast that comprised Papin's keel served to add strength to the bottom of his vessel.

"Fresh air will be attracted through the pipe CC, and the superfluous air expelled through the pipe *bb*;" This is another reference to Papin's ventilation system that, as with the *Turtle*, included inlet and outlet pipes. Bushnell's ventilation system is covered in more detail in chapter 9.

"The recurve barometer OO, open at both ends, whose lower part may be made of iron or wood, shews the depression of the ship very exactly." The depth gauge was a critical component on the *Turtle*. Papin very likely included one for the same reason, that being to ensure his submarine would descend only deep enough to come in contact with the target ship.

"It may be further depressed by letting in the external water by a cock. But to prevent the ship from being quite sunk by letting in too much water, two men ought always to be trying to depress it by the help of oars, and when they find it can be done without much stress, the cock is immediately to be shut, by which means the ship will for any space of time be kept lighter than water, and yet may, by means of the oars, be depressed more and more at discretion." Obtaining neutral (or nearly neutral) buoyancy was a major concern to both Papin and Bushnell. Both designs provided a method that enabled the pilot to control the addition of water ballast. Both vessels required that the pilot descend to the required depth with "oars," although of a much different design.

"The oars are to come through lateral holes, that are most exactly closed by leather bound about them, as we are told, was also practis'd in *Drebel's* ship." Both Papin and Drebbel used common oars as a mode of propulsion. Their vessels were designed to be rowed with the oars extending through openings in the side of the vessel and kept watertight with leather sleeves. Bushnell, while also using the term "oars" to describe both transiting and submerging, employed what was the genesis of the modern screw propeller. The concept of using oars to propel a submarine died hard. During the Civil War, the U.S. Navy contracted a Frenchman living in America, Brutus de Villeroi, to build a submarine. His vessel, the *Alligator*, was initially propelled by eight sets of oars later replaced with a single hand-cranked propeller. Its ten-month career was unremarkable and hardly noticed. On June 2, 1863, the *Alligator* sank while under tow. As described in chapter 1, even at the turn of the twentieth century, several authors writing about the history of submarine warfare included an illustration of the *Turtle* with oars extending from the sides of its hull.

"When we think fit to raise the ship, the thing is easily effected, partly by expelling the water by a pump contrived for that purpose." Both Papin and Bushnell provided bilge pumps that allowed the operators to regain positive buoyancy and return to the surface.

"As to the difficulty of breathing in such a ship, *Drebel* mentions that he had provided a certain quintescence of air, one drop of which emitted would render the vitiated air again fit for respiration, but Dr *Papin* imagines this rather a thing to be wish'd than a reality." Cornelius Drebbel had experimented with producing "air" from various chemical processes, including heating potassium nitrate (KNO_3). With limited air in the *Turtle*, Bushnell understood that it would be necessary for the underwater phase of his mission to occur in less than half an hour. Once submerged, he could not use his ventilation system, there being no hose extending to the surface as Drebbel and Papin had included.

Both Papin and Bushnell were responding to an imminent threat of war. The above comparisons are provided as an example of how one engineer's ideas might have found their way into a later engineer's design. We do not know, however, the extent to which Papin influenced Bushnell, if at all.

Bushnell must have encountered a great deal of skepticism about the whole business. He knew that to convince others to join him in his efforts, he first had to demonstrate that there was a scientific basis behind each idea. Bushnell's vision was a marriage of science and mechanics. His discussions with trusted professors, and certainly with his mentor Benjamin Gale, would have given him the credibility to take his ideas to the next stage, that being how and where to produce the critical parts for his submarine vessel.

Ideas led to designs that were then produced, installed, tested, and modified as necessary. It was a four-year process involving the unique talents of many individuals. It was not just two brothers. As Bushnell noted in the concluding remarks in his letter to Jefferson: "The above Vessel, Magazine, &c. were projected in the year 1771, but not completed, until the year 1775."

Figure 4.1. The "lower" Boston Post Road (there were three Boston Post roads that converged on New Haven) can be seen running along the Connecticut coast, passing through Saybrook, where it crossed the Connecticut River. The bend in the river near the "k" in "Seabrook" is Ayers Point, where we believe the *Turtle* was completed and launched in 1775. (*London Magazine, April 1758*)

Chapter Four

A SECRET "MACHINE"

David Bushnell was born and raised in Connecticut (figure 4.1), a colony where many were ardent supporters of American independence. Yet it was still a time when it could be difficult to know who was truly a patriot and who may have maintained loyalist sympathies. This uncertainty must have presented Bushnell with a dilemma.

The colonies were about to face the world's superpower, and his vision of a method to counter the British navy had to be kept secret in order to be effective. There was also concern that anyone who became actively involved with this form of clandestine warfare would be considered a traitor and terrorist by the British military courts. Hanging, the primary method of discouraging treason against the Crown, was a likely consequence.

While Bushnell's endgame was to sink British warships, his design would have to consider all of the tactical issues associated with accomplishing the mission. He had to design a method of propulsion, a system to ensure an adequate supply of air, some novel way to navigate by compass at night, and a procedure to control buoyancy and submerge to a specific depth in total darkness. If he could not overcome these design issues, there would be no mission. Not only was the entire concept of submarine warfare a novelty, each mechanical contrivance associated with it would require imagination and ingenuity. In a wartime environment, secrecy would be paramount, thus limiting the pool of talent that Bushnell could draw from.

Ezra Bushnell (1746–1786), David's brother, was the first person he knew he could trust. Ezra had continued with the business of farming the family homestead while David pursued his educational aspirations. They would have worked closely together, yet David knew that a project of this magnitude was something two brothers could not accomplish alone. A team had to be assembled that he could rely on and trust with his secret.

The process of turning a unique concept into reality in Revolutionary War America required the same attention to detail as it does today. In modern terms, Bushnell had to perform the functions of project manager and systems engineer. He had to be the scientist, the experimentalist, the engineer, the technician, the mechanic, the tactician, the logistician, and the salesman.

For his vision of submarine warfare to move forward, Bushnell became well versed in the science behind each technology that he incorporated. We don't know the extent that Bushnell confided in his Yale tutors, but they were there to direct and encourage their students' interests. He could, however, freely discuss the science with his trusted friend Benjamin Gale. Gale's credibility as a well-known doctor and accomplished scientist may have convinced other influential individuals to find ways of expediting Bushnell's project, including Ezra Stiles, who would soon become president of Yale College.

Bushnell would have tinkered with ideas on his own, as he likely did when demonstrating that gunpowder would detonate under water. Yet when it was time to create the final design and construct his complex vessel, he had to assemble a small team of artisans and mechanics. These individuals, Isaac Doolittle in particular, would have had the necessary insight to discuss specific requirements and determine the feasibility of producing the hardware. All of Bushnell's ideas, however, would stay on the drawing board until he solved the problem of access to scarce and expensive resources.

While developing the concept of a submarine with its many complex mechanisms, Bushnell also had to consider that someone would have to master their operation. That meant finding a suitable location with ample opportunities to train a pilot. But who would have the confidence in his design to take the helm? The first choice for this responsibility was his brother Ezra. It would be the Bushnell brothers who would tackle the issue of experimenting with and perfecting the vessel. But it would take more than two brothers to tackle the logistical issues associated with transporting and handling a vessel that would have weighed well over three thousand pounds.

If the *Turtle* were to be transported by land, Bushnell would need to have access to a wagon capable of carrying the weight across a miserable system of roads. Traveling on the Boston Post Road was no picnic, and the back roads around Saybrook would have been a challenge for even the healthiest team of oxen. A wagon, however, was an easy problem to solve. There may already have been one on the family farm easily adapted to carrying the *Turtle*.

Transportation by sea was a logical alternative. There is no mention of Bushnell owning a boat, however, and thus he would have had to rely on others to provide that resource. Employing a vessel large enough to handle the *Turtle* would require a crew consisting of more than just David and Ezra. The vessel would need to be rigged to move the *Turtle* into the water and then retrieve it once the testing or training was complete.

While some of the services that Bushnell was able to obtain may have been provided as a contribution to the patriot cause, much of the day-to-day operations had to be paid for. The owner and crew of a sailing vessel depended on their commercial enterprises for their livelihood. Most of the American "navy" consisted of privateers who were not paid by the Continental Congress, but received compensation from the sale of the cargo and vessels they captured. The British considered these privateers as pirates and when captured, they were treated accordingly. The crew of a vessel being used to support a weapon intended for a clandestine mission would also have been considered unworthy of honorable treatment.

"The shot heard 'round the world"[1]

By the spring of 1775, Bushnell was about to complete his final year at Yale College. The majority of his fellow students were profoundly patriotic and dedicated to American independence. Nearly one hundred students had organized a militia company and had been training on the New Haven Green in anticipation of what they perceived as an inevitable conflict with Britain. On the evening of April 21, 1775, a post rider from Massachusetts arrived in New Haven with news of a skirmish two days earlier between local militia and British regulars at Lexington and Concord. The reaction to this news is found in a brief entry in Yale student Ebenezer Fitch's diary: "Friday, April 21. To-day tidings of the battle of Lexington, which is the first engagement with British troops, arrived at New Haven. This filled the country with alarm, and rendered it impossible for us to pursue our studies to any profit." On April 26, Ezra Stiles, who was to become president of Yale College from 1778 to 1795, noted in his diary that within two days of hearing the news, the Yale contingent "marched from New Haven via Hartford for Boston."[2]

With half of the student body heading for Boston, the college closed until the end of May. Some students returned home and volunteered to serve in their local militia units while others remained at the college. Bushnell was anxious to complete his studies and his submarine. Marion Hepburn Grant suggested a possible scenario in her account of the *Turtle*:

> The Boston Massacre of 1770, the Boston Tea Party of '73, and the bloody battles between the Americans and redcoats at Lexington and Concord of '75, had already outraged numerous New Englanders. So, during his final college vacation, David Bushnell rushed back to Saybrook where he persuaded his husky younger brother, Ezra, to assist him with actually building both the mine and the submarine.[3]

We believe that *Turtle* construction had been underway for at least two years. By the spring of 1775 the hull would have been nearly complete. There is no way to know if Bushnell rushed home or spent his vacation working with Isaac Doolittle on the most critical elements of the *Turtle*. There were only a few individuals Bushnell trusted with his secret weapon; coordinating and directing the efforts of each would have occupied much of his free time.

With the *Turtle* nearly ready for launching, Bushnell needed to build and maintain enthusiasm among those he had enlisted to assist with this venture. Everyone on his team, however, would have had many other responsibilities. It was often the newspapers that inspired individuals to dedicate their energies to the impending confrontation. Newspaper "correspondents," who were rarely named, provided graphic descriptions of events intended to generate outrage among the local readers. Surprisingly, unedited American accounts also appeared in British publications. The following are excerpts from a Massachusetts newspaper article reprinted in the May 1775 issue of *The Universal Magazine*.[4]

From the Essex Gazette, Printed at Salem, in New England.
Salem, April 25

Last Wednesday, the 19th of April, the troops of his Britannic Majesty commenced hostilities upon the people of this province.

At Lexington, six miles below Concord, a company of Militia, of about 100 men, mustered near the meeting house; the troops came within sight of them just before sun-rise; and running within a few rods of them, the commanding Officer accosted the Militia in words to this effect: —'Disperse you Rebels—Damn you, throw down your arms and disperse:' Upon which the troops huzza'd, and immediately one or two Officers discharged their pistols, which were instantaneously followed by the firing of four or five of the soldiers, and then there seemed to be a general discharge from the whole body: Eight of our men were killed, and nine wounded.

It appeared to be their design to burn and destroy all before them; and nothing but our vigorous pursuit prevented their purpose from being put into execution. But the barbarity exercised upon the bodies of our unfortunate brethren who fell, is almost incredible: Not content with shooting down the unarmed, aged and infirm, they disregarded the cries of the wounded, killing them without mercy, and mangling their bodies in the most shocking manner.

We have the pleasure to say, that not withstanding the highest provocations given by the enemy, not one instance of cruelty, that we have heard of, was committed by our victorious Militia.

On June 10th the *London Gazette* printed General Gage's account of the events at Lexington and Concord. The article was reprinted in the June issue of *The Universal Magazine*, one month after the American account appeared. The following are excerpts:[5]

Lieutenant Colonel Smith finding, after he had advanced some miles on his march, that the country had been alarmed by the firing of guns and ringing of bells, dispatched six companies of light infantry, in order to secure two bridges on different roads beyond Concord, who, upon their arrival at Lexington, found a body of the country people drawn up under arms on a green close to the road; and upon the Kings troops marching up to them, in order to enquire the reason of their being so assembled, they went off in a great confusion, and several guns were fired upon the King's troops from behind a stone wall, and also from the Meeting-house and other houses, by which one man was wounded, and Major Pitcairn's horse shot in two places. In consequence of this attack by the rebels, the troops returned the fire, and killed several of them; . . . as soon as the troops resumed their march, they [the militia] began again to fire upon them from behind stone walls and houses, and kept up in that manner a scattering fire during the whole of their march of 15 miles, by which means several were killed and wounded; and such was the cruelty and barbarity of the rebels, that they scalped and cut off the ears of some of the wounded men, who fell into their hands.

General Gage says, that too much praise cannot be given to Lord Percy, for his remarkable activity during the whole day, and that Lieut. Colonel Smith and Major Pitcairn did everything that men could do, as did all the Officers in general; and that the men behaved with their usual intrepidity.

So . . . who to believe? Who actually fired "the shot heard 'round the world"? The Americans claimed that after a hearty "huzza!," the British regulars opened fire and only then did the militia fire back. The British claimed that after a polite inquiry as to why the Americans had assembled, the "rebels" scattered and began firing from behind stone walls; and only in consequence of this did the King's troops return fire.

Exaggeration of the facts was no different then than at any other time in history. Atrocities were likely committed by both sides. Each of the above accounts used the term "barbarity" to describe the actions of probably one or two individuals. The effect on the readers was also no different then than now. Passions were aroused on both sides; support for the impending crisis and the war that followed was the goal.

YANKEE DOODLE

Yankee Doodle went to town
Riding on a pony
Stuck a feather in his hat
And called it macaroni.[6]

A tune called "Yankee Doodle" was reportedly played by British troops when arriving in Boston in 1768. The words "Yankee" and "doodle" were intended to express British disdain for the American riffraff that the king was being forced to contend with. The original lyrics, however, took on many forms during the war and were quickly altered and adopted by New England troops in satirical defiance of the increasing British military presence. In his poem "M'Fingal" about a Tory who had been tarred and feathered, John Trumbull wrote a set of lyrics for a tune now very familiar to his compatriots. Trumbull included a reminder to the British of how "Yankies" had "instructed them in warlike trade," possibly a reference to the courageous stand by American militia at Bunker Hill.[7]

In the eyes of the British, the reference to a New Englander as a "Yankee" was not a kind gesture. Being a "Doodle" was even worse, but by far referring to someone as a "macaroni" held the most disparaging connotations. The following description is from the April 1772 issue of *Town and Country Magazine*.[8]

> The Italians are extremely fond of a dish they call Macaroni, composed of a kind of paste; and as they consider this as the *summum bonum* of all good eating, so they figuratively call everything they think elegant and uncommon macaroni. . . . the title of Macaroni was [considered] applicable to a clever fellow and [became] the standards of taste in polite learning, the fine arts, and genteel sciences; and fashion . . . became an object of their attention.

The article explained that this fashion statement had been carried to extremes whereby, "now Macaronies of every denomination from the colonel of the Train'd-Bands down to the errand boy . . . make a most ridiculous figure." The lengthy description of the Macaroni's clothing ended with this commentary: "In this free country every one has a right to make himself as ridiculous as he pleases . . . a Macaroni renders his sex dubious by the extravagance of his appearance, the shafts of sarcasm cannot be too forcibly pointed at them." American patriots simply smiled and redirected these "shafts of sarcasm" back to their antagonists. "Yankee Doodle" is now the state song of Connecticut.

THE AMERICAN CRISIS

It was a time when there were more loyalists than patriots and the status of your neighbor's sentiments was always a concern. In order for

Bushnell to proceed with his submarine concept, he needed a dedicated cadre of participants—people who were committed to the patriot cause and willing to make the required sacrifices. Bushnell had to choose wisely; the secret of his "machine" had to be kept among a few trusted individuals. Yet blatant arrogance was fanning the flames of discontent against British rule in the colonies, arousing and invigorating patriotism, and encouraging men to shift their loyalties to American independence.

Inspiration to choose sides came from many sources: patriotic, political, or straight from the pulpit. Thomas Paine published his popular pamphlet *Common Sense* in January 1776. In his introduction, Paine is quite clear as to the situation he and his countrymen faced: "The laying a Country desolate with Fire and Sword, declaring War against the natural rights of all Mankind, and extirpating the Defenders thereof from the Face of the Earth, is the Concern of every Man to whom Nature has given the Power of feeling . . ."[9]

When recalling Lexington and Concord, Paine's sentiments reflected the feelings of many Americans who lamented the break with England: "No man was a warmer wisher for a reconciliation than myself, before the fatal nineteenth of April, 1775, but the moment the event of that day was known, I rejected the hardened sullen-tempered Pharaoh of England forever."[10]

Paine's short pamphlet was published for the common colonial reader. With a population of between three and four million, enough copies were printed (including pirated editions) to reach nearly one in ten.[11] We can expect that Bushnell and his associates were all familiar with Paine's *Common Sense* and derived and maintained at least some inspiration from his words.

Beginning in December 1776, after Washington's defeats in New York, Paine began publication of a series of pamphlets titled *The American Crisis*. As before, he listed the author as "Common Sense." Paine began the first of this series with what some considered the most famous quote expressed during the war: "These are the times that try men's souls." The tenor of Paine's writings was to drive home the theme that "Tyranny, like hell, is not easily conquered; yet we have this consolation with us, that the harder the conflict, the more glorious the triumph."[12]

Fire and brimstone (and saltpeter and charcoal)

Other than his membership in the Congregational Church, we don't

know the role religion played in Bushnell's life, but American independence and English tyranny were common topics on Sundays. The fiery
rhetoric of the local charismatic pastors was inspirational to the devout
New England parishioners who were, by law, also members of the local
militias. Their sense of duty "to God and Country" would have encouraged those whom Bushnell relied on to support his activities behind the
scenes.

Many of the sermons that have survived were printed versions of
those preached before the various General Assemblies. Because the message in these sermons was considered of particular significance and had
been delivered to a limited audience, the General Assembly authorized
them to be published and therefore available to the public. After reviewing several contemporary sermons, one theme seemed to dominate the
passionate religious rhetoric: British tyranny. The words "tyrant," "tyranny," and "tyrannical" occurred sixty-five times in a sermon delivered May
29, 1776, by Samuel West, "Pastor of a Church in Dartmouth," on the
eve of the Declaration of Independence. In addition to declaring that
"we must beat our plow-shares into swords" and that tyrants will suffer
"fire and brimstone," West voiced a powerful message to the
Massachusetts General Assembly that "tyranny and arbitrary power are
utterly inconsistent with and subversive to the very end and design of
civil government" and later emphasized the necessity "of confirming our
selves as an independent state."[13] It would not be a surprise to the residents of Massachusetts when copies of the Declaration of Independence
arrived less than six weeks later.

Messages of patriotism and dedication continued to be heard
throughout the war, extolling the citizens to unite and maintain their
pursuit of freedom.

> Tyrants always support themselves with standing armies! And if
> possible the people are disarmed, or the militia neglected and
> kept low . . . And when it comes to this, it is difficult for them to
> unite in sufficient bodies to effect their deliverance. But if they
> would unite, nothing, nothing, could stand before them.
>
> When therefore we see in a manner, the whole world, except
> these American States, groaning under the most abject slavery,
> with so few successful attempts to deliver themselves, how stupid
> must we be, if we do not exert ourselves to the utmost to save

ourselves from falling into this remediless estate, this bottomless gulf of misery.[14]

Political and religious rhetoric inspired high ideals of American independence and patriotism, while British arrogance, fueled by reports of atrocities, bred anger and determination. In this environment, Bushnell would have little difficulty filling his need for human and material resources.

ISAAC DOOLITTLE (1721–1800)

Much of the equipment Bushnell needed to produce for his submarine was beyond the skills of a blacksmith. The valves and pumps; the compass, depth gauge and air ventilation system; the mechanism for attaching his explosive "magazine" and its timing device were all sophisticated designs requiring precision manufacturing. In the eighteenth century, the best artisans were to be found among the clockmakers, who acquired their skills after years as an apprentice. They also possessed the ability to visualize mechanical concepts and interactions between gears and levers. As clockmakers, they had the tools and machinery needed to convert Bushnell's ideas into reality.

Bushnell was careful not to include the names of any of his co-conspirators, although there is good evidence that he worked with two of Connecticut's skilled clockmakers. Benjamin Gale provides the only contemporary references to one of Bushnell's mechanics. The first is in his August 7, 1775, letter to Benjamin Franklin: "He is now in New Haven with Mr Doolittle an Ingenious Mechanic in Clocks & Making those Parts which Conveys the Powder, and secures the same to the Bottom of the Ship, and the Watchwork which fires it."

In his November 22, 1775, letter to Silas Deane, Gale noted that: "At the time of my last writing, I supposed the Machine was gone, but since find one proving the navigation of it in Connecticut River. The forcing pump [bilge pump] made by Mr. Doolittle, not being made according to order given, did not answer; which has delayed him."

An entrepreneur of many talents, Doolittle began operating his clock and watch business on Chapel Street in New Haven, Connecticut, in 1742. He also produced compasses and surveying instruments, and by 1770 had hired a watchmaker who had been trained in Europe.

His mechanical versatility is evidenced in the following article that appeared in the September 7, 1769, issue of the *Boston News Letter*, (reprinted in *Connecticut Clockmakers of the Eighteenth Century* by Penrose R. Hoopes, 1975). "We are informed that Mr. Isaac Doolittle, Clock and Watch maker, of New Haven, has lately completed a Mahogany Printing-Press on the most approved construction, which by some good judges in the Printing Way, is allowed to be the neatest ever made in America and equal, if not superior, to any imported from Great-Britain."[15]

In August 1774, Doolittle advertised that "having erected a suitable building and prepared an apparatus for Bell-Founding, and having had good Success in his first attempt, intends to carry on that Business, and will supply any that please to employ him, with any Size Bell commonly used in this, or the neighboring Provinces, on reasonable Terms."[16]

Isaac Doolittle, however, did not confine himself to making clocks and casting bells. In March of 1776, having received authorization from the Connecticut Council of Safety, he entered into a partnership with Jeremiah Atwater to build a powder mill in Westville, near New Haven. Over the next several months he supplied gunpowder to the nearby towns of Stratford, Milford, and of course, New Haven. By August, the Council of Safety had paid them £110 14s. for 4,100 pounds of this scarce and valuable commodity. They continued to operate their powder mill throughout the Revolutionary War. Doolittle, the man of many talents, was also appointed to a committee in June 1776 to search for viable lead mines within the colony.[17]

Doolittle's shop on Chapel Street was close to Yale (see figure 2.3), and readily accessible to Bushnell. Twenty years his senior and a well-known and respected mechanic, Doolittle would have been the ideal individual to discuss methods of producing Bushnell's complex designs. Doolittle was committed to American independence and there would have been little apprehension on Bushnell's part about sharing his vision of a submarine. When not in New Haven, however, Bushnell would return home to Saybrook where he could proceed with other aspects of the *Turtle*. There he apparently enlisted the help of another clockmaker and mechanical genius, Phineas Pratt. Pratt's association with Bushnell and the *Turtle* was not recorded until long after his death.

PHINEAS PRATT (1747–1813)

In 1864, the Reverend F. W. Chapman published *The Pratt Family: or the Descendants of Lieut. William Pratt, One of the First Settlers of Hartford and*

Saybrook. Chapman noted that: "He [Phineas Pratt] was a soldier of the Revolution. His name is found on the pay-roll of Captain Daniel Platt's company of the 7th Regiment, in the State of Connecticut, commanded by Colonel William Worthington, as one who marched as far as New Haven, on April 27, 1777. He served in the capacity of a soldier until solicited by Mr. David Bushnell, inventor of the American Turtle, to aid him in the construction of that machine. For his services in its construction he was released by the commander-in-chief . . ."[18]

Pratt opened his clockmaking business in the Potapaug section of Saybrook (now the town of Essex) about 1768, continuing long after the war.[19] He later worked with his son Abel developing the machinery used in the production of ivory combs. Chapman's book includes a description of Pratt's involvement with Bushnell during the attacks against the British. In 1870 Pratt's son, also Phineas, wrote an account of his father's association with the *Turtle* in a letter to the *Boston Journal of Chemistry*. This letter is in typescript format and may have been transcribed from an original handwritten version.[20] Excerpts from this letter and Chapman's book are included in chapter 21.

Another typescript document concerning Phineas Pratt mentions his early training as a ship's carpenter and goldsmith. The document is undated and is not signed but was written by one of his sons. From the information within the text, we feel that Pratt's son Phineas wrote this document at about the same time as the letter. The information about the elder Pratt's training sheds light on the skills he may have provided Bushnell.

It is our belief that while David Bushnell was at Yale, he entrusted the construction of the *Turtle* hull to his brother Ezra. Yet Ezra, the farmer, could not have accomplished the task alone, a task that would have to combine the skills of a shipwright and blacksmith. The Bushnell and Pratt families lived only about four miles apart, and with this proximity, likely had a long relationship, and in particular between Phineas Pratt and Ezra Bushnell, who were nearly the same age. In the undated document, the author noted that "he [the elder Phineas Pratt] followed the calking business in summer, and goldsmith business in winter. He served an apprenticeship at the ship carpenter shop of Dea. Abner Parker of Essex, Conn." Later in the document: "He served in the army near Boston till applied to by Mr. D. Bushnell to build the torpedo. He accepted and took part of the risk and expense and was promised a large reward if successful."[21] Our interpretation is that after Lexington and Concord, Phineas Pratt left Ezra Bushnell and the *Turtle* to join his militia unit heading to Boston. Once there, David Bushnell requested Pratt's commanding officer to allow him to return and help with the *Turtle*. Another Pratt family member, Phineas's father Aziriah, was a blacksmith and would have had the skills necessary to supply the metalwork for the *Turtle* hull.

KEEPING HIS SUBMARINE VESSEL A SECRET

Bushnell, who was constantly concerned with maintaining secrecy, relied on a few trusted friends to communicate his concepts to influential supporters. Tory sympathizers were everywhere, and even the local postmasters were not to be trusted. Correspondence was often intercepted and the contents diverted to the British. Bushnell's concerns were warranted, as evidenced in the various correspondence regarding the submarine in 1775.

Early in August, Reverend John Lewis, a tutor at Yale and familiar with Bushnell's interests, wrote a brief description of the vessel to Ezra Stiles, then Pastor of the Second Congregational Church in Newport, Rhode Island. In what is presumed to be an effort to obscure the details from unwanted eyes, Lewis penned the description in Latin, a second language to scientists and clergy, but likely to be dismissed by the local post rider as simply religious chatter from one reverend to another:[22]

> *Hic homo est machinae inventor, quae ad naves Bostoniae partu pulveris pyrii explosione destruendas, nunc est fabricata et fere perfecta. Machina ita est formata, ut 20 aut amplius pedes sub undas celeriter transeat, et pulveris pyrii 2000 lb. portare et navis carinae infigere possit. Statim vel post minuta decem vel semi-horam secundum operatoris voluntatem, horologium totam massam inflamabit.*

[This man is the inventor of a machine which is now made and almost perfected for the destruction of the fleet in the harbor of Boston, by the explosion of gunpowder. The machine is so constructed that it can move rapidly 20 or more feet under water, and can carry and attach to the hull of a ship two thousand pounds of gunpowder. A clock work will ignite the whole mass either immediately or in ten minutes or in half an hour, according to the will of the operator.]

During the summer and fall of 1775, Benjamin Gale, Bushnell's most ardent supporter, corresponded with his friends Silas Deane and Ben Franklin, two very influential members of the Continental Congress in Philadelphia. On August 7, the same time Lewis's letter was en route to Stiles, Gale sent a lengthy description of the submarine to Franklin, noting that "I have been Long Urging him for permission to Acquaint You with these facts. He at Length has Consented with this Condition that I request You would not Mention the Affair Until he has made the Experiment, when Completed, if Agreeable I will Acquaint You with the Experiments he makes with it down to Boston." It is apparent that Bushnell had intended to make his initial trials against the British fleet in Boston Harbor, and would only allow Gale to communicate with Franklin under the condition that secrecy would be maintained.

To honor Bushnell's wishes, Gale sought some degree of assurance from Franklin that there would be no leaks from members of the Congress. Gale, however, understood that the word might already have leaked, as evidenced in his letter to Franklin: "Your Congress doubtless have had intimations of the Invention of a new machine for the Destruction of Enemy Ships, but I sit down to Give you an Account of that Machine and what Experiments have been already made with it."

On August 10, 1775, while in Philadelphia, Silas Deane wrote to Gale telling him: "The Congress as you have heard make but a short recess; before the expiration of which, pray favor me with a line, and say what ground there is for the report of a certain new invention for destroying Ships. You are in the neighborhood, and therefore presume you can give me the particulars." Deane was telling Gale that word of "a certain new invention" had circulated among members of Congress. Their curiosity was building, but it would not be until November when Gale replied with the "particulars."

Bushnell had graduated in July and was increasing the tempo of his submarine construction. It would be a tight schedule if the *Turtle* was to be used against the British fleet off Boston, but that was his plan. However, by October this machine for the destruction of ships had not yet arrived on-site. Ben Franklin had traveled through Connecticut on his way to confer with Washington, arriving on October 18. While en route, Franklin visited Gale where they no doubt discussed Bushnell and the *Turtle*, and likely carried this information with him. Samuel Osgood, writing to John Adams from Roxbury, Massachusetts, on October 23, noted: "The famous Water Machine from Connecticut is every Day expected in Camp; it must unavoidably be a clumsy Business, as its Weight is about a Tun. I wish it might succeed [and] the Ships be blown up beyond the Attraction of the Earth, for it is the only Way or Chance they have of reaching St Peter's Gate."[23]

Benjamin Gale had assumed, however, that Bushnell had finally made the trip to Boston and in his November 9 letter to Silas Deane, he felt that he could "now write with greater freedom, as I conclude that by the time this reaches you the machine will be in the camp." This may have been a dangerous assumption, as it soon became evident that news of Bushnell's secret weapon had found its way to someone in the network of spies that New York Governor Tryon had established. Whether it was Gale's letter that had been intercepted or some other source, it was only a week later that Tryon dispatched a note to Vice Admiral Molyneux Shuldham. "The great news of the day with us, is now to Destroy the Navy. A certain Mr Bushnell has completed his Machine, and has been missing four weeks, returned this day week. It is conjectur'd that an attempt was made on the ASIA, but proved unsuccessful. Returned to New Haven in order to get a Pump of a new construction which will soon be compleated."[24] If Bushnell had, in fact, made an attempt against the *Asia*, the event would have occurred in New York. The *Asia* had arrived in Boston in December 1774, but was sent to New York the following May and remained there throughout 1775. The *Asia* was involved with the naval engagements around New York during the summer of 1776 and was there when the *Turtle* attacked the *Eagle*.

The brevity of Tryon's message may indicate that he and Vice Admiral Shuldham were already aware of the general nature of Bushnell's vessel, but were not taking the threat seriously. The information provided by Tryon's spy that the pump had failed only reassured the British that there was little to be concerned about.

Benjamin Gale also became aware of the problem with Bushnell's pump. On November 22, he updated his previous letter to Deane: "The forcing pump made by Mr. Doolittle [of New Haven], not being made according to order given, did not answer; which has delayed him." Issues associated with the British intelligence network continued to worry Gale. He was particularly suspicious of the local tavern keeper, where the post rider was required to leave the mail. In his December 7 letter to Deane, Gale vented his frustration that his letters "are all intercepted;

and if I send any, I am obliged to deliver them out of town or have them superscribed by some other person."

In spite of the obvious need to minimize public awareness of Bushnell's machine, rumors of a unique weapon being built to sink British ships were bound to spread. Delegates to Congress likely questioned the rumors and may have discussed the ideas with members of the military. Such rumors may have inspired Captain Daniel Joy to propose a similar vessel to the Pennsylvania Committee of Safety in January 1776, for use in the Delaware River.

> I think an Engin may be made with Copper and Glass Windows (Bulls Eys) in it, properly Ballast'd, so as a man may convey himself (by the help of a compass & a Candle lighted the first to direct him to his object, the other to let him know when he wants a supply of air), under a Man of War's Bottom, & to take with him two or three vessels, charged with 2 or 3 hundred wt of Gunpowder—These vessels to have machinery Locks that may be sett to any time, & on the upper part a pair of Jaws (to go by a spring) large enough to receive a Man of War's kiel.
>
> Then to proceed unto two other Men of War, leveing one at each. Then to proceed on shore—when the first blows up it will be looked upon as an accident, the second will cause doubts, and the third confusion.
>
> If any of the above hints proves of any advantage to the Publick, I shall be happy in executing it, and shall not think my time spent in vain.[25]

Joy's design bears a striking resemblance to Gale's early letters regarding the mine. To Benjamin Franklin (August 7, 1775), Gale writes: "It is so Contrived, as when it strikes the Ship . . . it Grapples fast to the Keils . . . the Powder is fired by a Gun Lock within the Cask which is sprung by a Watch work, which he can so order so as to have that take place at any Distance of Time he pleases." Then to Silas Deane (November 9, 1775): "Three magazines are prepared; the first, the explosion takes place in twelve,—the second in eight,—the third in six hours, after being fixed to the ship. He proposes to fix these three before the first explosion takes place."

Chapter Five

HOW TO FEED A TWO-TON TURTLE

FUNDING FROM THE CONNECTICUT COUNCIL OF SAFETY

On the eve of the Revolution, each colony created what was termed a "Committee of Safety" or, as with Connecticut, a "Council of Safety," to prepare, organize, and fund the limited resources in anticipation of war with Britain. Connecticut established an advisory committee in May 1775. By July the committee became the Council of Safety, that was "to assist his Honor the Governor when the Assembly is not sitting, to order and direct the marches and stations of the inhabitants inlisted and assembled for the special defense of the colony, or any part or parts of them, as they shall judge necessary, and to give order from time to time for furnishing and supplying said inhabitants with every matter and thing that may be needful to render the defense of the Colony effectual."[1]

While there is no record of the Pennsylvania Committee of Safety having accepted Joy's proposal to build a submarine, Bushnell had a bit more success with Connecticut's Council of Safety.

> Friday, 2d February [1776] A.M. . . . Mr. Bushnell was here, by request of the Governor and Council, and gave an account of his machine contrived to blow ships &c., and was asked many questions about it &c. &c., and being retired, on consideration, voted, that we hold ourselves under obligations of secrecy about it. And his Honor the D. [Deputy] Governor is desired to reward him for his trouble and expense in coming here, and signifie to him that we approve of his plan and that [it] will be agreeable to have him

proceed to make every necessary preparation and experiment about it, with expectation of proper public notice and reward.[2]

The following morning, the Council of Safety agreed that

some encouragement should be given to enable Mr. Bushnell to pay expenses incurred in preparing his machine for the design projected &c., and to carry forward the plan &c. &c., it appearing to be a work of great ingenuity &c., and a prospect that it may be attended with success, and being undertaken merely to serve the public, and of considerable expense to labor &c., its tho't reasonable that something should be done &c.[3]

What the Council considered reasonable was to order the Treasurer to provide Bushnell with £60. While £60 was a considerable amount of cash in 1776 (a year's pay for a lieutenant), it would by no means compensate Bushnell for his efforts. Patriotism no doubt enabled Bushnell to acquire help from many sources, but there were bills to pay. Years later, when writing to Thomas Jefferson, Bushnell lamented that after his unsuccessful attacks in New York: "I was unable to support myself, and the persons I must have employed, had I proceeded."

WAGE AND PRICE REGULATION

To gain an appreciation for the true cost of the critical components that Bushnell had to secure, a comparison can be drawn between these costs and the wages the average inhabitant of Connecticut would earn during the Revolutionary War period. Because of the scarcity of commodities and the demand placed on them during the war, wholesalers and retailers were tempted to engage in price gouging and monopolies.

In November 1776, the Connecticut General Assembly passed "an Act to prevent Monopolies and Oppression by excessive and unreasonable Prices for many of the Necessities and Conveniences of Life." The price regulation included beef, pork, wheat, rye, "Indian Corn," salt, sugar, tea, molasses, and of course, rum. The law also set the wages paid to farm labor: "That for the future the Price for Labor in the Farming Way, in the Summer Season, shall not exceed *Three Shillings per Diem*, and so in the usual Proportion at other Seasons of the Year."[4]

The wholesale price of a gallon of cheap New England rum was set at three shillings sixpence, slightly more than half the price of the best quality rum from the West Indies. By 1778, a gallon of New England rum had increased nearly fourfold to twelve shillings. Restrictions were also placed on retail prices, whereby "no Innholder in this State shall receive more than fifty per Cent. Advance on the wholesale price of any Liquors, or any foreign Articles herein stated, and by them sold in small Quantities."[5]

The General Assembly also set the wages for the regiments being formed in response to requests from the Continental Congress. In May 1776, as George Washington prepared for an assault on Boston, Connecticut passed the following resolution: "An Act for raising a Battalion or Regiment of Troops within this Colony to march to Boston or elsewhere in Pursuance of a Requisition now made by the Hon[orable] Continental Congress for the Defense of the United Colonies." This resolution included: "That the pay of each sergeant shall be 48s., each corporal 44s., each drummer 44s., each fifer 44s., and each private 40s. per calendar month." Officers fared much better, a captain receiving £8 and a lieutenant £5 per month.[6]

The military was also eager to encourage the troops by adding alcohol to the soldiers' diet. In May 1775, the General Assembly passed a resolution "that the allowance of said troops be as follows, viz.: three-quarters of a pound of pork or one pound of beef, and also one pound of bread or flour with three pints of beer to each man per day." The General Assembly also considered the troops' needs during tough times, adding "also one jill [or gill, four ounces,] of rum to each man upon fatigue per day, and at no other time."[7]

BUSHNELL'S NEED FOR GUNPOWDER

For a century and a half, the local militias had served alongside British Regulars during the colonial wars against the French, but were generally poorly supplied with military equipment. Most of the arms consisted of obsolete British muskets and equipment captured from the French. Weapons were also purchased abroad, but were typically old and in some cases unserviceable arms stored in European arsenals. The British had discouraged local manufacturing of any kind, preferring to export shiploads of raw materials from the colonies and return with British manufactured goods. As a result, there were few industries in America that produced military equipment.

Of particular note was the scarcity of gunpowder.[8] Bushnell had been experimenting with small quantities of gunpowder during his early days at Yale. The composition and handling of gunpowder was well-known to those involved with its production, and Bushnell would have had no need to study the science behind its explosive properties. Had it been of interest to him, he could have turned again to the scientific literature. In the *History of the Royal Society*, Thomas Sprat included two articles by a Society member describing the production and mixing of the components of gunpowder.[9]

Bushnell's intent was to prove his contention that gunpowder would detonate underwater and that the explosive forces would penetrate, or at least severely damage a ship's hull. He described these early experiments in his letter to Jefferson.

> The first experiment I made, was with about two ounces of gun powder, which I exploded 4 feet underwater, to prove to some of the first personages in Connecticut, that powder would take fire under water.
>
> The second experiment was made with two pounds of powder, inclosed in a wooden bottle, and fixed under a hogshead, with a two inch plank between the hogshead and the powder; the hogshead was loaded with stones as deep as it could swim; . . . A match was put to the priming, exploded the powder, which produced a very great effect, rendering the plank into pieces; demolishing the hogshead; and casting the stones and the ruins of the hogshead, with a body of water, many feet into the air, to the astonishment of the spectators.
>
> I afterwards made many experiments of a similar nature, some of them with large quantities of powder; they all produced very violent explosions, much more than sufficient for any purpose I had in view.

After demonstrating his idea of sinking ships with an underwater explosion to the "first personages of Connecticut" (Bushnell does not mention their names), it now became essential that he have access to this very valuable resource. Bushnell may have been able to acquire a few ounces of gunpowder from friends to support his initial testing. However, as he increased the magnitude of his demonstrations, these sources may have become reluctant to supply him with those "large quantities of powder." Regardless of Bushnell's reputation and influential friends, it is unlikely that the New Haven militia could have been convinced to spare any significant amount. Purchased from local merchants, gunpowder was an expensive commodity.

In a meeting between the governor and the Council of Safety on July 2, 1776, it was taken under consideration that "the price at which gunpowder at our mills should be sold when disposed of &c, and allowed and voted that for the present the price be fixed at five shillings and four pence [per pound]." At that same meeting, authorization was made to "receive and purchase at Elderkin and Wales's mill eight hundred pounds of gun-powder." In December 1775, Elderkin and Wales had been authorized by the General Assembly to build a powder mill in Windham and were well established in the business by the following spring.[10]

Prior to 1776, however, Bushnell would have had a difficult time acquiring his supplies of gunpowder. If we use the regulated price of 5s. 4d. per pound, his two pound experiment would have taken 10s. 8d. out of his pocket . . . more than three days pay for a local farmhand. At some point along his submarine warfare development, Bushnell needed government support as well as a few shillings from well-heeled friends. As noted in chapter 4, Isaac Doolittle, who was providing many of the mechanical components for the *Turtle*, was in the gunpowder business and could very well have set aside a sufficient quantity to supply Bushnell's needs. Yet for each of his mines, Bushnell would eventually need to acquire 150 pounds of gunpowder; valued at £40, this would have been a financially daunting task.

BALLAST LEAD FOR THE *TURTLE*

Bushnell to Jefferson: "The vessel was chiefly ballasted with lead fixed to its bottom; when this was not sufficient, a quantity was placed within, more or less, according to the weight of the operator; its ballast made it so stiff, that there was no danger of oversetting . . ." Bushnell also mentioned that about two hundred pounds of this could be jettisoned in an emergency. Benjamin Gale, writing to Silas Deane November 9, 1775, added: "His ballast consists of about 900 wt. of lead which he carries at the bottom and on the outside of the machine, part of which is so fixed as he can let run down to the Bottom, and serves as an anchor . . ."

After committing several hundred pounds of gunpowder, Bushnell also had to divert about half a ton of lead to his project that would otherwise have become cartridges for the Connecticut regiments. Lead was a bit more plentiful than gunpowder, and was found in at least three locations in Connecticut. Mines had been established in Farmington and New Canaan, but by far the most productive source was in Middletown, where a furnace had been built to refine the ore. On July 2, 1776, the Council of Safety "voted that the price of lead for the present be 6*d*. per pound." The Council also authorized the purchase "at the furnace at Middletown one thousand pounds of lead, at the price of 6*d*."[11] Middletown is located along the Connecticut River where lead could be readily transported to the coastal towns, including Saybrook, i.e., Bushnell.

At sixpence per pound, Bushnell would have been able to provide all of his ballast lead for about £25. This was no small piece of change and, as with the gunpowder, perhaps Bushnell would have had to "beg, borrow, or steal," and he and his associates may have used a combination of all three to satisfy this requirement.

CASTING THE "BRASS CROWN, OR COVER"

Copper had been used for weapons, tools, and decorative items for thousands of years. Eventually, it was discovered that combining copper and tin, readily available in Europe, would produce a much more durable alloy known as bronze. Also referred to as bell metal and gun metal, bronze was a metallic composition consisting primarily of copper alloyed with 8–10 percent tin. Lead was sometimes added to help the molten mixture flow into the molds.[12]

It wasn't until the thirteenth century that refinements in alloying technology led to the production of brass. The naturally occurring ore known as calamine was mixed with molten copper in various proportions, along with powdered charcoal, to create compounds with different properties. Calamine contains zinc carbonate (now known as the mineral smithsonite) often found in association with naturally occurring copper, but it wasn't until the late eighteenth century, that zinc was found to be the element in calamine that produced brass. By the turn of the century, metallic zinc had become the commodity of choice in brass foundries. The typical proportions were about two-thirds copper and one-third zinc. The copper to zinc proportions of modern brass range from 65% Cu and 35% Zn (yellow brass) to 85% Cu and 15% Zn (red brass). Modern bronze alloys typically contain 5%–10% tin.

Copper was mined in several of the provinces, including Massachusetts, Pennsylvania, New Jersey, and Connecticut. The copper mine in East Granby, Connecticut, then a part of Simsbury, produced a large quantity of copper ore. Yet, while there was a smelting operation near the mine, the majority of the copper ore was exported to England. The refined copper was then imported back to the colonies, along with zinc, tin, and brass. The operations at Simsbury did produce copper coins, but the copper mine is most well-known for having been converted to Newgate Prison in 1773. During the Revolutionary War, British captives mingled with criminals, but by far the majority of the prison population consisted of Connecticut citizens convicted of Tory activities. The prison reputation was such that George Washington, having convicted several "atrocious villains" under a court-martial, had "sentenced [them] to Simsbury in Connecticut."[13]

Although ironware was most often found in colonial homes, coppersmiths did produce some items for the kitchen, and a few brass foundries sprung up to cast pots, kettles, and an occasional church bell. With the approaching inevitability of war with Britain, a new urgency arose for a product that had always been imported and had not been produced in America: artillery. By the end of 1775, Pennsylvania's Governor William Penn observed that "the casting of cannon, including brass, which were cast in Philadelphia, had been carried to great perfection."[14]

While brass and bronze, because of their durability, were preferred for artillery, iron was readily available and foundries were established that could handle the large castings. One of the most productive was Connecticut's Salisbury Furnace. During its state-run operations, nearly 850 cannons were cast for the Continental army and navy as well as for privateers and the Connecticut militia. Mortars, swivels, hand grenades, and round shot of every caliber were also cast there during the war.[15] One might also recall that it was the Salisbury Furnace that received 326 gallons of rum in 1777.

The Continental Congress made every effort to encourage the establishment of brass foundries for producing cannons. After the British evacuated Boston, a foundry was established in New York under the supervision of Colonel Henry Knox. The founder, James Byers, soon began remelting the unserviceable brass cannons left behind by the British, but also used any old copper and brass that could be had, much of it appropriated from local sources. According to Knox, "Mr Byers . . . tells me that he has no doubt but in a little time he could procure Copper enough to cast 80 or 100 six pounders" and that "there are a great number of Stills which are only a pest to society which ought to change their form."[16] Knox may not have approved of the rum consumption at the Salisbury Furnace, in spite of its use to encourage cannon production.

Considering the scarcity of raw materials and that whatever could be had was converted into a cannon, Isaac Doolittle's bell foundry, established in 1774, may have had trouble staying in business. It was likely that he retained a supply of scrap metal in anticipation of the occasional order for a church bell. If, as we believe, Doolittle agreed to cast Bushnell's "brass crown," the hatch may very well have been produced with bell metal rather than brass. The distinction would have been irrelevant to Bushnell.

STEEL FOR "THE TIP OF THE SPEAR"

Located just in front of the pilot was the mechanism for attaching Bushnell's mine to the underside of his target. It was what he referred to as a wood-screw that by "pushing it up against the bottom of a ship, and turning it at the same time, it would enter the planks." The entire mission depended on the pilot being able to thread this screw into a ship's hull, thus it was absolutely critical that the wood screw maintain its sharpened point. Bushnell would not have resorted to the locally produced iron used for producing the average woodworking screws. High-quality steel would have been essential.

In the 1778 price regulations, the "best American manufactured steel fit for edged tools" was set at £2 per pound, probably a bit more than would have been the case in 1776.[17] A source of "free" steel was the British. The extremely high cost of imported steel forced gunsmiths to rely on "scrapped bayonets, swords, or other edged weapons and implements made of quality crucible steel."[18] Prior to the Revolution, militia units were often supplied with obsolete British weapons, including the bayonets that were of exceptional quality. The standard issue triangular bayonet blades were on average seventeen inches long. The socket that fit over the muzzle of the barrel and the first few inches of the blade were iron. The forward two-thirds of the blade, however, was made of steel. Figure 5.1 shows a standard Pattern 1768 British bayonet alongside another, shortened by American troops; the blacksmith would have reused the steel that had been removed.

Another source of steel may have come almost from Bushnell's backyard. Benjamin Gale, Bushnell's mentor and friend, had been working

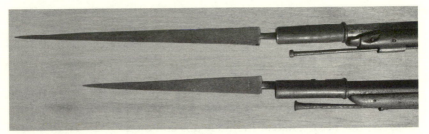

Figure 5.1. British bayonets used during the Revolutionary War. (*Roy Manstan*)

with his father-in-law, Jared Eliot, in the iron ore mining and smelting business. Both lived in the adjacent town of Killingworth. Eliot had received a prestigious gold medal award from the Royal Society of London for the book he published in 1762: *An Essay on the Invention of Art of Making Very Good, if Not the Best Iron, From Black Sea Sand.* In 1776, Gale described the production of high-quality steel in a letter to Ezra Stiles: "We carry on our own Steel manufacture very brisk, have made six tons of Good Steel since ye 25th last may. It is now in high reputation in New York, will sell preferable to English steel . . ."[19]

Cutting threads in a steel screw would have been a reasonably straightforward procedure for a skilled eighteenth-century mechanic. An early discussion of eighteenth-century manufacturing and metal trades can be found in Moxon (1703). Moxon described the craft of "Smithing," including the hardening and tempering of steel. He also included the use of lathes for turning wood, ivory and metals, and the procedures for creating screw threads. See also Bishop (1868), Lathrop (1909), and Mulholland (1981) for discussions of colonial metals and metalworking.

GLASS FOR THE *TURTLE*

To satisfy the popular interest in glass making during the mid-eighteenth century, a two-part article describing the composition of glass, the construction and operation of a glassworks, and the production of glass articles appeared in the September and November 1747 issues of *The Universal Magazine of Knowledge and Pleasure* (see figure 5.2). That same year, the magazine published articles about the development of the barometer by Torricelli, Boyle, Hooke, Huygens, and Des Cartes; electrical experiments by Hauksbee and Watkins; and the studies associated with the elasticity of air, including Hauksbee's vacuum-pump design.

The magazine brought to the attention of the general public the innova-
tive thinking that members of the Royal Society had been publishing in
the scientific journals.

Glass was of little consequence to the military during the
Revolutionary War, other than providing the general staff with a proper
container for serving imported Brandy. For Bushnell, however, its importance cannot be overstated. Without this commodity, the *Turtle* would
not have left the drawing board. Glass was used for two functions in his
submarine: windows and the depth gauge.

THE *TURTLE* HATCH WINDOWS

Bushnell: "There were likewise several small glass windows in the crown,
for looking through, and for admitting light in the daytime . . ."
Windows may seem irrelevant in a submarine designed to be used at
night, but they served an essential purpose. The majority of the training
that Bushnell provided to his pilots occurred during the day. Without the
windows, the pilot would have been in total darkness with the hatch
closed. It would have been extremely difficult to become proficient in
operating the complex assortment of mechanical systems. On the surface, he could have practiced with the hatch open, but when submerged,
the windows allowed enough light to enter the hull for the pilot to train.
Lee commented that "it had six glasses inserted in the head, and made
water tight, each the size of a half Dollar piece, to admit light—in a clear
day, a person might see to read in three Fathoms of water." The windows
were small but sufficient to illuminate the interior. Bushnell was fully
aware that repetitive training was essential to efficient operation of his
vessel.

Window glass was the bread and butter of colonial glassmakers. The
Crown had restricted the production of glassware in the colonies to util-
itarian objects such as window glass, cheap bottles for commercial use,
and some items for medical and apothecary applications. All other glass
was to be imported from British suppliers, including the high quality
items used for scientific study. To get around this restriction, the glassmakers produced much of their specialty glass after-hours, and rarely
advertised it for sale.

The glass used in the *Turtle* hatch would have to be thicker than the
household windows produced by colonial glassmakers. While glassblowing techniques were used to produce most glass items including window-

panes, thick glass was cast. For heavy plate glass, the molten glass was poured into a form and rolled flat.[20] For Bushnell, who installed his "windows" primarily to provide illumination during daytime training, optical purity was not essential.

William Liebknecht, archaeologist with Hunter Research Inc. (personal communication), suggested that a household article known as a "linen smoother" might have served Bushnell's purposes. He pointed out that an example of a linen smoother had been found at the site of the Germantown Glassworks, a Massachusetts manufactory destroyed by fire in 1755. The glassworks was rebuilt, but once again was burned to the ground, as evidenced in a notice placed in the February 20, 1769, edition of the *Boston Gazette and Country Journal* where the "Proprieter . . . is himself unable to renew that Manufacture . . ." The output from the Germantown Glassworks was primarily window glass, but a remarkable variety of bottles and household goods has been found at the site.[21]

The linen smoother from Germantown is a disc about three inches in diameter.[22] Although its thickness is not given, it is apparent from the illustration that it would have been sufficient to be adapted as a *Turtle* window. They were available in a variety of sizes and shapes; some from clear imported glass, others, like the Germantown example, were of green bottle glass. Such an item would have been available for sale at a local shop, and readily ground into a size adaptable to Bushnell's purposes.

For a free source of glass, however, he might have approached his Yale professors or fellow students. Bushnell and his friends would have brainstormed the topic, finding creative ways of providing light to the interior of the hull. Optical glass used and abused by students for sixty years may have been scratched and chipped beyond its useful life. This is truly speculation, but "borrowing" a half dozen thick glass discs from among old and obsolete optical instruments would have enabled Bushnell to move on to the next issue. No one would have allowed this to become a showstopper.

GLASS FOR THE DEPTH GAUGE

More obvious to the operation of a submarine was a requirement for a depth gauge. Bushnell adapted the idea of a "water barometer" as described in the pages of the *Transactions of the Royal Society*, the details of which are discussed in chapter 10. The glass for this depth gauge would have been more difficult to obtain than his windows. Bushnell described it as "a glass tube eighteen inches long, and one inch in diameter, standing upright, its upper end closed, and its lower end, which was open, screwed into a brass pipe, through which the external water had passage . . ." A piece of cork would rise and drop in proportion to the water depth. In order for the pilot to observe the depth in darkness, the cork was topped with bioluminescent "foxfire."

Our assumption is that it would have been difficult to have a glass tube produced exactly to his specifications. There were no glass producers in New England when Bushnell was designing his submarine. As mentioned above, the Germantown Glassworks in Braintree, Massachusetts, closed in 1769 after having burned down a second time. The nearest glassworks was in New Jersey. One might speculate, therefore, that Bushnell and his friends returned to the basement of Yale College to peruse obsolete scientific instruments in storage.

The article "An Abstract of M. Abbé Nolet's Essay on the Electricity of Bodies" that appeared in the August 1747 issue of *The Universal Magazine of Knowledge and Pleasure,* provides a hint as to an apparatus that might have been adapted for Bushnell's depth gauge. The following is an extract from that article.[23]

> To electrify a body, Dr. Hauksbee, an English physician, introduced the use of the glass tube, and afterwards of a ball or globe of the same metal [glass was often referred to as a metal]; which has been the most common instrument in electricity ever since.
>
> This tube, says Abbé Nolet, must be three feet long, or thereabouts, an inch or 15 lines diameter, and a full line thick.[24] It must be very straight, stopped close at one end, and made to open with a stopper at the other, to keep out all damp and dust, and at some times to let in fresh air.
>
> Your English or Bohemia chrystal glass is accounted the best to make these tubes and globes: but, says the Abbé I have observed that the glass of which wine bottles are made, will electrify very well.

It was a keen interest in science that inspired Ben Franklin to pursue many experiments that he would have read about as a young man. In 1747, Abbé Nollet wrote to Franklin about a device to measure electricity that he referred to as an "electrometer" or more appropriately an "electroscope."[25] Why was this of importance to Bushnell's glass?

Franklin was a supporter of the most successful glass-producing enterprise in the colonies. His friend, Caspar Wistar, established a glass manufactory in 1738 in the town of Alloway, New Jersey. In the 1740s Franklin, and various colleges, obtained glassware from the Wistarburgh Glassworks to conduct scientific experiments. According to Arlene Palmer, "Franklin corresponded with a number of prominent colonists about electrical science and undertook to provide individuals as well as institutions with apparatus, including tubes. In one of these letters, Franklin mentions that the Wistar-made tubes have a 'greenish cast, but [the glass] is clear and hard, and, I think, better for electrical experiments than the white [colorless] glass of London which is not so hard.'"[26] Fragments of glass tubes of varying diameter have been found during archaeological excavations at the site of Wistar's glass manufactory (Liebknecht and Tvaryanas 1999; Liebknecht, et al., 2004).

Figure 5.2. The long cylinder Bushnell adapted for his depth gauge would have been manufactured using the glassblowing techniques illustrated here. (*Universal Magazine 1753*)

Franklin also corresponded with Peter Collinson in England. Collinson was responsible for having Franklin's discoveries published in a pamphlet, *Experiments and Observations on Electricity* (London, E. Cave, 1751). In addition to his source of Wistar glass, Franklin obtained many instruments for his scientific investigations from England including a "glass tube for generating static electricity."[27]

What can be inferred here is that a long glass tube was the experimental instrument of choice until glass spheres became popular. While the abbé preferred the expensive crystal glass available in England, tubes produced from common glass of the type used for wine bottles would also suffice. By the 1770s, Yale may very well have retired its cylindrical apparatus (if it had one) to the basement in favor of the spherical variety for carrying on electrical experiments.

If one of these tubes, built in accordance with Abbé Nollet's description, was being used at Yale, it would have been the ideal solution to

Bushnell's depth gauge design. The tube would have been one inch in diameter, necessary to accommodate the cork float, and would already have had one end closed. Other tubes created for scientific purposes, barometers for example, were much smaller in diameter. Even if Nollet's three-foot long tube had broken, only a sufficient length (eighteen inches for Bushnell) was needed to enable the pilot to read changes in depth. As we found with our replica, an eighteen-inch long depth gauge was ideal for installation in the *Turtle*.

Transport and handling logistics

When completed, Bushnell's vessel would have weighed well over three thousand pounds, yet he never describes the logistics that were required to handle the *Turtle*. The submarine was likely cradled on its side, making it easily concealed, whether transported overland by wagon, or in the cargo hold of a ship.

During the summer of 1775 after Bunker Hill, Washington and the hastily assembled Continental army occupied all of the land approaches to Boston. Without an American navy to blockade the harbor, the British had unrestricted access to the sea. Bushnell originally intended that his submarine would be carried to Boston where it would harass the British fleet. The most direct route was the Boston Post Road, crossing the Connecticut River at Saybrook, only a mile from where the *Turtle* was built and tested.

Transporting a nearly two-ton "experiment" over land would have been slow—but safe. British sympathizers were everywhere, but the army was essentially bottled up in Boston. Besides, had British regulars on patrol inspected a wagon carrying a submarine, none of them would have had any idea what they were looking at.

As a colony well entrenched in the maritime trades, handling heavy items was a routine activity in Connecticut. Merchant vessels would have been rigged to load and offload cargo, including quarried rock, pig iron for the forges, lead from the mines at Middletown, heavy cannons for coastal defenses, and the ubiquitous hogsheads filled with West Indian rum.

Salvaging valuable cargo was a lucrative business, encouraging industrious minds to invent a variety of "diving machines." Diving bells, around since the sixteenth century, had always been supported by vessels rigged for that purpose. The need to provide a sufficient volume of air to

Figure 5.3. The *Turtle* would have been deployed and retrieved over the side of the sloop with rigging similar to that used by John Lethbridge for his "diving engine." (*Universal Magazine 1753*)

the divers resulted in the "machine" being very buoyant. In order to submerge, the buoyancy had to be counteracted by an equal weight in ballast. This made the diving bell extremely heavy in air. Bushnell's submarine was no exception.

A description and illustration of the rigging used to handle what the author referred to as a "diving engine," invented by John Lethbridge in 1714 (see also the discussion in chapter 9), appeared in 1753 in the supplement to volume 13 of *The Universal Magazine*. The illustration (figure 5.3) shows a single masted vessel we estimate to be about forty feet long. The mast is rigged with a boom crane that can swing out over the port beam. Lines to an anchor and to the diver accommodate operations in a current. The whole assembly is raised and lowered with the block-and-tackle arrangement that was standard rigging used for load handling. The vessel shown here may have been a bit small to move the two-ton *Turtle* out over the port gunwale, but the handling procedures would have been the same.

Transport by sea would not have presented serious logistical problems. The risk to Bushnell, however, was that all vessels were susceptible to search and capture by the British, regardless of whether it was a warship or a merchant ship. The intercepted vessel, or "prize," would be brought into port and the cargo sold. Some vessels, if not sold, were converted for use by the British navy. Early in the war, however, Bushnell was apparently able to find a vessel of sufficient size to handle the *Turtle* with a crew willing to accept the risk.

PART TWO

REPLICATION

Our replica *Turtle* was christened by Connecticut Senator Eileen Daily on November 10, 2007, on the waterfront of the Connecticut River Museum, less than two miles from where David Bushnell first launched the original *Turtle* in 1775. (*Warner Lord*)

After a bit of time travel helped uncover and interpret the conditions that Bushnell would have faced during his years at Yale, we set about the task of replicating his vision of submarine warfare. To create a replica with any degree of credibility, we were obliged to think critically as to what we would have done in response to an inevitable war with the greatest military power in the world.

We pondered how he engineered each aspect of his submarine in accordance with the anticipated mission, how he selected and motivated his team, how he obtained the critical materials and components, how he managed to solve the logistical hurdles. Tackling these must have consumed all of his mental energy, yet while struggling with each issue, he managed to complete his education and ultimately create a working submarine.

The following chapters in part 2 describe the Old Saybrook High School *Turtle* Project, and the aspects of a replica that we considered critical to maintaining the form and function of the original vessel. They include the propulsion system; the hull shape; the ballasting of the vessel; the hatch; the ventilation and air supply; the navigation instruments and illumination; and the weapon with its attachment mechanism. The rationale used by project participants when replicating each component is described within its relevant historical context and according to the level of technology available in the eighteenth century.

The "we" in the previous paragraphs does not just refer to the authors. Over the life of the project, dozens of students were involved; several volunteers brought their insight and physical help to the classroom and waterfront; many other individuals donated time and resources; the Navy supported the project mission throughout; and feedback from each of the test pilots was essential. This may suggest that we had a solid advantage over Bushnell. We suspect, however, that he also interacted with fellow students, had patriotic individuals who volunteered their time and resources, had support from government sources (including George Washington), and depended on feedback from his pilots to understand and perfect his *Turtle*.

Chapter Six

THE
TURTLE PROJECT
Not Just Another Birdhouse

Replicating David Bushnell's sub-marine vessel

Our assumption was that with war with England on the horizon, all participants were prepared to turn David Bushnell's ideas into a reality. The local artisans and mechanics that Bushnell partnered with, including Isaac Doolittle of New Haven and Phineas Pratt of Essex (then a part of Saybrook), were known for their mechanical skill and ingenuity, skills that endured beyond the Revolutionary War. Their patriotism and dedication to American independence also contributed to their motivation to make the *Turtle* a success.

The concepts that Bushnell incorporated into his submarine were truly unique. His use of what was the world's first screw propeller is extraordinary. Among his many innovations were a depth gauge, an air circulation system, pumps and valves for ballasting, and check valves, in case one of his pumps failed. He incorporated a sophisticated timing mechanism for detonating his underwater mine, and a simple but effective way to attach the mine to the underside of the targeted warship. Because the *Turtle* was to conduct covert night missions, he applied a naturally occurring bioluminescent material to illuminate his compass and depth gauge.

From the beginning, the *Turtle* replica was intended to be capable of being launched and operated in actual at-sea conditions (figure 6.1). This requirement drove the decisions as to the design and materials used in its construction. All of the mechanical systems necessary to operate the original *Turtle* were included. They were built, however, in such a way that there would be no compromise in ensuring the safety of the pilots. For example, we used O-rings for all of the hull penetrations, modern pumps and check valves, scuba tanks for the pilot, and the impact-resistant polycarbonate "Lexan" for the windows and depth gauge instead of glass as used in the original *Turtle*.

The goal of the *Turtle* Project was to interpret the eyewitness accounts in light of the technical and scientific knowledge available in the eighteenth century, and create a replica true to the submarine as envisioned and implemented by its inventor. Once completed, the intention was to investigate the capabilities of the original vessel by subjecting the replica to a series of operational tests. The testing of the replica *Turtle* was primarily associated with propulsion, transiting, and maneuvering. Using a replica to evaluate the performance of Bushnell's vessel requires that the replica represent as closely as possible the original in its form and function. For example, using O-rings to ensure a watertight seal on the propeller shaft will not measurably affect the performance of the propeller. On the other hand, testing the *Turtle* with a modern three-bladed constant-pitch propeller, which bears no resemblance to the form and function of Bushnell's rudimentary two-bladed propeller, would give totally unrealistic results.

STUDENTS

The *Turtle* Project was one of many that Old Saybrook High School had created under the direction of Fred Frese, to expand the educational opportunities available to their students. The school's technical arts program is open to college-bound students and those considering entering the trades. The only criteria are motivation and creativity.

The students participated in all aspects of the *Turtle* Project, from laying the keel in 2003, to its christening in 2007, and operational testing in 2008. Each year new students would sign on to participate, and over the four-year program, dozens had contributed their energy to the *Turtle* Project. They cut and installed the *Turtle* hull planking, welded and fit the steel, and designed and built many of the mechanical components. Several students who began working on the *Turtle* in their freshman year continued throughout their four-year high school career. Figures 6.2 and 6.3 show the *Turtle* hull partially completed and one of the students finishing its steel deck and upper skirt.

Figure 6.1. The *Turtle* viewed from above during tests at the Mystic Seaport Museum. The upper skirt is just below the hatch and deck and above the upper barrel band. (*Jerry Roberts*)

The authors, working with one of the students, created the *Turtle* logo (figure 6.4). As a template, we used examples of the "dolphins" worn by navy submariners worldwide. When a sailor trains and qualifies to serve onboard a navy submarine, he is eligible to wear a pin designating one of the various categories of submarine service. In particular, we selected the "dolphins," associated with deep submergence vessels. We then created our project "dolphins" by adapting a drawing of our *Turtle*, along with illustrations of sea monsters from the sixteenth century.[1]

VOLUNTEERS

A complex project like this requires the energy and commitment that comes from volunteers. Some worked with students in the classroom,

Figures 6.2 and 6.3. *Turtle* under construction in 2004, left. (*Old Saybrook High School*) Old Saybrook High School student Joe Flammang, right, fitting the *Turtle*'s steel deck and upper skirt to the hull in spring 2007. (*Roy Manstan*)

while others designed and built critical components necessary for the *Turtle* to operate safely. Volunteers Ken and Bonnie Beatrice helped with the unique historical research necessary to build a truly representative replica. Their help was also critical during the operational testing.

One of the students who had worked four years on the *Turtle*, Joe Flammang, continued on with the project after graduating in June 2007. His contribution, along with that of his father, Paul Flammang, was essential to enabling the *Turtle* to meet its scheduled November launch.

Other volunteers were involved with the handling and transportation of the *Turtle* to the various venues. The *Turtle*, weighing well over three thousand pounds, requires significant effort and the proper equipment to be lifted, transported, and launched.

TEST PILOT SELECTION

The replica *Turtle* had from the start been intended to be seaworthy. There would, therefore, be a requirement to have individuals trained and qualified to operate the vessel. The pilot, in addition to being familiar with the *Turtle*'s mechanical systems, must be competent to use the safety equipment installed in the replica, in the event of an emergency situation. The project director, Fred Frese, agreed with the suggestion from the Navy coordinator, Roy Manstan, that pilots should be Navy-trained divers.

It was also decided that whenever the *Turtle* would operate in the water, there would be a team of safety divers onsite who would accompany the *Turtle*. The Naval Undersea Warfare Center (NUWC), as a part of the education partnership with the high school, agreed to provide divers from the Engineering and Diving Support Unit (EDSU) for the initial trials, for the christening and launch, and for the test program.

Figure 6.4. The *Turtle* project logo. (*Old Saybrook High School*)

The *Turtle* pilots were selected from this pool of EDSU divers. They included Roy Manstan, Paul Mileski, and David Hart. The test pilots, in addition to being Navy-trained divers, have dedicated their careers to working with military technology, and on projects where teamwork is essential.

SAFETY TEAM

Ken Beatrice, who has been a *Turtle* Project volunteer from the beginning, served as the project safety coordinator, and interfaced with the dive team during all operations. Beatrice, a NUWC retiree, had been a member of the EDSU dive team for over twenty-five years, serving as a diving supervisor until retirement (figure 6.5).

The EDSU dive team has been a component of NUWC since the 1960s. It is currently under the direction of Jack Hughes. Members of the EDSU are primarily civilian engineers and technicians on the NUWC staff, or drawn from the active duty military, who have been assigned to NUWC as mission specialists. There are also Navy Reservists who support the EDSU as part of their Reserve duties. Each individual on the EDSU staff, whether civilian or military, has been trained at one of the Navy's diving schools, primarily at the Naval Diving & Salvage Training Center (NDSTC), Panama City, Florida.

The EDSU provided a minimum of five divers onsite during each at-sea operation. Each team included a diving officer (either Jack Hughes or Vic Marolda), a diving supervisor, and three divers, two being in the water with the *Turtle* at all times. Occasionally, the situation did not require that divers be equipped with scuba, and they supported the testing as surface swimmers.

Figure 6.5. *Turtle* project safety officer Ken Beatrice briefs the navy dive team just prior to its christening and launch on November 10, 2007. The Connecticut River Museum is in the background. (*Bruce Greenhalgh*)

Chapter 1 sought to emphasize the many interpretations of the *Turtle* made by authors over two centuries. At the beginning of the twenty-first century, Rick and Laura Brown of Handshouse Studio, working with students from the Massachusetts College of Art, studied in detail the contemporary accounts written by and about Bushnell and the *Turtle*. Their interpretation resulted in a full-size, working replica (figure 6.6) that was launched into Duxbury Harbor near Boston in January 2003. The following March, their *Turtle* was brought to the Hydrodynamics Laboratory at the U.S. Naval Academy in Annapolis, Maryland where it underwent a series of tests.[2]

A quick comparison with our replica (figure 6.1) will find several distinct and important differences. In particular are the design and construction of their copper hatch and the vessel's nearly egg-shaped hull. Also significant is their placement of the vertical propeller, the wood screw mechanism, and the ventilation pipes above the profile of the hatch. Interpretation is simply a matter of what perspective the interpreter brings to a subject. The following chapters serve to explain the current authors' perspective and how we resolved various eighteenth-century design questions during the development and construction of the *Turtle* Project replica.

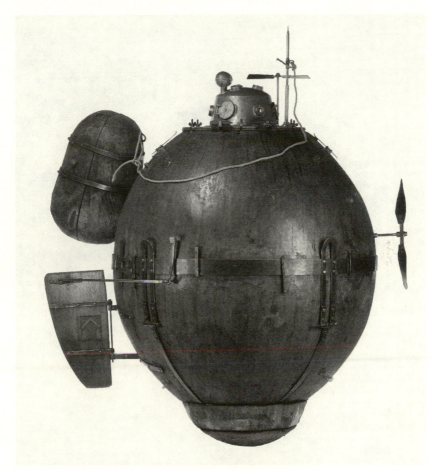

Figure 6.6. Profile of the *Turtle* replica built by Rick and Laura Brown of Handshouse Studio with students from Massachusetts College of Art. (*Cary Wolinsky*)

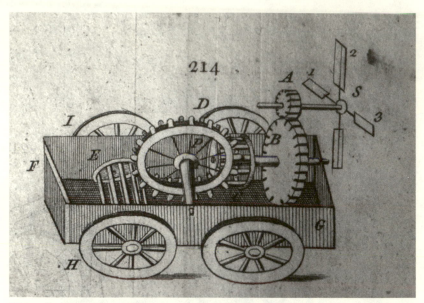

Figure 7.1. The improbable (or impossible) wind-driven "sailing chariot" illustrated in *The Principals of Mechanics* by Emerson (1773). (*Connecticut River Museum*)

Chapter Seven

PROPULSION

How to Move a Two-ton Turtle

Replicating the world's first propeller

The reliability of the *Turtle* Project test results relies on replicating the vessel as accurately as possible. The *Turtle*'s transiting speed, maneuvering, submerging, propeller thrust, and pilot endurance tests are all based on the propulsion system design, and the replica propeller in particular. As with hull shape, the form and function of our propeller had to be derived from the few written descriptions and from various assumptions on our part.

The bladed propeller that Bushnell conceived was his most unique contribution to the development of a practical submarine vessel. For that reason, we have attempted to trace the genesis of his idea by understanding the evolution of devices that converted linear forces (wind and water) to rotational motion and torque. From that we can make the jump, as we believe Bushnell did, to a device that would convert torque back into linear motion, i.e., a propeller.

Bushnell's propeller

Bushnell understood that to successfully apply his ideas of using underwater explosives, the mission would have to be covert; that his vessel carrying the mine would need to "approach very near a ship, in the night, without fear of being discovered." There were only two methods of propulsion available to him at the time: sails and oars. A practical man, he was faced with the dilemma of having an effective weapon without a means of delivery.

We will not delve into the history of submarines here, only to mention three previous attempts to produce a submarine vessel. The Reverend John Wilkins in his book *Mathematical Magic* described Cornelius Drebbel's use of oars in the 1620s. "As for the *progressive* motion of it, this may be effected by the help of several Oars, which in the outward ends of them, shall be like the fins of a fish to contract and dilate."[1] Giovanni Borelli in the 1670s and Dr. Denis Papin in the 1690s also employed oars. Sails, the other common mode of propulsion in the eighteenth century, were certainly not a consideration for submerged mobility, although thirty years after Bushnell's *Turtle*, Robert Fulton included sails on his *Nautilus* for transiting on the surface.

For Bushnell, propulsion was certainly one hurdle he had to overcome before proceeding, and he must have dismissed oars and sails early. There is no record of when he conceived the idea of a screw propeller for his submarine, or what inspired the concept. He undoubtedly discussed the dilemma with his scientific mentors, who would have been well-read and familiar with many unusual mechanical "engines."

A unique propulsion system was patented in 1661 by Toogood and Hayes. They proposed a hydraulic jet system, where water would be drawn in from a ship's bow and forced out aft using some form of pump or bellows, the details of which are unknown, although it was likely an adaptation of the Archimedian helical screw. Such a device would have been impossible to incorporate into Bushnell's small vessel, and it is doubtful that he was aware of this obscure British patent. It wasn't until 1787 when James Rumsey successfully ran a steam-powered water jet propulsor on the Potomac River.

The "natural philosophers" of the seventeenth and eighteenth centuries were fond of proposing a multitude of contraptions consisting of gears, levers, screws, and pulleys that had little chance of actually working. Figure 7.1 is from Emerson's 1773 *Principals of Mechanics* showing "a *chariot* or waggon to sail against the wind [using] the sails of a windmill . . . This waggon will always go against the wind, provided you give the sails power enough, by the combination of the wheels."[2]

This was, however, also an era when many clever individuals created equally clever, and functional, devices. Members of the Royal Society published their studies of natural phenomena in the Society's *Transactions* and included detailed descriptions of the unique scientific instruments they had devised. These scientists also published their own volumes relating their theories and experiments. Robert Hooke was among the most prolific and, as with so many individuals from the age of enlightenment, his followers continued to publish his scientific accomplishments after he died. In 1705, two years after his death, Richard Waller published *The Posthumous Works of Robert Hooke.* Waller included Hooke's "Lectures concerning Navigation and Astronomy," where he had proposed "an Instrument to keep an exact account of the way of a Ship through the Water . . . 'tis of great use to know the true Velocity of a Ship . . .'"

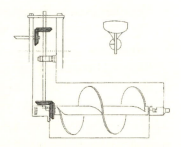

Figure 7.2. The helical propeller patented by F. P. Smith in 1836. (*Stevens 1893. Stevens Institute of Technology*)

Waller found documents written by Hooke among the Royal Society's unpublished papers: "which I have here publisht, hoping they may give the Ingenious some hints of improving them, which indeed has been my chief aim in Printing many of the foregoing Discourses." One of the discourses Waller referred to described Hooke's method for determining a ship's velocity, written in November 1683: "I shew'd an Instrument I had contrived [to] some of the Society above twenty Years since, by which the way of a Ship through the Sea might be exactly measur'd, as also the velocity of any running Water or River, and thereby the comparative velocity of it in its several parts; by this also the quantity of the Water vented by any River into the Sea, or any other River, might be found . . . [The] part of the Engine now shewn was the Vane, Fly, or first mover of the whole . . ."[3] By the "Vane or Fly," Hooke was referring to the blades of a propeller that would turn as it was pulled through the water when the ship was underway. This device became known today as a Taffrail Ship Log.

Most early mechanical propulsion concepts, however, were adaptations of the Archimedian screw, a helical pumping system originally designed to move water upwards. Propeller designs that called for a blade helically wound around a central shaft were proposed and patented into the second quarter of the nineteenth century. Examples include Lyttleton in 1794, Shorter in 1800, Cummerow in 1828, Woodcroft in 1834, and Smith in 1836 (figure 7.2).[4] Notable exceptions were the bladed propellers used by Robert Fulton in France on his submarine *Nautilus*, and by Col. John Stevens on his steam-powered small craft in the Hudson River, at the very beginning of the nineteenth century.

Bushnell described his propulsion system where "an oar, formed upon the principal of the screw, was fixed in the forepart of the vessel." He also noted that the pilot "could row upward, or downward . . . with an oar, placed near the top of the vessel"—this oar also being based on the principal of the screw. It is not surprising that early interpreters of Bushnell's letter to Jefferson simply drew an Archimedian screw as his means of propulsion. Because the *Edinburgh Encyclopaedia* illustration was published before 1836, when John Ericsson had patented a bladed propeller, Lieutenant Barber (see chapter 1) probably assumed that Bushnell had not conceived of a modern propeller.

In designing the replica *Turtle* propulsion system, the assumption was made that Bushnell's propellers, both for transiting and for vertical depth control, were of a two-bladed screw propeller design. The size of the blades comes from Lee: "It had two oars, of about 12 inches in length, & 4 or 5 in width, shaped like the arms of a windmill, which led also inside through water joints, in front of the person steering, and were worked by means of a wrench (or crank)."

"Shaped like the arms of a windmill"

Next was to decide what shape to make the replica blades. Bushnell only referred to them as oars, while Lee and Gale described the oars as resembling windmill blades. Another source was Charles Griswold's 1820 letter retelling Lee's account of the *Turtle*, where the terms windmill blades, oars, and paddles were used.

Windmills in the American colonies during the eighteenth century were derived from either English or Dutch designs. Both designs used elongated rectangular blades attached to arms radiating from the central hub. Typically, English designs had the blades secured along the central axis of the blade, while Dutch designs were attached along the leading edge of the blade. Windmills were common in New England and Bushnell would have been familiar with their operation. Figure 7.3 shows a Rhode Island windmill built in 1787.

Windmills may have inspired the concept of a bladed propeller, but we were not convinced that the *Turtle* propeller would have had rectangular blades. A possible source of the shape of Bushnell's windmill-like

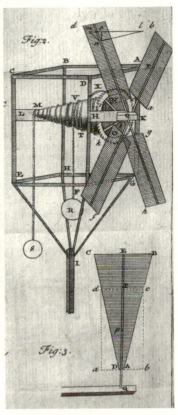

Figure 7.3, above, the Jamestown Windmill, Jamestown, Rhode Island. Originally built in 1787 to replace one destroyed by the British when occupying Narragansett Bay, this windmill was restored in 2001. (*Roy Manstan*) Figure 7.4, right, the Anemoscope, a device designed to measure wind-generated torque as described by Benjamin Martin in his *Philosophia Britannica*. (*Martin 1747*)

propeller is found in the *Philosophia Britanica* by Benjamin Martin. There is no record of what textbooks were at Yale and available to Bushnell while a student there. This particular text, however, was published in many editions during the last half of the eighteenth century, and is found on the 1791 list of Yale library holdings. The volume referred to during the *Turtle* Project is the first edition published in 1747. In volume 2, lecture 7, "The Doctrine of Winds and Sounds," there is a description of the "Anemoscope," a device designed in the eighteenth century to determine the torque applied to a windmill shaft by winds of varying velocity. An illustration of Martin's anemoscope is in figure 7.4.

In the description of the Anemoscope experiment, typical rectangular blades were used. Martin, however, proposed that windmill blades should be "in the Form of an Isosceles Triangle in each Sail . . . in stead of the equal Parallelogram Sail in common use."[5] He described how the force of the wind acting on the sail area, provides more torque when the "Center of Gravity" (centroid of the area) is at its greatest distance from the center of the shaft: "for in the Triangular Sail the Center of Gravity is at P [2/3 the length of the triangle, see figure 7.4] whereas in the Rectangular Sail the Center of Gravity is at the middle Point."

Martin suggested that the concept had inspired another author "to propose Sails in the form of Elliptic Sectors, for the Centers of Gravity in them are also removed to about two Thirds of their length, and are moreover better adapted to fill a circular Space when placed oblique to the Wind, so that no Wind be lost when you would take in all that falls on a given Space or Area."

Martin also discussed the calculations that determine the optimum angle for his windmill blades. He understood that the force of a fluid on an object was proportional to the mass density of the fluid, the square of the velocity of the fluid, and the projected area of the object. "Now in order to determine the absolute force of the wind, we must compare it with that of Water, as follows. Since Air and Water are both Fluids, if they move with equal velocities, their Effects in a given Time will be as [proportional to] the Quantities of Matter."

Martin had shown that the torque generated by any "fluid," for example, wind driving a windmill, was dependent on the projected area of the blades. If the blades were set perpendicular to the wind, all of the force is directed parallel to the windmill shaft and the shaft would not turn. By setting the blades oblique to the wind, a portion of the wind's force is directed tangential to the shaft generating rotational torque. Using vector math, Martin demonstrated that an angle of 54°44′, with respect to the axis of the shaft, would provide the maximum torque.

Windmill theory was applied to other mechanical devices during the eighteenth century. The current of heated air and smoke that rises in a chimney was harnessed by installing a multi-bladed wheel in the flue. This wheel, known as a "smoke jack," was designed to turn a roasting spit in a fireplace hearth. Col. John Stevens described his propeller in a letter written in 1805 to Philadelphia scientist Dr. Robert Hare: "To the extremity of an axis passing nearly in a horizontal direction through the stern of the boat, is fixed a number of arms with wings like those of a windmill or smoke jack."[6] Robert Fulton, in his patent application for a steam-powered vessel, also used this analogy to describe propeller blades that he referred to as "flyers like those of a smoke jack."[7]

Figure 7.5 illustrates this "Smoak Jack" as described in Emerson's 1773 edition of *The Principles of Mechanics*: "BE is a smoak jack. AB is a horizontal wheel, wherein the wings or sails are inclined to the horizon. The smoak or rarified air moving up the chimney at B, strikes these sails, which being oblique, are therefore moved about the axis of the wheel, together with the pinion C, of 6 leaves. C carries the toothed wheel D of 120 teeth, all of these are of iron. E a wooden wheel 4 or 5 inches diameter; this caries the chain or rope F, which turns the spit. The wheel AB must be placed in the first part of the chimney, where the motion of the air is swiftest; and that the greatest part of it may strike upon the sails. The force of this machine is so much greater, as the fire is greater. The sails B are of tin, 6 or 8 in number, placed at an angle of 54 1/2 degrees [essentially the same as Martin's anemoscope]."[8]

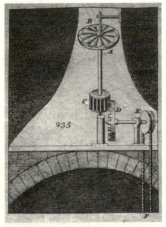

Figure 7.5. Many eighteenth-century households had a "smoke jack" installed in their chimneys to harness the rising current of hot air and smoke for turning a roasting spit. From William Emerson's *The Principals of Mechanics* (1773). (*Connecticut River Museum*)

Bladed "reaction wheels" had also been used to harness the energy from flowing water. Bushnell, however, was applying the "windmill" approach in reverse, i.e., rotating the shaft to generate thrust rather than using water to generate torque. Optimizing the angle of the blades for propulsion is not as straightforward as windmill theory. For any particular blade angle, propeller thrust (or wind if designing a fan) will increase as shaft rotation increases. Increasing the blade angle will also increase thrust. However, the torque, or power, needed to rotate the shaft also increases. Bushnell would have wanted a propeller that would maximize the thrust, but not at the expense of the individual. He knew that his pilot would need to maintain sufficient torque to the shaft for an extended period of time. He may have tried various blade designs, but as a starting point, would likely have considered Martin's optimal windmill-blade angle.

REPLICA PROPELLER

Specifying the blade angle for the replica propeller presented the same dilemma that Bushnell must have faced. The convention used to specify modern propeller blades requires that the blade face angle be measured with respect to the plane of rotation, rather than the axis of rotation used by Martin and Emerson. Thus, the 54º44´ angle of Martin's windmill blade is equivalent to a modern 35º16´ propeller blade.

The decision was made to consider multiple designs, including one with a blade angle of 25º, three variations with blade angles of 35º (Martin's windmill), and one with a blade angle of 45º. Other decisions about the propeller were more straightforward and included Martin's suggestions of an "isosceles triangle" for four of the propellers, and one with an "elliptic sector." Lee's description of Bushnell's propeller blades, as being twelve inches long with a maximum width of five inches, provided dimensional limits.

Unlike modern propellers, where the blade angle varies with respect to the radial distance from the hub (figure 7.6), the assumption was made that Bushnell (as with windmills) would have produced a blade where the blade angle remained constant over its entire length.

Initially, a single bronze propeller was produced with the triangular blade permanently set at 25º (figure 7.7).[9] A second propeller hub was then built with extensions set at 35º that would facilitate blade changes. Three pairs of triangular blades and one pair of elliptical "paddle-shaped" blades were produced for the second hub (figures 7.8 and 7.9).

Two pairs of these interchangeable blades (figure 7.9) were modified to evaluate propeller variations. A 10º wedge was bonded to one pair of the triangular blades and, when attached to the 35º hub extensions, produced a net blade angle of 45º. A strip about a half-inch wide along the leading and trailing edges of the second pair of triangular blades was bent approximately 10º, producing a slightly more helical blade contour when attached to the hub.

Table 1 provides dimensional parameters associated with the five replica propellers. Note that the diameter of all five propellers, measured along the centerline of the blade, is twenty-seven inches (each blade being twelve inches as specified by Lee).

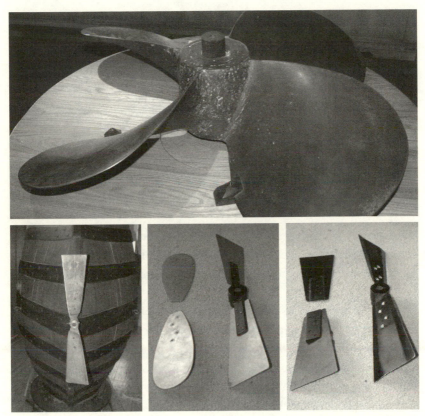

Figure 7.6, top, the propeller from the steamboat *Sabino* in the collection of the Mystic Seaport Museum, Mystic, Connecticut. Figure 7.7, bottom left, the bronze propeller that we believe resembles the size and shape of Bushnell's propeller. This is propeller No. 1 in Table 1. Figure 7.8, bottom center, the *Turtle* Project propulsion tests were designed to evaluate a variety of propeller concepts that we believe Bushnell may have considered. To facilitate this, we designed a hub with interchangeable blades. Each of their characteristics are shown on Table 1. Shown here are the paddle-shaped blades on the left (propeller No. 4) and the triangular blades (propeller No. 3), right, installed on the hub. Figure 7.9, bottom right, showing the blades used for propeller No. 2, left, and propeller No. 5, right with its modified leading and trailing edges. (*Roy Manstan*)

INTERNAL DRIVING MECHANISM

Turning the propeller to drive the submarine would depend solely on the physical stamina of the pilot. When the pilot rotates the propeller, the blades force water backward. This force, called thrust, is a function of the mass of water displaced and its aft acceleration. It is this thrust that

Table 1. Replica Propeller Parameters

Propeller	Material	Weight	Shape	Blade Angle
1	Bronze	9 lb. 9 oz.	Triangular	25°
2	Aluminum	4 lb. 1 oz.	Triangular	45°
3	Stainless Steel	5 lb. 7 oz.	Triangular	35°
4	Stainless Steel	5 lb. 7 oz.	Elliptical*	35°
5	Aluminium	3 lb. 15 oz.	Triangular (modified edges)	35° (45°)

*paddle-shaped

Source: The *Turtle* Project

causes the vessel to move forward. There is, of course, an equal and opposite force that the pilot overcomes through physical effort. In other words, he must be able to generate sufficient horsepower to maintain the propeller's rotation.

In referring again to Bushnell's description of the propeller: "An oar, formed upon the principal of the screw, was fixed in the forepart of the vessel; its axis entered the vessel, and being turned one way, rowed the vessel forward, but being turned the other way rowed it backward; it was made to be turned by hand or foot," it is the last phrase that provides a clue to the driving mechanism. Lee only mentioned that the propellers "were worked by means of a wrench (or crank)." Gale, in his November 9, 1775, letter to Silas Deane, added another clue regarding the operation of the *Turtle* propellers: "which are turned by foot, like a spinning wheel." Figure 7.10 illustrates a spinning wheel with its treadle that would have been a common sight in an eighteenth-century home.

From these sources, we have gathered that the driving mechanism included a combination of a hand-operated crank and foot-driven treadle. This is logical in that the pilot would have the option of using either set of muscles to keep the propeller in motion, and would be able to alternate hand or foot as required. It would also free up his hands to run other mechanical systems, including making course corrections with the rudder.

Possible use of a flywheel

There is no mention of a flywheel in any of the source material. However, the use of a flywheel type device to provide rotational momentum was common during the eighteenth century. Potters used a heavy

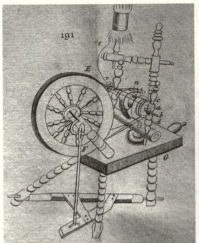 

Figure 7.10, left. According to Benjamin Gale, the *Turtle*'s propeller was "turn'd by foot, like a spinning wheel." This illustration from William Emerson's *The Principals of Mechanics* (1773) shows the common spinning wheel and its foot-operated treadle. (*Connecticut River Museum*) Figure 7.11, right, the treadle and flywheel as installed in the *Turtle* replica construction. (*Roy Manstan*)

kick wheel to maintain rotation when throwing pots. A heavy stone wheel, driven by a foot treadle, was used for sharpening the edges of axes and scythes. The spinning wheel (figure 7.10) employed a large diameter wheel that served two purposes. First, it allowed the operator to turn the smaller diameter spindle at a much higher speed. Secondly, when the operator pushed down on the treadle, the rotational momentum of the heavy wooden rim caused the wheel to continue to rotate and return the treadle to its upper position, where the operator would repeat the cycle.

Figure 7.11 shows the treadle and flywheel installation during early stages of the replica construction. Additional illustrations of the internal drive mechanism are in chapter 12 (figures 12.5 and 12.6). The propeller tests and feedback from all of the replica test pilots provide a compelling argument that Bushnell's pilot would not have been able to sustain any significant propeller rotation without a flywheel. A discussion of the flywheel as used in the testing, the test pilot's observations, and our conclusions are provided in chapters 13 and 15.

Figure 8.1. The keel of our replica *Turtle* was made as a flat disc to allow it to stand upright while working on the hull. Most early interpretations show the keel to be a continuation of the hull curvature ending in a point, a feature we believe to be very unlikely on the original vessel. (*Roy Manstan*)

Chapter Eight

THE HULL

WAS THE TURTLE MORE THAN A BARREL?

THE HULL TAKES SHAPE

Critical to determining the vessel's operational capabilities and limitations is creating an accurate representation of the hydrodynamics associated with hull shape. Beyond Bushnell and Gale referring to the *Turtle* as resembling two tortoise shells "joined together," details as to the size and shape of the hull had to be developed from an interpretation of verbal descriptions within the text of their letters. The interior of the replica *Turtle*, however, was designed to accommodate the pilot, the mechanical systems, and the safety features necessary to allow the at-sea testing.

Most underway operations would occur with the *Turtle* surfaced, i.e., with the upper portion of the hull and hatch exposed. In Bushnell's words: "The skillful operator could swim so low on the surface of the water, as to approach very near a ship, in the night, without fear of being discovered." There would be no doubt in his mind that these operations would require an extended transit. If the operator were exhausted when he finally gained the target ship, the mission would be less likely to succeed. Bushnell would need to minimize the physical demands on the pilot, creating a hull shape that would maximize propulsion efficiency when transiting through the water.

Many small craft, whaleboats for example, were double ended to improve rowing efficiency. This hull shape was common knowledge and would certainly have been a consideration. A hint that the bow and stern were not rounded comes from Bushnell: "At one edge which was directly before the operator, who sat upright, was an oar for rowing forward or backward. At the other edge, was a rudder for steering." The need for rowing efficiency and the use of the term "edge" to describe the bow and stern governed our interpretation of the general double-ended hull shape.

Hull dimensions

The only dimensions that are available come from Gale's observations in 1775. Our replica corresponds to his November 9, 1775 letter to Silas Deane where he writes: "In length, it doth not exceed 7 1/2 feet from the stem to the higher part of the rudder: the height not exceeding 6 feet."

The width of the replica is only an estimate based on Bushnell's description of the interior of the hull: "The internal shape of the vessel, in every possible section of it, verged towards an ellipsis, as near the design would allow, but every horizontal section, although elliptical, yet as near to a circle, as could be admitted. The body of the vessel was made exceedingly strong; and to strengthen it as much as possible, a firm piece of wood was framed, parallel to the conjugate diameter, to prevent the sides from yielding to the great pressure of the incumbent water, in a deep immersion."

Based on the above, our hull was dimensioned at six feet from the deck to the keel, six feet stem to stern, plus about a foot and a half for the rudder. It was decided that the maximum width of the hull (Bushnell's "conjugate diameter") would be four feet, creating an elliptical curvature to the exterior of the hull and ample room inside for the pilot requirements. The streamlined hull shape with its distinct stem is apparent in most of the *Turtle* illustrations in this book.

Hull thickness and construction

The only discussion of the type of construction and the thickness of the hull comes from secondary sources, including a set of notes (author uncertain) appended to Ezra Lee's 1815 letter to David Humphreys, claiming "its sides were at least six inches thick." Bushnell only thought it necessary to indicate that the "the body of the vessel was made exceedingly strong . . .", and that he took precautions "to prevent the sides from yielding to the great pressure of the incumbent water, in a deep immersion."

In Charles Griswold's subsequent retelling of Lee's account, published in 1820, he claimed that the hull was "composed of several pieces of large oak timber, scooped out and fitted together." Griswold also noted that "it was bound around thoroughly with iron bands." He is the only author that mentions the use of the iron bands, a construction detail that may have been taken for granted and omitted by other contemporary writers.

Our replica was built from one-inch thick white oak planks. Each plank was laid over a frame that established the elliptical contour. The joints were shiplapped and caulked, and we used several wide steel bands to reinforce the hull. Based on our replica dimensions and the initial testing, we determined that fully submerged, our *Turtle* displaces about two tons. Being comfortable with our external dimensions and shape, we expect that Bushnell's submarine also displaced about the same two tons. Considering that we required significantly more lead than what was used in the original *Turtle*, the only way to compensate for the greater ballasting would be to have used thicker planking on our replica.

As we considered the construction of our *Turtle*, we were concerned that Lee's claim of a six-inch thick hull would have reduced the internal space available to accommodate the multiple mechanical systems. The interior of our replica was a tight fit, even for our one-inch thick hull. So was the original *Turtle* hull six inches thick or was it built from one-inch planks? Our feeling is that the truth lies somewhere in between. The following are possible scenarios that Bushnell may have considered.

James Walsh, in his book *Connecticut Industry and the Revolution*, discusses shipbuilding in the eighteenth century, and in particular, the use of naturally curved or "compass" timber for a ship's framing. He notes that "in a ship of about 300 tons, with a keel of eighty feet, there would be approximately 560 pieces of compass timber needed just for the frames and each piece would be about a foot square." Walsh also mentions that timber merchants would sell compass wood to shipwrights.[1]

Bushnell, when considering his hull design, may have discussed its construction with one of the local shipwrights. The shipyard of Uriah Hayden was only about five miles from Bushnell's home. There would have been little concern about Hayden's politics, as his shipyard was contracted by the Connecticut Council of Safety to build one of the state's finest naval vessels, the *Oliver Cromwell*, launched in 1776. It is conceivable that Bushnell convinced Hayden to provide lengths of compass wood that would conform to the curvatures of his elliptical submarine hull.

In order to build a vessel of reasonable size while retaining some degree of exterior streamlining, a six-inch thick hull would have severely limited the interior volume. The cross section of a compromise hull could have incorporated a common arched bridge concept (see figure 8.2), where the thinner center span increases in thickness toward the end supports. In this way, the interior volume of the *Turtle* would have been shaped, such as noted by Bushnell, where "every horizontal section, although elliptical, yet as near to a circle, as could be admitted," whereas the exterior would maintain a more streamlined contour. A hull of this design will provide resistance to the forces of water pressure, in a similar fashion to how an arched bridge reacts to the vertical loads it is subjected to.

BUOYANCY AND BALLASTING

Reverend John Wilkins, in his *Mathematical Magic*, explained the concept of submarine buoyancy: "If this Ark be so ballast as to be of equal weight with the like magnitude of water, it will then be easily movable in any part of it."[2]

Bushnell adopted a tactic requiring the *Turtle* to transit afloat, with only enough freeboard to allow the pilot to navigate with his deadlights open, without fear of waves washing over the hull and into the *Turtle*. Submerging required the pilot to add small quantities of water into the bilge until the vessel attained neutral buoyancy. At that point, the pilot would rotate the vertical propeller to adjust his depth.

Buoyancy of a vessel is a function of the volume of water that the hull displaces when afloat. The greater the weight of the vessel, including any added ballast weight, the greater will be its draft, or depth to the keel. We estimate that the *Turtle* weighed about three thousand five hundred pounds with an empty bilge and without its pilot. The *Turtle* was, as Bushnell stated, intended to remain afloat when approaching the enemy, but with the majority of the hull below water. The *Turtle's* draft could be adjusted by adding water to its bilge, but there would have to be sufficient volume remaining to enable the pilot to add more water until the hull became neutrally buoyant in preparation for submerging.

It is important to remember that the density of seawater is about 64 pounds per cubic foot (lb./ft.[3]), whereas the density of pure freshwater is 62.4 lb./ft.[3]. Thus, a three thousand five hundred pound *Turtle*, when afloat in freshwater, will displace more freshwater and thus float with a deeper draft than it would when afloat in saltwater.

Figure 8.2. This arched bridge design illustrates the ability of a structure to support heavy loads in spite of the thinner section along the top of the central arch. (*Gentleman's Magazine*, March 1747)

Salinity and the effect on buoyancy

The differences in water density are caused by the added weight of dissolved salts. The "saltiness" or salinity of the water is a function of the proportion of these salts, primarily sodium chloride (NaCl), measured in "parts per thousand" (ppt). The salinity of seawater may typically be as high as 32 ppt (water temperature also affects the solubility of the salts and thus the salinity), whereas pure freshwater will have no dissolved salts, i.e., 0 ppt. The proportion of salts can also be viewed in terms of percent rather than parts per thousand, i.e., 32 ppt is equivalent to 3.2%.

When fully submerged, the *Turtle* hull displaced about sixty-two cubic feet of water. In pure freshwater, the submerged displacement would therefore be 62.4 lb./ft.^{3} multiplied by the displaced volume of 62 ft.^{3}, or 3,869 lb. In saltwater (at 64 lb./ft.^{3}), the displacement would be 3,968 lb. The maximum difference in displacement would then be about one hundred pounds.

In reality, however, the *Turtle* operated in coastal environments where these extremes are rarely encountered. Where a river empties into a harbor, the fresh river water will ride over the heavier saltwater entering from the ocean. While there is mixing at the interface between these two bodies of water, there will likely be some salinity gradient from the surface to the bottom. When preparing to submerge, the pilot will adjust his ballast to become neutrally buoyant, according to the density of the surface water. He then rotates the vertical propeller to drive the vessel downward.

As the pilot descends from the surface, he will enter the slightly more saline ocean water. In an estuary environment, depending on the amount of mixing, the salinity may be 15 to 20 ppt on the surface, increasing to 20 to 25 ppt near the bottom. With each increase in salinity of 1 ppt, the density of the water increases by 0.05 lb./ft.^{3}. For the *Turtle*, with a volume of 62 ft.^{3}, this means a 3.1 lb. increase in buoyancy.

While this may seem inconsequential, this small increase will result in the *Turtle* becoming slightly more buoyant as it descended. If the difference in density becomes significant, the vertical propeller will not have sufficient thrust to drive the vessel down. To compensate, the pilot will have to add water to the bilge and reestablish neutral or near-neutral buoyancy, allowing him to use the vertical propeller.

BALLASTING AND HULL STABILITY

The ballast was concentrated at the keel and in the bilge. This ensured that the center of gravity would be well below the center of buoyancy, providing a stable hull. Bushnell noted that "its ballast made it so stiff, that there was no danger of oversetting." We noticed the inherent stability during the early testing of our replica in the Connecticut River, when winds and waves began to increase. The pilot could feel the hull move slightly up and down, but there was no perceptible rocking motion.

There were four aspects to ballasting Bushnell's submarine, including about seven hundred pounds of lead permanently installed at the keel, another two hundred pounds at the keel that could be jettisoned in an emergency, an unspecified quantity that could be added internally to account for differences in the weight of the pilot, and the critical water ballast that the pilot would use to attain neutral buoyancy when submerging. Each of these are discussed below.

KEEL BALLAST

The density of lead ensures that it will provide the greatest amount of ballast in the smallest volume. Bushnell was able to acquire a sufficient supply for the *Turtle* in spite of its expense, and that it was in demand for supplying ammunition to the Continental army.

There are no contemporary descriptions of the shape and method of attachment of the keel ballast. Most drawings from the nineteenth and early twentieth centuries show a pointed keel simply following the general curvature of the hull. When designing the replica, we decided on a flat disc, allowing the *Turtle* to be freestanding. It enabled us to climb into the hull during construction, without fear of tipping over. We felt that the original *Turtle* would have also been easy to handle and work on in an upright position, although it could have been transported lying on its side in a cradle. A forklift was used to facilitate installation of the keel (figure 8.1), but with the heavy keel as a pivot point, we were able to lower the replica to the horizontal and raise it again with relatively little effort, without requiring the forklift.

EMERGENCY BALLAST

Bushnell was very concerned about safety, particularly during training: "I never suffered any person to go under water, without a strong piece of rigging made fast to it, until I found him well acquainted with the operations necessary for his safety." He also provided the pilot with the ability to jettison a significant amount of ballast in an emergency: "About two hundred pounds of the lead, at the bottom, for ballast, would be let down forty or fifty feet below the vessel; this enabled the operator to rise instantly to the surface of the water, in case of accident."

Ezra Lee also noted the purpose of this additional ballast: "Seven hundred pounds of lead were fixed on the bottom for ballast, and two hundred weight of it was so contrived, as to let it go in case the pumps choked, so that you could rise at the surface of the water."

In Gale's letter to Franklin, however, he indicated that "he has an Anchor, by which he Can remain in Any Place . . . and Weigh it at Pleasure . . . about 1000 wt of Lead is his Ballast, part of which is his Anchor." Gale also mentioned an anchor in his November 9 letter to Deane.

It is not surprising that some modern interpretations have concluded that Bushnell's two hundred pound lead ballast was not simply a weight that could be dropped in an emergency, but served as an anchor that the pilot could lower and haul back up at will. Barber's illustration (see figure 1.2) shows a crank and line extending below the pilot's seat to the ballast. This is a highly unlikely, and in fact impossible, interpretation. First, raising a two hundred pound clump mired in the mud would have been a Herculean task. It is more likely that if he had a powerful enough winch, the pilot would have cranked the *Turtle* down to the anchor, rather than hauling it up.

Second, running a lift line through the hull would have provided a path for water to enter, and the *Turtle* would have sunk as soon as it was launched. Even with the hatch closed, the check valves that supplied the ventilation system would be open, and the hull interior would be at atmospheric pressure. Any opening into the vessel below the waterline would allow water to enter the hull, forcing the air inside to exit through the ventilation pipes (or the deadlights if they were open).

Bushnell took pains to ensure that all hull penetrations were water-tight: "Wherever the external apparatus passed through the body of the vessel, the joints were round, and formed by brass pipes, which were driven into the wood of the vessel, the holes through the pipes were very exactly made, and the iron rods through them, were turned in a lathe to fit them, the joints were also kept full of oil, to prevent rust and leaking." This system would have been used for the two propellers, the dogs for securing the hatch, the wood screw mechanism, and the pin that secured the mine to the back of the vessel.

Using the same watertight brass pipe and an iron rod, Bushnell could have secured the emergency ballast to the keel. If needed, the pilot would rotate the iron rod in such a way as to allow the ballast to fall free of the keel. Bushnell had provided about "forty or fifty feet" of line that would anchor the distressed *Turtle* until it and the pilot could be retrieved. This system would be particularly useful during training, when equipment still being developed and tested was most likely to fail. In a combat mission, the pilot could quickly drop the weight, exit the *Turtle* floating on the surface, and hopefully swim to safety.

INTERNAL LEAD BALLAST

Whenever the Turtle was launched, Bushnell would have adjusted the trim to what he considered the optimum level. Draft, and thus freeboard, will depend on the salinity of the water as well as weight of the pilot, all of the other contributing factors remaining constant. This additional ballasting must have been significant, as Bushnell specifically mentioned it in his letter to Jefferson: "The vessel was chiefly ballasted with lead fixed to its bottom; when this was not sufficient, a quantity was placed within, more or less, according to the weight of the operator." Taking the time to note the need to accommodate the weight of the pilot implies that several individuals (see chapter 21) of different size had operated the *Turtle* during its brief time afloat (from the summer of 1775 to the fall of 1776). The only mechanical issue with internal ballast is securing the loose lead blocks inside the hull if the *Turtle* was laid on its side when being transported.[3]

Water ballast and pumps

The most critical aspect of ballasting the *Turtle* occurred when submerging. Bushnell: "When the operator would descend, he placed his foot upon the top of a brass valve, depressing it, by which he opened a large aperture in the bottom of the vessel, through which the water entered at pleasure; when he had admitted a sufficient quantity, he descended very gradually; if he admitted too much, he ejected as much as was necessary to obtain an equilibrium, by the two brass forcing pumps, which were placed at each hand."

Ezra Lee added details about this valve: "It was sunk by letting in water by a spring near the bottom, by placing your foot against which, the water would rush in and when sinking take off your foot & it would cease to come in & you would sink no further, but if you had sunk too far, pump out water until you got the necessary depth."

As water entered the bilge, air inside the hull was forced out through the ventilation pipes, and the hull slowly moved downward. Equilibrium was obtained when the top of the hatch was just awash. The pilot lifted his foot, allowing the spring to automatically close the seawater valve. In this position, and as the pilot submerged using the vertical propeller, the check valves in his ventilation system closed.

This process was tested on the replica *Turtle*, and seemed to be fairly straightforward. Our testing, however, was done in daylight when the pilot could see the water passing the deadlights along the side of the hatch. Combat missions in 1776, and possibly much of the training, occurred at night when the water level, and thus equilibrium, would have been difficult to determine. The pilot may have been able to watch his depth gauge, but the *Turtle* likely had to descend less than eighteen inches before the hatch was awash. Bushnell would have needed to place a specific mark on the depth gauge, indicating the exact depth when equilibrium was obtained. Regardless of the inherent difficulty, all of Bushnell's pilots managed this critical maneuver.

We determined that we had to add about 250 to 300 pounds of water (about four and a half cubic feet) to bring the hatch awash. There was

Figure 8.3, left. Bushnell had installed a foot-operated valve to allow water to enter the bilge for additional ballast. We used a hand-operated ball valve, in part because it was the most reliable modern method to safely and efficiently add water to our bilge. Figure 8.4, right. This student is installing a brass screen over the opening into the hull for the hand-operated bilge pump. (*Roy Manstan*)

plenty of room in the bilge below deck to accommodate this much water. Bushnell had installed a foot-operated valve, while our replica incorporated a hand-operated ball valve (figure 8.3). In either case, we could readily control the flow rate through the valve, such that flooding the bilge became a simple task. As with the original *Turtle*, our replica was equipped with two hand pumps to adjust the water ballast, one of which is visible just forward of the ball valve in figure 8.3. Bushnell also covered the openings "with a plate, perforated full of holes to receive the water, and to prevent any thing from choking the passage, or stop the valve from shutting." Likewise, we installed screens over the openings (figure 8.4).

Once neutrally buoyant, it only takes a few pumps to create enough positive buoyancy to remain safely at the surface. In this position, the check valves in the ventilation pipes of Bushnell's *Turtle* would open, allowing the pilot to continue pumping out the water without drawing a vacuum in the hull. However, it takes a lot (and I repeat "a lot") of pumps to evacuate the entire four cubic feet of water in order to return to the original freeboard. In order to accommodate our aging pilot (and in the pursuit of safety), we also installed an electric bilge pump.

For Bushnell, bilge pumps would not have presented a major problem. Hand-operated "forcing pumps" were common in the eighteenth century. Efficiencies of these pumps were measured in the number of hogsheads of water a man could pump in an hour. Bushnell only needed a small "one hogshead per hour" pump (about a gallon a minute). His only issue would be finding or producing two that would fit into the confined interior of the hull. With two small capacity pumps, his pilot could easily remove a couple gallons (sixteen pounds of water ballast) within a minute.

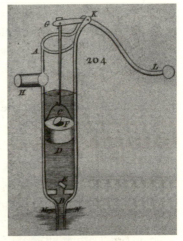

Figure 8.5. Bushnell had installed two hand-operated "forcing pumps" to enable the pilot to remove water ballast from the bilge. This illustration of a common eighteenth-century water pump is from William Emerson's *The Principals of Mechanics* (1773). (*Connecticut River Museum*)

The mechanism was not complicated, and Bushnell turned once again to Isaac Doolittle to produce the "two brass forcing pumps" custom fit for the *Turtle*. That Doolittle produced the pumps comes from Gale's November 22, 1775, letter to Deane, explaining that a problem with "the forcing pump made by Mr. Doolittle" had delayed the *Turtle* from being deployed against the British in Boston.

Figure 8.5 is from Emerson (1773), figure 204, and shows the cross section of an eighteenth-century hand pump that Bushnell and Doolittle would have been very familiar with. The mechanical components in the modern hand-operated bilge pumps used in our replica are essentially identical in design and operation as that described by Emerson. The defect that Bushnell discovered in Doolittle's pumps was likely an easy problem to correct, there being few moving parts as is evident in Emerson's description:

GB is a common sucking pump; GKL the handle; CD the bucket; E,F two clacks opening upwards. When the end L is put down, the end G raises the sucker or bucket CD, and the valve or clack F shuts; . . . the water [rises] up the pump, opens the valve E, and ascends thro' the hole B into the body of the pump DB . . . . when the handle L is raised, the bucket CD descends, the valve F opens and lets the water ascend through it, and the pressure of the water shuts the valve E, so that the water cannot return through B. Then whilst the end L is put down again, the sucker CD is raised again, together with the water above it, whilst more ascends through B. So that at every stroke of the handle, water is raised into the pump, till at last it flows through the pipe H.

The bucket, sucker, or piston, is to be surrounded with leather to fit exactly, and must move freely up and down in the barrel, and must also exactly fill it. Of valves or clacks, some are flat, made of leather; others are conical: and they must all fit very close, and move freely.[4]

Chapter Nine

HATCH AND LIFE SUPPORT

How to Survive Inside a Turtle

Bushnell's hatch

Bushnell provided a lengthy description of his submarine hatch, yet it has resulted in as many interpretations as the *Turtle*'s propulsion system. First, he notes: "The entrance into the vessel was elliptical, and so small as to barely to admit a person. This entrance was surrounded with a broad elliptical iron band . . ." Bushnell doesn't define how elliptical this entrance was, only mentioning that "every horizontal section [of the hull] although elliptical, yet as near a circle, as could be admitted." Our *Turtle* replica was built with an elliptical upper deck but included a circular entrance.

He then described the hatch:

> Above the upper edge of this iron band, there was a brass crown, or cover, resembling a hat with its crown and brim, which shut water tight upon the iron band: the crown was hung to the iron band with hinges so as to turn over sideways, when opened. To make it perfectly secure when shut, it might be screwed down upon the band by the operator, or by a person without.

Contemporary accounts provide little additional insight into the shape of Bushnell's hatch. Lee only noted that it was "a composition head hanging on hinges." Benjamin Gale, in his November 9, 1775, letter to Silas Deane, described it as "a brass top or cover, which receives the person's head as he sits on a seat, and is fastened on the inside by screws." This implies that the brass hatch is tall enough to accommodate the pilot's head.

The *Turtle* hatch has seen many interpretations by writers, historians, and other replica designers. Lieutenant Barber's 1875 illustration (see figure 1.2) shows a hatch that is parallel sided and rounded fore and aft, simulating a very elongated ellipse. Barber's illustration also implies that Bushnell's "broad, elliptical iron band" was wide enough to include the "three round doors," and that the hatch was secured above the pilot's head.

We decided not to try to guess the shape of the original vessel's hatch, but designed ours in such a way that the final casting could accommodate all of the components that Bushnell had included. The interior of our hatch was about twenty-two inches in diameter and eight inches high. While this may have been slightly larger than Bushnell's hatch, we found that the interior of our hatch was sufficient to accommodate the piping used for our ventilation and air supply, and still provide room for the pilot to move his head and manipulate the air system controls. In this way, we could replicate the essential functions but not compromise safety during the subsequent operational testing.

There is no record of where Bushnell obtained his hatch. However, when Isaac Doolittle established his bell foundry in New Haven in 1774, Bushnell may have found the perfect partner. Casting bells was a technology well understood in the eighteenth century. Doolittle had advertised that he could provide "any Size bell commonly used in this, or the neighboring Provinces"[1] and must have employed individuals familiar with the foundry business, including experienced pattern makers. The pattern maker could have given Bushnell guidance in how to include provisions in the initial casting to accommodate his windows, doors, and air pipes. For example, Bushnell's description of the "three round doors . . . large enough to put the hand through," also included that they "were ground perfectly tight into their places with emery, hung with hinges and secured in their places when shut." This was not an afterthought; his rough casting would have been produced with the ability to mount hinges and create watertight surfaces.

CREATING THE REPLICA HATCH

The design and construction of our replica hatch was a partnership between the students, engineers and technicians at the Naval Undersea Warfare Center (NUWC), and the Mystic River Foundry, Mystic, Connecticut. To create our replica, we followed the same process that

Figure 9.1, upper left. Sketches of the *Turtle* hatch were supplied to the Naval Undersea Warfare Center (NUWC), where they were entered into a Computer Aided Design (CAD) program and from which the casting pattern was produced. (*NUWC*) Figure 9.2, bottom. The rough casting of the hatch from the Mystic River Foundry was brought to NUWC, where the critical surfaces were machined. The *Turtle* Project students visited NUWC, hosted by Scott Boyd (center), Donald Cressman (far left), and Steve Tobiaz (far right). Also present are authors Roy Manstan (back row left) and Fred Frese (front row right). (*Dave Stoehr, NUWC*) Figure 9.3, upper right. Old Saybrook High School student Will Tucker helps fit the hatch to its hinge on what will become the *Turtle*'s deck. (*Roy Manstan*)

Bushnell must have experienced. Our first visit was with Sharon Hertzler, the owner/operator of the foundry. She explained the entire process and how we would need to design our pattern. The foundry uses casting procedures that have changed little from what occurred at Doolittle's bell foundry. She explained that in order to produce a thin casting of the dimensions we had wanted, the overall thickness of our hatch would need to be about one-quarter inch. We also had to include five-eighths-inch thick raised "bosses" that would be machined later, to allow us to provide our windows with watertight seals. Based on her recommendations, we created a final set of drawings.

As with Bushnell, our next step was to produce a pattern. Under the education partnership with the school, NUWC converted our design drawings to a computer model (figure 9.1), compatible with their CNC (computer numerical controlled) milling machine, and produced the pattern from high-density machineable foam. The pattern was then hand-finished and painted, in order to facilitate creating the casting mold. The pattern was then brought to the foundry, where the sand mold was produced and the casting poured.

The casting was then returned to NUWC, where the critical surfaces were machined. The students were able to visit the foundry where the owner, Sharon Hertzler, described eighteenth-century casting processes. They also visited NUWC (figure 9.2), where they could see their hatch during the final machining process. Figure 9.3 shows one of the students helping fit the hatch and its hinge to the steel deck.

The foundry provided the casting as a donation to the *Turtle* Project. The value of this contribution to the success of the Project cannot be overemphasized. The quality of the casting and the attention to detailed machining ensured a safe and reliable watertight hatch. The precision machining needed to accommodate the windows, also produced at NUWC, can be seen in figure 9.4. While certainly speculative, Isaac Doolittle likely understood Bushnell's financial limitations and may have provided the original hatch at no cost.

A WATERTIGHT SEAL

Keeping the *Turtle* watertight was no simple task for Bushnell, or our replica for that matter. His propeller shafts, rudder shaft, mine attachment system, and the various hatch components all required precision machining to ensure a tight fit. The rotating shafts were "kept full of oil, to prevent rust and leaking." There is no doubt that some water did manage to seep across these joints and along the seams of the hull planking. Bushnell noted that his bilge pumps were designed to enable the pilot to control his ballast and "whenever the vessel leaked . . . he also made use of these forcing pumps." These hull penetrations were relatively small, and leakage would not have presented a major problem. However, the junction between the hatch and the hull was another matter.

The replica hatch was cast with a five-eighths-inch thick flange that was designed to include an O-ring groove. A flat ring, two inches wide, was welded to the steel deck and served as the mating surface for the

Figure 9.4. Detail of the hatch window. (*Roy Manstan*)

hatch O-ring. When lowered into position, the hatch was secured with two dogs operated by the pilot, or by the support team. When the dogs were fully engaged, they forced the hatch flange against the flat ring on the deck, compressing the O-ring about a sixteenth of an inch. This compression ensured a watertight seal.

Bushnell's dependence on well-machined and oiled metal-to-metal surfaces was not an option for the hatch seal. He described his hatch as "resembling a hat with its crown and brim." His "brim" may have been similar to the flange on the replica. When Bushnell added that the hatch "shut water tight upon the iron band," it is likely that he used a leather gasket between the "brim" and the "iron band."

An example of this type of seal was described for a "diving engine" invented by John Lethbridge in 1714. Lethbridge developed a one-atmosphere suit to compete with the diving bells commonly used during the eighteenth century. A description of the device was published in the September 1753 issue of *The Universal Magazine of Knowledge and Pleasure*. The illustration that accompanied the article is provided here in figure 9.5.

The diver would enter the copper cylinder, extending his arms through two openings kept watertight with leather sleeves. The article describes "a cover to fit the engine, fastened down with screws . . . and leather between the borders, so as to prevent leaking in any depth of

water."[2] With the individual secured inside, the "diving engine" was lowered to the bottom, where the occupant could retrieve lost objects. As
with the *Turtle*, keeping the vessel watertight was essential, and a leather
gasket was the solution. Lethbridge initially experimented with a wood-
en barrel. He later partnered with Thomas Rowe, using the improved
version illustrated here to salvage vessels along the coast of Europe.

VENTILATION

In 1648, Reverend John Wilkins first described the problem of the limited supply of air inside a submarine: "It is observed, that a barrel or cap,
whose cavity will contain eight cubical feet of air, will not serve a
Urinator or Diver for respiration, above one quarter of an hour." Yet, as
a technological visionary, he speculated about the ability of humans to
adapt to breathing underwater. "I will not say that a man may by custome
(which in other things doth produce such strange incredible effects) [i.e.
evolution] be inabled to live in the open water as fishes do, the inspiration and expiration of water serving instead of air, this being usual with
many fishes that have lungs; yet it is certain that long use and custome
may strengthen men against many such inconveniences of this kind . . ."[3]

Bushnell's mission plan was to require that his vessel make its
approach to the target "low on the surface of the water," only submerging to attach his mine or "magazine." When submerged, the pilot would
have to rely on the air contained within the hull. Bushnell concluded that
"the inside was capable of containing the operator, and air, sufficient to
support him thirty minutes without receiving fresh air." Benjamin Gale,
in his August 1775 letter to Benjamin Franklin, was a bit more generous,
indicating that during tests, the pilot "continued about 45 Minutes without any Inconveniency as to Breathing."

Even during the transit, unless the sea was unusually calm, the hatch
would need to remain closed. Waves washing over the hull would otherwise flood the *Turtle*. Under these conditions, Bushnell had to provide a
means of supplying air to the pilot (who had not yet evolved to "live in
the open water like fishes" as Wilkins had projected) during extended
surface transits. Bushnell described his solution: "There were in the brass
crown, three round doors, one directly in front, and one on each side,
large enough to put the hand through—when opened they admitted
fresh air; their shutters were ground perfectly tight into their places with
emery, hung with hinges and secured in their places when shut . . ." Lee

also mentioned the doors that served as his primary air supply: "Three round doors were cut in the head (each 3 inches diameter) to let in fresh air untill you wished to sink, and then they were shut down and fastened."

These "three round doors" could be kept open during transit, allowing sufficient fresh air to circulate into the area where the pilot sat, yet small enough that little water would enter if an occasional wave rolled across the deck and over the hatch. We believe that three of the small windows Bushnell included to illuminate the interior during daylight training were installed in these doors (see chapter 5). We therefore installed windows in each of the replica hatch deadlights, one looking directly forward and one on each side. We decided to secure the windows with retaining rings rather than using hinged doors. When we performed the pilot endurance tests

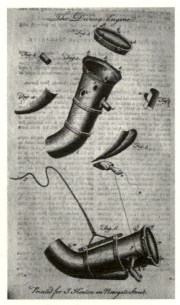

Figure 9.5. The cover of John Lethbridge's "diving engine" was kept watertight with a leather gasket secured with screws. (*Universal Magazine, September 1753*)

described in chapter 17, we removed the retaining rings and pulled out the windows to simulate the same source of fresh air that Lee would have had with the doors opened (as shown in figure 9.4).

Once the *Turtle* set out on its mission, it would be several hours before it would return. Weather and sea state could quickly change from benign to malevolent. In the event that conditions precluded keeping the three doors open, the pilot would need to continue to resupply the interior with air. In a mission of this nature where the life of the pilot is at stake, it is necessary to provide primary and secondary sources of air. Bushnell added a sophisticated ventilation system that would serve his need for this secondary or back-up air supply.

There were two air pipes in the crown. A ventilator within drew fresh air through one of the pipes, and discharged it into the lower part of the vessel; the fresh air introduced by the ventilator, expelled the impure light air through the other air pipe. Both air pipes were so constructed, that they shut themselves whenever the water rose near their tops, so that no water could enter through them, and opened themselves immediately after they rose above the water.

Lee does not mention its use and it is likely that the conditions never required him to operate the ventilator. Benjamin Gale, however, does include a brief description in his November 9, 1775, letter to Silas Deane: "In the same brass head are fixed two brass tubes, to admit fresh air when requisite, and a ventilator at the side to free the machine from the air rendered unfit for respiration." Adding to the complexity was the need to provide check valves in the external brass piping that, according to Bushnell, automatically "shut themselves whenever the water rose near their tops, so that no water could enter through them, and opened themselves immediately after they rose above the water."

Denis Papin, in 1695, included "the Hessian rotary sucker and forcer, which attracts the external air through the pipe" as the air supply for his submarine (see chapter 3 and figure 3.3). While no additional details are given, the brief description implies that Papin employed what was later referred to as a "centrifugal bellows." Figure 9.6 is an illustration of a "Machine for changing the Air" devised by Dr. J. T. Desaguliers in 1734, and described in the *Philosophical Transactions of the Royal Society*. A seven-foot diameter version was designed to circulate air in hospital rooms, prisons, and mines. Desaguliers also noted that "The Machine may also serve in a Man of War, to take away the foul Air between Decks, occasioned by the Number of Men in the Ship, and to give them fresh Air in a few Minutes."[4] His initial design that was demonstrated before the Royal Society, however, was seven inches in diameter and ideal for Bushnell's application in the *Turtle*.

It is a credit to Bushnell's concern for the safety of the pilot that he included a secondary air source that would have been a complex and costly custom-built system. The *Turtle* replica also included primary and secondary air supplies. A scuba tank installed inside the hull provided primary air. Figure 9.7 shows the internal scuba tank mounted just behind the pilot. Note the T-bar handle above the tank used for disengaging the mine from inside the hull. In the event that the primary air supply failed, a secondary source was available from another scuba tank

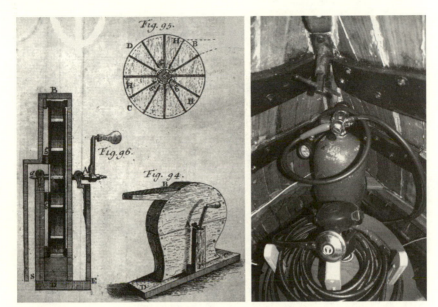

Figure 9.6, left, a rotary "centrifugal bellows" from the eighteenth century, similar to what Bushnell described as "a ventilator within drew fresh air through one of the air pipes. . ." (*Martyn 1747*) Figure 9.7, right, a scuba tank was mounted just behind the test pilot. In the event of an emergency, a seventy-five-foot umbilical hose connected the tank to the pilot's regulator, allowing the hatch to be opened underwater and enabling him to ascend to the surface. (*Roy Manstan*)

set pierside, or in a support boat, that could be connected via a long umbilical to one of the brass pipes in the hatch. The secondary source was available whenever the *Turtle* was afloat and the hatch closed (see, for example, chapter 17).

Depths in Water.		Air compress'd, to		Depth in Water.		Air compress'd, to	
Feet.	Fathoms	parts.	Inches.	Feet.	Fathoms	parts.	Inches.
00	00	1	60	32	0	$\frac{11}{65}$	$30\frac{18}{67}$
1	0	$\frac{11}{34}$	$58\frac{4}{12}$	33	$5\frac{1}{2}$	$\frac{1}{2}$	30
2	0	$\frac{11}{35}$	$56\frac{4}{9}$	66	11	$\frac{2}{3}$	20
3	$\frac{1}{2}$	$\frac{11}{36}$	55	99	$16\frac{1}{2}$	$\frac{3}{4}$	15
4	0	$\frac{11}{37}$	$53\frac{19}{37}$	132	22	$\frac{4}{5}$	12
5	0	$\frac{11}{38}$	$52\frac{2}{13}$	165	$27\frac{1}{2}$	$\frac{5}{6}$	10
6	1	$\frac{11}{39}$	$50\frac{10}{13}$	198	33	$\frac{6}{7}$	$8\frac{4}{7}$
7	0	$\frac{11}{40}$	$49\frac{1}{2}$	231	$38\frac{1}{2}$	$\frac{7}{8}$	$7\frac{1}{2}$
8	0	$\frac{11}{41}$	$48\frac{4}{41}$	264	44	$\frac{8}{9}$	$6\frac{2}{3}$
$8\frac{1}{2}$	0	$\frac{4}{5}$	48	297	$49\frac{1}{2}$	$\frac{9}{10}$	6
9	$1\frac{1}{2}$	$\frac{11}{42}$	$47\frac{3}{7}$	330	55	$\frac{10}{11}$	$5\frac{5}{11}$
10	0	$\frac{11}{43}$	$46\frac{4}{43}$	363	$60\frac{1}{2}$	$\frac{11}{12}$	5
11	0	$\frac{11}{44}$	45	396	66	$\frac{12}{13}$	$4\frac{8}{11}$
12	2	$\frac{11}{45}$	44	429	$71\frac{1}{2}$	$\frac{13}{14}$	$4\frac{2}{7}$
13	0	$\frac{11}{46}$	$43\frac{5}{6}$	462	77	$\frac{14}{15}$	$4\frac{4}{15}$
14	0	$\frac{11}{47}$	$42\frac{9}{47}$	495	$82\frac{1}{2}$	$\frac{15}{16}$	$3\frac{3}{4}$
15	$2\frac{1}{2}$	$\frac{11}{48}$	$41\frac{1}{2}$	528	88	$\frac{16}{17}$	$3\frac{9}{16}$
16	0	$\frac{11}{49}$	$40\frac{10}{49}$	561	$93\frac{1}{2}$	$\frac{17}{18}$	$3\frac{3}{7}$
$16\frac{1}{2}$	0	$\frac{1}{2}$	40	594	99	$\frac{18}{19}$	$3\frac{3}{18}$
17	0	$\frac{11}{50}$	$39\frac{3}{4}$	627	$104\frac{1}{2}$	$\frac{19}{20}$	3
18	3	$\frac{11}{51}$	$38\frac{41}{51}$	660	110	$\frac{20}{21}$	$2\frac{6}{7}$
19	0	$\frac{11}{52}$	$38\frac{10}{13}$	693	$115\frac{1}{2}$	$\frac{21}{22}$	$2\frac{8}{11}$
20	0	$\frac{11}{53}$	$37\frac{1}{3}$	726	121	$\frac{22}{23}$	$2\frac{14}{17}$
21	$3\frac{1}{2}$	$\frac{11}{54}$	$36\frac{2}{7}$	759	$126\frac{1}{2}$	$\frac{23}{24}$	$2\frac{1}{2}$
22	0	$\frac{11}{55}$	36	792	132	$\frac{24}{25}$	$2\frac{2}{7}$
23	0	$\frac{11}{56}$	$35\frac{5}{14}$	825	$137\frac{1}{2}$	$\frac{25}{26}$	$2\frac{2}{11}$
24	4	$\frac{11}{57}$	$34\frac{42}{45}$	858	143	$\frac{26}{27}$	$2\frac{2}{3}$
25	0	$\frac{11}{58}$	$34\frac{13}{19}$	891	$148\frac{1}{2}$	$\frac{27}{28}$	$2\frac{1}{4}$
26	0	$\frac{11}{59}$	$33\frac{33}{59}$	924	154	$\frac{28}{29}$	$2\frac{2}{3}$
27	$4\frac{1}{2}$	$\frac{11}{60}$	33	957	$159\frac{1}{2}$	$\frac{29}{30}$	2
28	0	$\frac{11}{61}$	$32\frac{28}{31}$	990	165	$\frac{30}{31}$	$1\frac{29}{11}$
29	0	$\frac{11}{62}$	$31\frac{36}{31}$	1023	$170\frac{1}{2}$	$\frac{31}{32}$	$1\frac{1}{11}$
30	5	$\frac{11}{63}$	$31\frac{7}{9}$	1056	176	$\frac{32}{33}$	$1\frac{1}{17}$
31	0	$\frac{11}{64}$	$30\frac{45}{10}$	1089	$181\frac{1}{2}$	$\frac{33}{34}$	$1\frac{13}{17}$

Figure 10.1. Robert Boyle, in the late seventeenth century, discovered the relationship between water depth and pressure. He devised a sixty-inch long tube to demonstrate this relationship, the tabulated results of which were published in the *Philosophical Transactions of the Royal Society.* (*Lowthorp 1716*)

Chapter Ten

NAVIGATION

A Compass and a Depth Gauge
Guided the *Turtle*

Navigation at night

The tactical problem that faced Bushnell was that any attack would have to be attempted at night. Bushnell also understood the physical demands that would be placed on his pilot, and would have attempted to minimize the distance required to navigate to the target. Lee begins his account with, "We set off from the City—the Whale boats towed me as nigh the ships as they dared to go, and then cast me off." Before Lee was set free of the tow, he would have immediately established his compass bearings, then closed the hatch and confirmed his course heading by making a visual sighting through the open deadlight.

We know from Lee that the weather was favorable. According to the U.S. Naval Observatory Web site, on the night of September 6, 1776, the moon was in a waning crescent phase (23% of disk illuminated).[1] If there was minimal cloud cover, moonlight may have been sufficient for Lee to make course corrections based on visual sightings. Once at the target, he submerged and had to rely on his depth gauge and compass to position himself under the hull.

Regardless of how much moonlight was available, the interior of the *Turtle* was pitch-black. A standard mariner's compass and his custom designed depth gauge would have been impossible to read. Bushnell, however, found that he could illuminate these critical instruments with the bioluminescent fungi "foxfire."

COMPASS

Bushnell only mentions that "a compass marked with phosphorous directed the course, both above and under the water." From Bushnell's perspective, this device was a common navigation instrument, not requiring any elaboration. Ezra Lee did provide some additional information: "A pocket compass was fixed on the side, with a piece of light wood on the north side, thus +, and another on the east side thus -, to steer by while under water." Isaac Doolittle sold compasses in his New Haven shop, and may have adapted one to carry the foxfire.

Our replica was not intended to transit long distances, and certainly only during daylight. We did, however, include a compass as well as a GPS (Global Positioning System) device. The compass allowed us to view the effects of the surrounding metal on compass performance. We noticed that when installed, apparent north was far removed from magnetic north. This deviation remained constant, however, and we were able to maintain our course during the November 2007 launch and sea trials in the Connecticut River.

The GPS was intended to be used only when operating on the surface. Its antenna was mounted in one of the deadlight windows on the top of the hatch, through which we were able to receive a strong signal. Our plan was to monitor direction and speed during the sea trials. Unfortunately, the pilot had left his glasses on shore and was unable to read the small GPS display. Our assumption is that Ezra Lee, at age twenty-nine, probably had much better eyesight than the test pilot, aged sixty-two, and would have loved GPS technology.

DEPTH GAUGE

When Lee came alongside HMS *Eagle*, his next task was to submerge and attach his mine to the bottom of the ship. In order to position his submarine, Lee had to descend to the proper depth and transit a sufficient distance to contact the flat portion of the hull. This maneuver required both his compass and depth gauge. A compass was almost an off-the-shelf item. The depth gauge, however, was a different matter.

Because of its uniqueness, Bushnell included (as have we) a more detailed description of this device:

A glass tube eighteen inches long, and one inch in diameter, standing upright, its upper end closed, and its lower end, which was open, screwed into a brass pipe, through which the external water had a passage into the glass tube, served as a water-gauge or barometer. There was a piece of cork with phosphorous on it, put into the water-gauge. When the vessel descended the water rose into the water-gauge, condensing the air within, and bearing the cork, with its phosphorous, on its surface. By the light of the phosphorous, the ascent of the water in the gauge was rendered visible, and the depth of the vessel under water ascertained by a graduated line.

The use of a depth gauge for a submarine was proposed by Denis Papin in the late seventeenth century (see chapter 3). The article about Papin's "diving ship" in *The Gentleman's Magazine* (1747), described his depth gauge as: "the recurve barometer . . . open at both ends, whose lower part may be made of iron or wood, shews the depression of the ship very exactly." The author of the article unfortunately was a bit inaccurate in his description of the "barometer" in that if it were "open at both ends," the submarine would have sunk. Papin, a scientist and member of the Royal Society, would have provided, as described by Bushnell, a depth gauge with "its upper end closed."

During the early years of the Royal Society of London, "the Barometer or baroscope was first made publick by that notable Searcher of Nature, Mr. [Robert] *Boyle*, and employed by him and others [including Edmund Halley], to detect all the minute Variations in the Pressure and Weight of Air." The properties of air and the concept of atmospheric pressure evolved from experiments by these and other seventeenth-century scientists. Halley predicted that the height of the atmosphere could be as high as fifty-three miles, but felt it was more likely limited to forty-five miles.[2]

Subsequent experiments by Boyle, Halley, and others determined the relation between water depth and the compression of air. A cylinder devised with a check valve was "let down 33 Foot into the Water, the Mouth [and check valve] downwards, and after a little stay drawn up, was found to be so very near half full of Water." After many trials, the experimenters "concluded that the Quantity of the Air, that filled the Bottle before it was immersed in the Water, was at the Depth of 33 Feet, compressed into half the space it took up before, and so proportionately at other Depths."[3]

Experiments with the mercurial barometer led to what is referred to as "Boyle's Law," or that the ratio of the initial and final volume of air is equal to the reciprocal of the ratio of the initial and final pressure. In the experiment described above, the initial volume of air was at one atmosphere. At a depth of thirty-three feet, pressure had increased to two atmospheres, and the air had compressed to one-half the original volume. To illustrate the relationship, the Royal Society published a table of "Depths from the Surface of the Water to the Bottom of the Air included in a Cylinder of 60 inches, closed at one End, and having the open end downwards."[4] A portion of this table is provided in figure 10.1.

This relationship was well established in the literature by the time Bushnell was studying at Yale. An early application of this theory to a "Sea-Gage" for measuring ocean depths was described in the *Philosophia Britannica*, one of the books available to Bushnell. The device was initially designed to measure up to a depth of one-eighth of a mile, but "since 'tis reasonable to suppose the Cavities of the Sea bear some Proportions to the mountainous Parts of Land, some of which are more than three Miles above the Earth's Surface . . . ," the author suggests a redesign to "explore such great Depths."[5]

The depth relationship in figure 10.1 is for a cylinder sixty inches long. Bushnell reduced the length of the cylinder for his "water-gauge or barometer" to eighteen inches. He then recalculated the compression of the entrained air using Boyle's Law and marked the cylinder with a "graduated line" in such a way that "the depth of the vessel under water [could be] ascertained" by the pilot. For example, at a depth of thirty-three feet, the air in Bushnell's gauge would be compressed to nine inches.

The table was produced from tests in seawater, and we presume that Bushnell's depth gauge was based on similar calculations. The seventeenth-century experimental philosophers, however, understood that freshwater weighed less than saltwater, and that difference would be important when studying freshwater environments:

> Some members of the R. Society did, with two different sorts of Instruments, make divers Experiments for finding the Proportion of the Compression of Air under Water, . . . in the Mouth of the River of Medway, at the time of high Water, where the Depth was then about 19 Fathom, and the Proportion of the Weight of the Saltwater to that of the same Quantity of Fresh

Water, taken out of the River Thames, was 41 to 42. . . . The Proportion of the Weight of Salt Water to that of Fresh, was found by weighing some Ounces of both in a Bottle, whereof the Weight was exactly known, and which was made with so small a Neck, that the Addition or Diminition of one single Drop in it was discernable.[6]

This "41 to 42" proportion of fresh- to saltwater measured by these members of the Royal Society equals 0.9762. Using modern measurement techniques, the density of freshwater (distilled) has been determined to be 62.4 pounds per cubic foot (lb./ft^3), whereas saltwater, with a salinity of 32 parts per thousand (ppt), is 64.0 lb./ft^3. The ratio of these densities is 0.9766. We should not dismiss the pursuit of accuracy and precision by the natural and experimental philosophers of the seventeenth century.

In keeping with Bushnell's design, we produced an eighteen-inch long and one-inch diameter depth gauge for the replica. Figure 10.2 shows the gauge held up alongside the *Turtle* prior to installation. Figure 10.3 shows the lower section as installed and the "graduated line" markings for the various depths. Note that our cylinder was made of an impact resistant polycarbonate material [Lexan] rather than glass.

By the early nineteenth century, the depth gauge was a standard feature on diving bells. An illustration of this gauge (figure 10.4) appeared in the *Edinburgh Encyclopaedia* (Brewster 1832), (see also chapter 1). Their description is similar to the gauge designed by Bushnell:

> It will be satisfactory to the divers to know at what depth they are from the surface; and for this purpose a gauge, represented in Plate CCXXXI [231]. Fig. 6. should be fixed withinside the bell. It is a glass tube a b, hermetically sealed at the top, and the bottom cemented into a metal tube b, which turning at right angles, has a screw to fix it into the side of the bell. To defend it from injury, the tube is bedded in a piece of board, which has divisions and feet marked upon it, to show how high the water rises in it; . . . the water entering freely into the lower end of the tube, condenses [compresses] the air in the glass tube into a space proportional to the intensity of the pressure. Thus at 33 feet deep, the water will rise up half way to the top of the tube . . .[7]

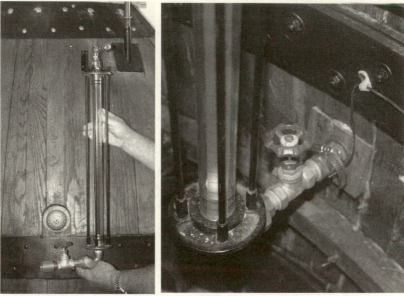

Figure 10.2, left. Our replication of Bushnell's original depth gauge is shown alongside the partially completed *Turtle* hull. Bushnell described his scaled down version of Boyle's experimental depth gauge as "a glass tube eighteen inches long, and one inch in diameter, standing upright . . ." Figure 10.3, right, ". . . its lower end, which was open, screwed into a brass pipe through which the external water has passage into the glass tube." (*Roy Manstan*)

Foxfire

With Bushnell's entry into the world of clandestine warfare, navigating to his target in the dark presented yet another obstacle. Benjamin Gale, in his November 9, 1775, letter to Silas Deane, made specific reference to Bushnell's solution to the problem of illumination. "On the inside is fixed a Barometer, by which he can tell the depth he is under water; a Compass, by which he knows the course he steers. In the barometer and on the needles of the compass is fixed *fox-fire*, i.e. wood that gives light in the dark."

On December 7, 1775, Gale wrote again to Deane: "He proposes going in the night, on account of safety. He always depends on fox-wood, which gives light in the dark, to fix the points of the needle of his compass, and in his barometer, by which he may know what course to steer and the depth he is under water, both which are of absolute necessity for personal safety of the navigator." In the same letter, Gale also noted: "He has tried a candle, but that destroys the air so fast he cannot remain under water long enough to effect the thing."

Bushnell was never specific about the material he used for illumination, only mentioning that he applied "phosphorus" to his compass and depth gauge. Ezra Lee added that both instruments were illuminated with "light wood," while Gale refers to "fox-fire" and "fox-wood" in his correspondence.

A material that glowed with no flame or heat and could sustain its illumination without any apparent consumption of air was a mystery that begged for an answer. It had been a curiosity among scientific minds throughout history. The mysteries of natural phenomena drove many seventeenth-century philosophers to publish their observations in spite of potential religious and political consequences (Galileo was excommunicated for publishing his scientific inquiries).

Sir Francis Bacon, a contemporary of Galileo, was a pioneer in the field of empirical and experimental science and documented his experiments and observations in his *Silva Sylvarum, or a Natural Philosophy in Ten Centuries*, first published posthumously in 1628.

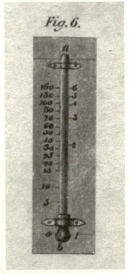

Figure 10.4. By the early nineteenth century, diving bells had become a common sight on the waterfront and were frequently used for salvage and underwater construction. Many were equipped with a depth gauge similar to that used on Bushnell's *Turtle*. (*Brewster 1832*)

Note that the term "Centuries" refers to the one hundred numbered experiments described in each of "Ten" chapters. The following quotes are from his "*Experiment Solitary touching Wood Shining in the Dark*":[8] "The Experiment of Wood that shineth in the dark, we have diligently driven and pursued . . . for that of all things that give light here below [i.e. not in Heaven], it is the most durable, and hath least apparent motion."

Bacon also noted that this strange wood "that did first shine, and being laid dry in the house, within five or six days lost the shining; and laid abroad again, recovered the shining." The experiments included his notion that the light was associated with rotting of the wood. "There was the shining part pared off, till you came to that, that did not shine; but within two days the part contiguous began also to shine, being laid abroad in the Dew; so as it seemeth the putrefaction spredeth."

Toward the end of the seventeenth century, members of the Royal Society were experimenting with the properties of air, or "pneumatics." Robert Boyle published a paper in the *Transactions of the Royal Society* comparing the properties of glowing coal to "shining wood." Boyle noted that a live coal is extinguished by crushing it underfoot, whereas shining wood when crushed, retains its light among the fragments; that coal produces smoke, whereas the shining wood does not; and that "a *quick Coal* is actually and vehemently *hot*; whereas I have not observed *shining Wood* to be so much as *lukewarm*."

The quality of foxfire of most interest to Bushnell, however, was its ability to sustain the illumination. Boyle emphasized that quality: "a *live Coal*, being put into a small close Glass, will not continue to *burn* for very many *Minutes*; but a Piece of *shining Wood* will continue to *shine* for some whole *Days*. The assertion . . . may be easily made out by what I have tried upon *shining Wood* sealed up *hermetically* in very small Glasses, where the *Wood* did for several Days retain its *Light*."[9]

Foxfire was (and still is) prevalent in New England forests, and certainly a phenomenon familiar to Bushnell. It wasn't until the early nineteenth century that the luminescence of rotting wood became associated with fungi. The most likely candidate for Bushnell's "phosphorous" is the honey mushroom, *Armillaria mellea*.[10] There are several related species, including *A. gallica*, that possess bioluminescent properties. The common name derives from the tan or golden color of the mushroom caps and not from its taste, although some species are edible.

The luminescence does not occur in the fruiting body (i.e., the stalk and cap associated with the reproductive process) among the *Armillaria*. It is only associated with its mycelium and rhizomorphs. The fibers of these branching root-like structures infiltrate dead or dying trees feeding on (and contributing to) the rotting wood. The production of light is a chemical process that occurs when the mycelium is growing and feeding, and there are environmental conditions essential to maintaining this process. As Sir Francis Bacon discovered, the wood must remain moist in order for the mycelium to feed and produce light.

Bushnell had planned to attack the British fleet at Boston during the fall of 1775, but found that his "phosphorous" ceased to produce light.

Bioluminescence in *Armillaria* is temperature dependent, occurring below 86°F and, much to the chagrin of Bushnell, above freezing. In the December 1775 letter to Silas Deane cited above, Benjamin Gale described the problem Bushnell faced: "He now finds that the frost wholly destroys that quality in the wood, of which he was before ignorant, and for that reason and that alone he is obliged to desist." Bushnell had to put his plans on hold until spring.

Figure 11.1. The total destruction of the brig *Dorothea* by Robert Fulton's clockwork torpedo on October 15, 1805. (*Pesce 1906*)

Chapter Eleven

FIREPOWER

THE TURTLE AND ITS INFERNAL MACHINE

According to Benjamin Gale's November 9, 1775, letter to Silas Deane, the mission of the *Turtle* as initially conceived was to secure one mine to a ship with the timing mechanism set for twelve hours; return to another ship with a second mine set for eight hours; and finally a third one set for six hours. Under this scenario, Bushnell must have planned to produce a minimum of three mines, each rigged with its tether line and wood screw. The quantity that Bushnell actually produced is unknown.

In September and October 1776, the *Turtle* was used in three missions against British ships. After the initial and unsuccessful attempt against HMS *Eagle*, Ezra Lee made a hasty retreat back to Manhattan. With the British in pursuit, Lee jettisoned the mine. It detonated in the harbor just as expected. With two subsequent attempts conducted in the Hudson River, we know that Bushnell certainly had carried at least one additional mine with him to New York, and probably had a third in reserve.

Bushnell described his mine as "a large powder magazine . . . made of two pieces of oak timber, large enough when hollowed out to contain one hundred and fifty pounds of powder." If he had three weapons, each containing 150 pounds of powder, he had enough gunpowder to load thousands of cartridges for a standard infantry musket (one pound of musket powder was sufficient to produce about fifty cartridges). Those supporting Bushnell's venture had to be convinced that committing a valuable resource for this "engine of devastation" was worthy of the cost.

Bushnell had tested his concept of detonating gunpowder underwater and found two important results: that gunpowder would, in fact,

ignite and explode when submerged, and that the force would propagate
upward toward the surface, disintegrating any structure in its way. In his
letter to Jefferson, Bushnell described the success of his early experiments: "A match put to the priming, exploded the powder, rending the
plank into pieces; demolishing the hogshead; and casting the stones and
ruins of the hogshead, with a body of water, many feet into the air, to the
astonishment of the spectators."

Convincing his scientific mentors and supporters was essential. If
these influential individuals remained skeptical, they would be unwilling
to continue their enthusiasm for the idea. Benjamin Gale expressed his
misgivings in his letter to Benjamin Franklin, wondering whether: "100
wt of Powder will force its way through the ship I fear the Water will give
way before the Bottom of the Ship, and the force of the Explosion
Eluded." Yet Gale remained an ardent supporter, ending his letter with,
"His reasoning so Philosophically and Answering every Objection I ever
made that In truth I have great relyance upon it." It was obvious that
Bushnell and Gale had debated the effectiveness of using underwater
explosives against a British warship. Their discussion no doubt also covered other unique technologies that Bushnell proposed for his complicated delivery system. Had he not provided convincing arguments, Gale
would have been unwilling to risk his own reputation when seeking the
interest and support of Ben Franklin and Silas Deane.

If, after the war, there was any lingering doubt about the explosive
potential of an underwater mine, all uncertainty was removed with the
successful experiments carried out by Robert Fulton during the first
decade of the nineteenth century. The woodcut (figure 11.1) was originally published in 1810 in Fulton's *Torpedo War and Submarine
Explosions,* illustrating the destruction of the brig *Dorothea* on October
15, 1805, as witnessed by members of the British political and military
establishment. Fulton noted that he had "filled one of the Torpedoes
with one hundred and eighty pounds of powder, and set its clockwork to
eighteen minutes." Fulton described the destruction: "At the expiration
of eighteen minutes, the explosion appeared to raise her bodily about six
feet; she separated in the middle, and the two ends went down; in twenty seconds, nothing was to be seen of her except floating fragments."[1]
Two years later Fulton conducted a similar, and successful, demonstration in America.

THE MINE

The weapon Bushnell planned to use consisted of three components: a container to hold the gunpowder, a clockwork detonator, and the hull attachment mechanism. When developing an overall plan for our replica of the *Turtle*, we concentrated on the vessel itself and the mechanical systems that the pilot was required to operate. We made an estimate of the dimensions of the mine and designed a method of securing it to the back of the submarine, such that the test pilot would be able to jettison the mine as Ezra Lee had done.

Bushnell did not describe the shape or dimensions of his mine or "magazine," only that it was produced from hollowed-out pieces of oak sufficient to contain the powder charge. He also did not provide clues as to the location of the clockwork and flintlock firing mechanism, nor did he explain how it was all kept watertight.

To gain insight into the dimensions of the mine, we turn to a view of the gunpowder used by the military, the composition of which remained essentially unchanged until well after the Civil War. Gunpowder is a combination of saltpeter (potassium nitrate), brimstone (sulfur), and charcoal.[2] The process of mixing and creating the "powder" allowed the mills to manufacture their explosives in several grades of fineness, depending on the application. During the Revolutionary War, there were essentially two primary grades of gunpowder produced for military applications. A coarse grade, with the largest granules or "grains," was for use in artillery. Musket powder was midgrade and used to manufacture the paper cartridges that the infantry carried in leather covered wooden "cartouche boxes" on their belt. The soldier would prime their flintlocks with a small quantity of the powder from their pre-made cartridges.

Access to the primary ingredient, saltpeter, created a cottage industry among the farming community. It could be isolated from its source, cow (or other mammalian) urine, through filtration and distillation.[3] A premium was paid to anyone who could bring their product to the powder mills being established throughout the state, including that of Isaac Doolittle of New Haven, the same Doolittle that ran a bell foundry and produced the clockworks for Bushnell's mines.

An early, although also unsuccessful, attempt to use underwater mines was by Cornelius Drebbel. In 1626 Drebbel was commissioned by Charles I to develop a "water petard," also referred to as a "water myne."

Figure 11.2. The petard was a form of ordnance well known in the seventeenth century. Cornelius Drebbel is said to have adapted this device for underwater use at the behest of Charles I in 1626. A year later Drebbel's "water petards" were used, although unsuccessfully, against the French. (*Elton 1668*)

Charles I must have been impressed with Drebbel as he was also interested in his "boates to goe under water."[4]

A petard (figure 11.2) was an explosive device designed to detonate against a structure, such as a bridge abutment or castle portcullis. The "charges for these Petards are to be the finest powder that can be got, beaten hard into the Petard."[5] The cone shape and the air space between the packed powder and the structure created what is now referred to as a shaped charge, where the explosive force is initially directed toward the structure. There is no mention of how Drebbel intended to adapt the petard to underwater warfare. Denis Papin apparently also intended to use an underwater explosive with the submarine he had built during the last decade of the seventeenth century for Charles, Landgrave of Hesse–Cassel. Papin's submarine (see chapter 3) included a cylindrical compartment with an opening "through which the man inclined in the cylinder will be able to destroy the enemy's ships."[6]

When Bushnell designed his mine, he had estimated that 150 pounds of gunpowder would be sufficient to destroy a British warship. He then needed to build a watertight magazine to carry that quantity of powder, plus the timing and firing mechanisms. Assuming that Bushnell didn't simply fill the cavity with loose powder, the total volume that would hold 150 pounds of gunpowder depended on its compaction. As with all of his components, the details of his mine were a well-guarded secret.

Submarine designer Denis Papin, working with Christian Huygens in the late seventeenth century, presented the results of his experiments with gunpowder in a paper published in 1674. Papin stated that "a *Cubic Foot* can hold much more than 72 pounds of Powder."[7] In keeping with the seventeenth-century petard, and what Papin likely used as well, we believe that Bushnell would have preferred musket powder to the coarser artillery powder. From Papin's "more than 72 pounds," we used a density of seventy-five pounds per cubic foot to simplify our estimate that Bushnell's mines were produced with a cavity of two cubic feet to contain the 150 pounds of gunpowder. As a comparison, however, we were able to calculate the density of artillery powder to be about fifty-five to sixty pounds per cubic foot based on information in Barnes (1869) and a publication from the United States Navy Bureau of Ordnance (1874).[8]

If we consider Bushnell's two cubic foot capacity mine to have a cylindrical cavity 24 inches long, the diameter will be approximately 13 1/2 inches. It is likely that his mine, built from two pieces of hollowed-out oak timber, may have had 3-inch-thick walls, making the outside dimensions at least 2 1/2 feet long by 1 1/2 feet in diameter.

A size comparison can be made with one of the many barrel-shaped containers in which most eighteenth-century commodities were shipped, stored, and sold. For example, a barrel of wine was thirty-one and a half gallons, a barrel of ale contained thirty-two gallons and a barrel of beer held thirty-six gallons. Other common terms (see glossary) included "hogshead" (thirty-three gallons), "kilderkin" (eighteen gallons), and "firkin" (nine gallons). With a volume of two cubic feet (~sixteen gallons), Bushnell's mine would have been about the same size as a kilderkin.

THE CLOCKWORK TIMING DEVICE

Another design dilemma that Bushnell faced was how to detonate his mine. With an initial strategy that would have the *Turtle* deliver three mines in succession, it would be necessary to provide timing devices on each mine set such that all of the explosions would take place concurrently, or approximately so, with ample allowance for the *Turtle* to escape. While Bushnell later considered his initial idea to plant three mines too ambitious and opted for a single mine, he would still need a timed detonation. The solution was to modify a clockwork that could be set to trigger a flintlock mechanism, the sparks from which would ignite a charge of priming powder that would in turn detonate the 150 pounds of powder contained in his mine.

There is, however, an inherent problem with using a flintlock. If the mine tips at an angle sufficient to spill the priming powder, the sparks from the triggered lock will not ignite the charge. Fulton, in his 1810 publication, described exactly this problem when it occurred while

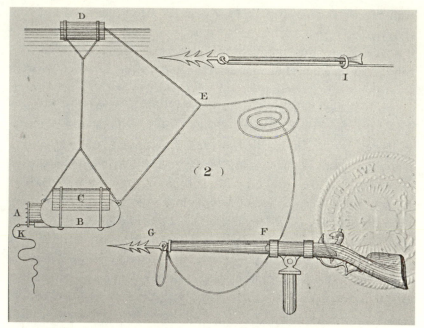

Figure 11.3. Robert Fulton's clockwork torpedo was designed to be secured to the targeted ship with a harpoon fired from the attacking vessel. (*Barnes 1869*)

demonstrating his clockwork mine to an American audience in 1807. "The brig was anchored, the Torpedoes prepared and put into the water in the manner before described; the tide drove them under the brig near her keel, but in consequence of the locks turning downwards, the powder fell out of the pans and they both missed fire."[9] Fulton later corrected the problem and eventually completed a successful demonstration.

Bushnell would have faced the same issue. We believe that his clockwork was mounted at the top of the mine, such that the position of the lock and priming powder remained upright. The center of buoyancy of the mine was designed to be sufficiently far above its center of gravity that when disengaged from the *Turtle*, the mine would float vertically up against the underside of the target. A device consisting of a crank and gears in the collection of the Connecticut Historical Society (see plates following page 50 in Wagner 1963) is purported to be part of Bushnell's firing mechanism. Having studied this item, we are unconvinced that it was a component of the original clockwork, although it may have served in some other of Bushnell's "infernal machines."

Figure 11.4, left. This example of a Confederate clockwork torpedo is in the collection of the West Point Museum. It was among those used by Southern forces against Union gunboats in the St. John's River, Florida. Figure 11.5, right. Detail of the Confederate torpedo timing mechanism and percussion detonator. Note that the clockwork was produced in Bristol, Connecticut. (*West Point Museum: Roy Manstan*)

Fulton's clockwork torpedo (figure 11.3) was designed to be deployed by a harpoon gun carried on an attacking vessel. The harpoon would be fired into the targeted hull, allowing the torpedo to be suspended below the water line. A pin attached to a lanyard would be pulled from the torpedo, engaging the clockwork set with sufficient time for the small craft to escape.

It would be more than fifty years before another attempt was made to create a clockwork torpedo. Early in the Civil War, the Confederacy fully understood that the Union navy would be capable of overwhelming the southern coastline and isolating the South from its dependency on imported military supplies. In response, a Torpedo Bureau was established in Richmond with responsibility for the design and manufacture of a weapon that served the Confederacy well. Confederate torpedoes were built in a wide variety of designs; some were moored and floating just below the surface, others were mounted at the ends of timber frames set into river bottoms. There were also spar torpedoes carried by the "David" class semi-submersible boats and the *Hunley*, the first submarine to successfully sink a vessel in combat.[10]

Among the most unique, however, were several clockwork designs, also called horological torpedoes.[11] One type intended as a floating torpedo has survived and is in the collection of the U.S. Military Academy museum at West Point (figures 11.4 and 11.5). According to information accompanying the torpedo, it was used in the St. John's River in Florida and sent to the Academy shortly after the war. The device is cylindrical, 18 1/2 inches in diameter and 12 1/2 inches high. The clockwork, set into a circular recess flush with the top surface, is designed to trigger a percussion hammer that detonates a percussion cap near the end of a gun barrel. The barrel is positioned vertically, such that its muzzle sits in the middle of the torpedo's seventy-five pound charge of gunpowder. It was designed to have the timer engaged by pulling a spring-loaded pin just prior to either being set adrift upstream, or when placed in direct contact with its target, if used during a clandestine operation by swimmers, or with a small-craft. Bushnell may have used a similar pin that would be pulled free by the buoyancy of the mine when released by the *Turtle* pilot.

From an examination of the West Point museum specimen, it appears that when the torpedo was disarmed, the clockwork was also disassembled and some of the firing mechanism removed to ensure that it could not be rearmed. Nonetheless, the clockwork was an off-the-shelf item produced, according to an address stamped on the clockwork frame, by Wm. J. Hill of Bristol, Connecticut. We can surmise, therefore, that Bushnell and Isaac Doolittle may have also simply adapted a clockwork from Doolittle's inventory.

The attachment mechanism

In replicating the *Turtle*, we concentrated our interest on the mechanism Bushnell would have used to secure the mine to the underside of the target ship. He faced an interesting problem. First, he had to devise a method to guarantee that the mine would remain in place. Threading a wood screw into the hull was a logical answer. The mine, once disengaged from the *Turtle*, would float up against the hull, held in place by a rope running to the screw. This seems straightforward enough, but there was also the problem of how to thread the screw into the wood and then release it from the threading mechanism. Bushnell devoted a lengthy portion of his letter to Jefferson describing the entire weapon system. The following is an extract dealing with the wood screw attachment mechanism that we read and reread many times before designing what we felt was an acceptable replica.

In the forepart of the brim of the crown of the submarine vessel, was a socket, and an iron tube, passing through the socket; the tube stood upright, and could slide up and down in the socket, six inches: at the top of the tube, was a wood-screw (A) fixed by means of a rod, which passed through the tube, and screwed the wood-screw fast upon the top of the tube: by pushing the wood-screw up against the bottom of a ship, and turning it at the same time, it would enter the planks; driving would also answer the same purpose; when the wood-screw was firmly fixed, it could be cast off by unscrewing the rod, which fastened it upon the top of the tube.

This mechanism consisted of four primary components: the socket, the iron tube, the iron rod, and the wood screw. Our first assumption was that the socket was permanently fixed to "the forepart of the brim of the crown." The tube was able to rotate and slide vertically within the socket. The iron rod was likewise able to slide and rotate within the tube. Finally, this rod was connected to the wood screw, in such a way that the screw could be secured at the top of the tube and later released.

When in position under the target hull, the pilot would slide the tube up until the tip of the screw contacted the wooden hull. The pilot then rotated the tube, forcing the wood screw to thread into the planking. Once in place, the pilot then had to hold the tube with one hand while unscrewing the end of the iron rod from the back of the wood screw with his other hand. Once the iron rod was detached from the wood screw, the pilot could slide the tube down and out of the way. The pilot was then free to jettison the mine by turning another rod from inside that held the mine in a cradle against the back of the vessel. It was the process of casting off the mine that activated the clockwork timing mechanism.

Keeping the attachment mechanism with its multiple moving parts watertight would have been a challenge. Bushnell's answer was that, "wherever the external apparatus passed through the body of the vessel, the joints were round, and formed by brass pipes . . . the holes through the pipes were very exactly made, and the iron rods, which passed through them, were turned in a lathe to fit them; the joints were also kept full of oil, to prevent rust and leaking."

For the solid shafts such as used with the two propellers, there was only one sliding surface. In this case, however, the outside surface of the iron tube had to be machined and polished to fit inside the socket. The interior of the iron tube also had to be bored out with precision in order to provide a sliding fit to the iron rod. When installed on the *Turtle*, the crank handle, used by the pilot to thread the wood screw into the targeted hull, was added to the tube. Another handle, used by the pilot to release the wood screw, was placed on the iron rod. The whole system had to be disassembled occasionally to allow it to be cleaned and oiled. Such was the talent and ingenuity of Bushnell and Isaac Doolittle who, according to Benjamin Gale in his letter to Franklin, made "those Parts which Conveys the Powder, and secures the same to the Bottom of the Ship . . ."

THE REPLICA

When designing our replica, we were not constrained to eighteenth-century manufacturing methods, and employed a generous number of O-rings for all of our hull penetrations. We did, however, also ensure that the mechanism could be removed to allow us to clean and grease the O-rings and the mating surfaces prior to each launch.

Figure 11.6 shows the disassembled replica wood screw mechanism. From left to right are Bushnell's "iron tube" designed to rotate and slide within the socket, the "rod, which passed through the tube," and the wood screw that was held "at the top of the rod." The crank handle and the wing nut on the rod can be removed to allow the tube to be installed or removed by sliding it through the socket.

Figure 11.7 is the assembly prior to installation into the hull, including the "strong piece of rope [that] extended from the magazine to the wood-screw." A close-up view of the *Turtle* replica wood screw as modified to insert into the tube (figure 11.8), is shown alongside a wood screw taken from a British musket produced around 1770. The thread pitch of both screws is ten threads per inch.

The "socket," as described by Bushnell, was installed on the replica *Turtle* in a location we believe corresponds to what he referred to as the "brim," just forward of his hatch or "brass crown." The final installation with the screw visible at the top of the tube and socket is seen in figure 11.9. Note that the wood screw, the "tip of the spear," lies below the profile of the hatch. This position protected its sharp and vulnerable point when the *Turtle* submerged and maneuvered into position, a process requiring the pilot to slide along or bump up against the underside of the

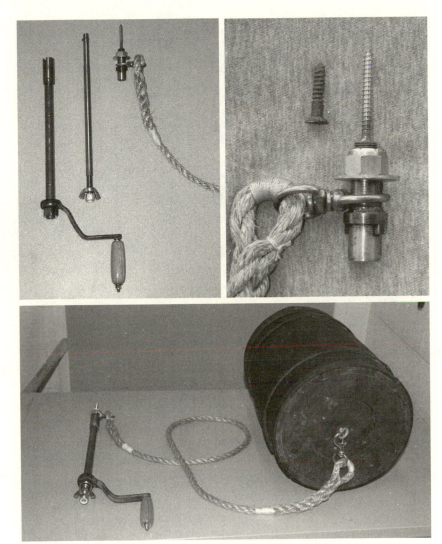

Figure 11.6, upper left, components of the replica mine attachment mechanism. Figure 11.7, bottom, The assembled components with the rope tether running to the mine. Figure 11.8, upper right, detail of the replica screw alongside a wood screw from a British musket, ca. 1770. (*Roy Manstan*)

ship's hull. The hand crank was just forward of the pilot and, as specified by Bushnell, within easy reach "so that every thing might be found in the dark."

We leave this chapter with a mystery. Recently, *Turtle* Project volunteers Bruce and Janna Greenhalgh (see aerial photography of the

Figure 11.9. The "socket" holding the wood screw mechanism was located just forward of the hatch. (*Roy Manstan*)

Connecticut River, figure 19.2) found an article by H. Edward Moore in the Autumn 1934 issue of *Mechanics and Handicraft* titled "The First Submarine Ever Built in America." The article describes a relic carrying an old yellowed label: "Portion of David Bushnell's submarine torpedo used in the Revolution."[12]

This certainly grabbed our attention. Accompanying the article were photos of this "torpedo" and the flintlock firing mechanism (figures 11.10a and 11.10b), and a caption that read: "This ingenious and unique torpedo model consists of a heavy cylindrical tin powder-magazine 10 1/2 inch long by 4 1/2 inch in diameter (tapering at both ends)—on the side of which is fitted a block of wood whittled out by hand and hollowed out on the inside to contain a flint-lock gun lock, with a trigger arrangement on the outside operated by a rod of wood."

According to Moore, the "rediscovery of what is said to be the first experimental working model made and used by David Bushnell" had been in the collection of the Peel Museum in New York City. After the museum closed, Moore noted that the "torpedo" was sold at auction, and after many years was bought by a Midwestern collector from "a relic dealer's shop in Chicago."

So, was Moore's "relic" actually used by David Bushnell? The "tin powder-magazine" was of a size that could contain about five pounds of

Figures 11-10a (left) and 11-10b (right). A museum tag reads "Portion of David Bushnell's submarine torpedo used in the Revolution" reportedly "rediscovered" in 1934 (left). Detail of the flinlock firing mechanism (right). (*Mechanics and Handicraft, Autumn 1934*)

gunpowder and certainly would have created what Bushnell described as a "very violent explosion" that he occasionally demonstrated "for the satisfaction of the gentlemen," including his mentor Benjamin Gale. Gale made note of this in his letter to Ben Franklin, assuring him that Bushnell "has made the Experiment of firing Powde[r] Under Water." When Bushnell wrote to Thomas Jefferson in 1787, he was very specific about having "made many experiments . . . some of them with large quantities of powder."

The photos, however, show no sign that this "torpedo" would have been watertight. Yet if the whole mechanism was set into a submerged barrel "loaded with stones" (as Bushnell mentioned), the flintlock detonator could be fired remotely by pulling a long cord attached to the wooden rod shown on the photo. There is a tantalizing possibility that Moore has described and photographed a demonstration model made by Bushnell. Maybe a "Midwestern" collector will read this and the "relic" with its tattered label will be rediscovered again.

PART THREE

INVESTIGATION

The *Turtle* is launched into the Connecticut River, November 10, 2007. (*John Nilson*)

By the summer of 1775, Bushnell was ready to demonstrate the capabilities of his "submarine vessel" to a very limited audience. His primary mentor during the evolution of the *Turtle* was Benjamin Gale who, in a letter to Benjamin Franklin, related his eyewitness account of this "new Machine for the Destruction of Ships of War." Gale then described how "in the most private manner he conveyed it on Board a Sloop in the Night and Went out into the Sound. He then sunk under Water, where he Continued about 45 Minutes without any Inconveniency as to Breathing."

While Gale's letter does not distinguish whether David or his brother Ezra piloted the *Turtle* during this trial, it is very likely that both of the brothers were involved with the initial testing. David was keenly aware of his own physical limitations and realized that he would depend on his brother to conduct most of the testing, as well as the intended mission against the British.

The goal of the *Turtle* Project was not simply to determine if the *Turtle* was an operational submarine—that question was answered with the Bicentennial replica—but to uncover the mysteries of how it operated. We believe that the care and attention to detail that led to the design of our replica allowed us to investigate with confidence the performance of the original vessel.

The testing described in the following chapters provided physical data regarding the hydrodynamics of the vessel and the characteristics of its propulsion. We concentrated our efforts on the performance of our concept of Bushnell's propeller, and in particular, the interaction between the pilot and the overall propulsion system. We found it necessary to consider multiple propeller designs, all of which fall within constraints associated with the limited descriptions of participants and eyewitnesses. Our test results enabled us to document what we believe is the most likely design that propelled the *Turtle* into history.

Yet the picture would be far from complete were it not for the attention paid to the feedback from our test pilots. For that reason, we have included all of the commentary provided during the testing, and have used these responses in our final evaluation of the *Turtle*. The operation of our replica would subject our pilots to the same physical, and to some extent mental, stress experienced by Bushnell's pilots in 1775 and 1776. We are confident that the voices of our pilots echo the voices of Ezra Bushnell, Ezra Lee, and Phineas Pratt.

Chapter Twelve

WALKING IN BUSHNELL'S SHOES

OPERATIONAL TESTING OF THE *TURTLE*

THE REPLICA *TURTLE* GOES TO SEA

The *Turtle* replica has been operated in the water on three occasions. The first occurred on October 22, 2007, at the Essex Boat Works, located along the Connecticut River, in Essex, Connecticut. The *Turtle* was lowered into one of the slips where the test pilot, Roy Manstan, could evaluate the mechanical and safety systems. After several hours, the *Turtle* was determined to be seaworthy and plans for the official christening and launch could proceed.

On November 10, the *Turtle* was transported to the Connecticut River Museum, also in Essex, Connecticut. The museum had agreed to host the event and at 10:30 a.m., State Senator Eileen Daily christened our vessel the submarine "*Turtle*." All in attendance watched as the *Turtle* was set into the water alongside the museum and headed out into the Connecticut River with author Roy Manstan at the controls (figure 12.1). After completing a series of maneuvers, the *Turtle* returned to the pier.

MYSTIC SEAPORT

Early in 2008, the Mystic Seaport Museum in Mystic, Connecticut, was approached about the possibility of conducting the final series of operational tests at their facilities. After meetings between the *Turtle* Project and Seaport staff, a date was agreed upon and preparations began.

The Seaport had recently completed construction of a shiplift facility, designed to allow the docking and launch of any of the museum vessels scheduled for hull maintenance and restoration. Figure 12.2 shows

Figure 12.1. The *Turtle* underway in the Connecticut River after its christening and launch in 2007 (*Warner Lord*)

the shiplift with a series of carriages on which a vessel's hull is blocked and secured. The lift platform is lowered by electric winches located along the length of the facility. A vessel is then brought into position above the carriages and secured. When the platform is raised, the carriages can be rolled along tracks to bring the vessel ashore. Figure 12.3 shows the Seaport's historic ship, *L. A. Dunton,* in the lift.

The shiplift was available to support the *Turtle* Project from May 19 through 21, 2008. The facility, located at the south end of the museum waterfront, is run by the museum shipyard staff and operates between 7:00 a.m. and 3:30 p.m. While it is located at a fair distance from the main channel of the Mystic River, there was some concern that tidal currents at the shiplift would affect the testing. After reviewing the tide charts for the Mystic area, it was found that there would be three periods of slack water during the time available to the Project. This provided ample opportunity to conduct the two primary propulsion tests that required minimal current conditions.

The *Turtle* was delivered the morning of the 19th and set into position on the shiplift. High winds during the afternoon precluded launching the *Turtle,* and one of the slack tides was lost. Weather predictions indicated that conditions would be improving and preparations began for

Figure 12.2, left, the shiplift facility at the Mystic Seaport Museum where the replica *Turtle* was tested in May 2008. The rolling cradles are used to secure a vessel being brought ashore for maintenance or restoration. Figure 12.3, right, the museum's historic ship *L. A. Dunton* ashore on the shiplift cradles. (*Daniel Manstan*)

an early start on the 20th. The test team assembled the following morning and testing commenced on schedule.

OPERATIONAL TESTING

The tests described below were primarily designed to provide data for evaluating the overall performance of the *Turtle* replica and the pilot. The results were then used to speculate as to the performance of the original vessel and its pilots. The initial testing considered the propulsion characteristics of the *Turtle* and covered: (Test 1) static thrust from the propeller; (Test 2) hull resistance (towing load vs. speed); (Test 3) transit speed with the *Turtle* operating on the surface; (Test 4) maneuvering (backing down and S-turns); and (Test 5) submerged operations (including the mine attachment mechanism). The testing then turned to the pilot and included: (Test 6) the pilot's endurance, i.e., his ability to sustain sufficient power to the internal drive mechanism; and (Test 7) the ergonomics associated with the position of the pilot with respect to the operational controls.

TEST INSTRUMENTATION

The majority of the testing related various parameters to propeller rotation, measured in revolutions per minute (rpm). The pilot monitored rpm using an EXTECH Model 461940 Panel Tachometer (figure 12.4),

Figure 12.4. The VHF radio, GPS, and tachometer used during the 2007 launch and sea trials and the 2008 operational testing. (*Roy Manstan*)

modified to run directly from a 12-volt battery source mounted inside the *Turtle*. An EXTECH Model 461955 Proximity Sensor was used in conjunction with a metallic stud mounted on the flywheel. The tachometer provides a resolution of 0.1 rpm in the range of 5 to 1000 rpm. Our test plan specified that propeller operation would be limited to a range of 30 to 100 rpm.

The photo also shows the West Marine Model "Submersible" VHF radio used throughout the testing. The *Turtle* was also equipped with a Garmin Model 48 GPS unit that was only used during the launch and transit demonstrations in November 2007. Not shown is a compass mounted just to the right of these instruments and directly in front of the pilot.

TURTLE COCKPIT

The interior of our replica was arranged as close to the original *Turtle* as we could envision. The circular entrance is shown on figure 12.5, early in the construction phase. Looking down into the hull, the partially completed drive mechanism is visible. The rectangular opening in the floor behind the treadle is for access to the bilge.

The completed *Turtle* interior is shown on figure 12.6. The pilot would slide in through the opening and sit on the heavy wooden bench that was, as Bushnell described, positioned "parallel with the conjugate

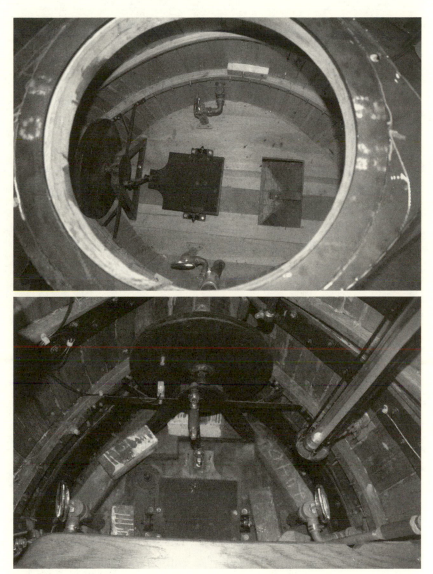

Figure 12.5, top. The interior of the *Turtle* during construction. Figure 12.6, bottom. The completed *Turtle* cockpit just prior to the November 2007 launch. (*Roy Manstan*)

diameter." In front of the pilot was the propeller with its crank and treadle. Note that the crank handle is near top dead center of the shaft rotation. The EXTECH proximity sensor for the tachometer can be seen mounted on the shaft support bracket. Also visible in this photograph is the additional lead ballast set on the floor along the hull perimeter. The hand-operated bilge pumps are shown as installed port and starboard, and just forward of the pilot's bench. This was a convenient location for the pilot. The tiller can be seen "hard to starboard" along the forward edge of the bench. The eighteen-inch-long cylindrical depth gauge is shown mounted on the starboard hull, its graduated lines clearly visible along the pilot's line of sight.

TEST PILOTS

The propulsion and maneuvering tests were accomplished by Paul Mileski, who had been selected as the pilot for the bulk of the testing described in the following sections of this book. Roy Manstan, the *Turtle* Project test director, being the most familiar with the ballasting and other onboard mechanical systems, performed the submerging and wood screw tests. David Hart most closely represented the age and presumed physical capabilities of the original pilots (Ezra Bushnell, Ezra Lee, and Phineas Pratt), and was selected from the NUWC dive team for the pilot endurance tests.

Documenting the interaction between the pilot and various operations of the *Turtle*, while subjective, is a critically important element in these tests. Having the majority of the operational testing performed by the same individual provided consistency among the parameter observations. All of the pilots' comments are transcribed at the end of each set of test results, giving the reader a sense of what the original *Turtle* pilots might have described to Bushnell when they were training in the Connecticut River. The terminology may be modern, but the sentiments are as human today as they would have been in 1776.

BUSHNELL'S PROPELLER

When Bushnell decided to employ an entirely new concept in vessel propulsion, his "oar for rowing forward or backward," there was no precedent to fall back on to design the most efficient "oar." We have already discussed what may have been the origin of the idea of a propeller and the blade angle that he may have considered. Because the mission of his vessel depended on efficient propulsion, he may have experimented with more than one design. Whether he had the opportunity to experiment is certainly a question, considering all of the other mechanical systems he had to design and incorporate into his submarine.

There is no indication of whether his propeller was wood, forged out of iron, or cast in brass. Building a wooden hub with spokes designed to accommodate blade variations would have been a relatively straightforward task. A forged iron propeller would be more difficult to alter, although changes in blade angle could be accomplished in the forge. A cast brass propeller would be the most difficult to modify.

Regardless of what the original *Turtle* carried for its propeller, it certainly worked. Ezra Lee described his experiences with propulsion: "With hard labor, the machine might be impelled at the rate of 3 [k]nots an hour for a short time." He further claims to have "rowed for 5 glasses [2 1/2 hours, see glossary], by the ships' bells before the tide slacked."

The glossary describes the time-keeping process onboard ships, and how the ship's bell was used to specify half-hour intervals during a typical four-hour watch. (See the terms "Bells," "Glasses," and "Watch.") If Lee had approached his target at midnight, he would have heard the *Eagle*'s bell rung eight times, denoting the end of the 8:00 p.m. watch and the beginning of the midnight to 4:00 a.m. watch. Each time the bell was rung, a half-hour interval had passed. One "bell" would have been rung at 12:30 a.m., two "bells" at 1:00 a.m., three "bells" at 1:30 a.m., and so on. When Lee indicated that he "rowed for 5 glasses," he meant five half-hour intervals, or a total of two and a half hours, had passed.

MODERN PROPELLER DESIGN

A thorough discussion of modern screw propellers is beyond the scope of this book. Naval architecture texts typically cover this topic in detail (see, for example, Barnaby (1967), Comstock (1967), Gillmer and Johnson (1982), and Gerr (2001)). The concepts of pitch and slip, however, are important in evaluating and understanding propeller performance, and are summarized below.

Propeller "Pitch"

Propellers are typically defined in terms of "pitch." Propeller pitch is similar in concept to screw threads. The threads lie along a helical path spiraling along the length of the screw. When you drive a screw into wood, it will move into the material a specific distance with every 360° rotation of the screwdriver. Screws are often specified as "threads per inch." For example, ten threads per inch will enable a screw to travel one-tenth of an inch per revolution. That distance (i.e., one-tenth of an inch), is referred to as the "pitch." Viewing a modern propeller hub from the side

Figure 12.7. This side view of the *Sabino* propeller at the Mystic Seaport Museum illustrates the change in the angle of the blade from the hub out to its tip. (*Roy Manstan*)

(figure 12.7), the blade is seen at an angle with respect to the centerline of the hub that varies (becomes steeper) from the root to the tip. Propeller blades are essentially short segments of a helical Archimedean screw. If you trace a path along 360° of this helix (as with a screw thread), the horizontal distance between the starting point and ending point is the pitch.

The pitch of the helix is defined by the equation:

$P = 2\pi R \tan \Theta$

Where:

P = Pitch (inches)

R = Radius to a location on the blade (inches)

Θ = Blade Face Angle at "R"

For most applications, the blades of modern propellers, such as shown in figure 12.7, are termed "constant pitch." In this situation, "P" in the equation above will remain constant over the length of the blade only when "tan Θ" decreases (i.e. when the blade face angle Θ becomes steeper) at the same rate that the radius "R" increases. If, however, the blade face angle (and thus "tan Θ") remains constant, then its pitch "P" will increase at the same rate as the radius "R." The *Turtle* propeller blades meet this condition and as such are considered to have a "variable pitch." Figures 12.8a and 12.9b illustrate the helices and associated "pitch" for the variable pitch (*Turtle*) propeller and the constant pitch (*Sabino*) propeller.

Propeller "Slip"

As a propeller rotates through 360°, it will drive a vessel horizontally. If water behaved like wood, one rotation of the propeller would drive the

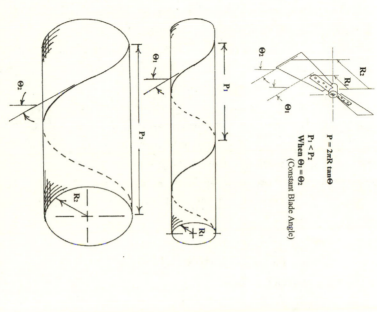

Figure 12.8a. *Turtle* replica, variable pitch propeller. Helical path of a point at "R" along the surface of the propeller as it rotates and transits through the water. (*Roy Manstan*)

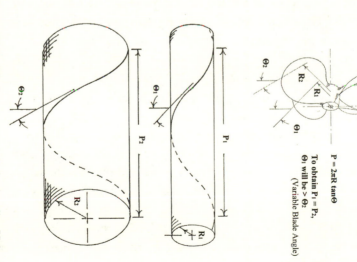

Figure 12.8b. Modern constant pitch propeller. Helical path of a point at "R" along the surface of a modern constant pitch propeller as it rotates and transits through the water. (*Roy Manstan*)

vessel a distance corresponding to this pitch. Where wood, however, remains stationary with respect to the motion of the screw threads, water will flow in response to propeller rotation. This means that in water the theoretical advance, or the lateral distance traveled per rotation if water behaved like wood, will always be greater than the actual lateral distance the vessel travels in one rotation. The difference between theoretical and actual advance is referred to as "apparent slip."

For typical vessels with the propeller mounted in the stern, the slip is affected by the velocity of the water within the wake, making the actual slip greater than the apparent slip. The Turtle propeller, however, is mounted at the stem and out of the influence of the wake. Under these conditions, apparent slip is the same as actual slip.

Slip "S" is often shown in terms of percent as defined by the equation:

S = (1 - V/Pn)x100
Where:
V = Speed of advance of the vessel (fps)
P = Propeller pitch (ft.)
n = revolutions per second (rps)

In the static thrust tests, where the speed "V" in the above equation is zero, propeller slip is 100%. Barnaby (1967) states that "the maximum thrust of a propeller is exerted at zero speed of advance, a condition that is termed 100% slip."[1] When transiting, however, as the pilot starts rotating the propeller from the "at rest" position, the propeller initially experiences 100% slip (i.e., maximum thrust) until the hull and its water boundary layer begin to accelerate. As this initial thrust overcomes the inertial forces, and the vessel begins to move, the "V" in the above equation is no longer zero and slip begins to drop from 100%.

When the vessel has reached a constant speed, i.e., no longer accelerating, thrust is now only required to overcome the hydrodynamic forces (resistance) associated with motion of the hull through the water. When these forces are balanced, slip remains constant and the vessel will continue at that speed as long as the pilot can sustain the propeller rpm. As far as the pilot is concerned, the higher the slip, the greater the effort that he must sustain to keep these forces (resistance and thrust) in balance.

Chapter Thirteen

PROPELLER THRUST

HORSEPOWER = MANPOWER

TEST 1: STATIC THRUST

When Col. John Stevens began experimenting with steam-driven screw propellers in 1802, he devised an adjustable propeller hub that allowed him to determine the optimum blade angle. In a letter written in 1805, Stevens noted that these windmill-like "arms are made capable of ready adjustment, so as that the most advantageous obliquity of their angle may be attained after a few trials."[1] He understood that blade angle was critical to efficient propulsion, thus minimizing the required steam generated horsepower. Because the *Turtle* depended on manpower, an efficient propeller was essential. Our testing simply included a variety of propeller designs that we believe represent a reasonable approximation of what Bushnell may have tried.

Gillmer and Johnson (1982) define propeller thrust as "the force produced by the ship's propeller that overcomes the resistance of the ship."[2] It is therefore essential to determine both the "resistance" of the *Turtle* hull and the maximum thrust that the pilot can deliver and sustain. The first tests were designed to measure propeller thrust as a function of rpm.

Our experiment follows Gerr (2001) who states that "thrust at maximum power with the boat tied to a dock is called *static thrust* or *bollard pull*."[3] For a modern vessel, maximum power as provided by the ship's engine is sustainable. In the *Turtle*, however, propulsion was in the hands (or feet) of Ezra Lee who could only deliver maximum power "with hard labor . . . for a short time."

Figure 13.1. The diver changing the propeller during the static thrust measurements. (*Jerry Roberts*)

With limited test time, we selected four of the five propeller designs that we had built for the replica *Turtle*, and that we felt were the most likely to represent Bushnell's options. The support divers were able to quickly change the propeller between tests (figure 13.1).

Our goal was to evaluate each design as to its effectiveness in transiting and maneuvering our, and consequently Bushnell's, *Turtle*. Propeller thrust, and thus the vessel's speed through the water, is a function of propeller speed of rotation (rpm) and the ability of the pilot to maintain that rpm, i.e., the amount of horsepower that he can sustain. Horsepower produces propeller thrust; propeller thrust produces forward motion.

TEST PROCEDURES

The *Turtle* was lowered in the shiplift until afloat. The test pilot added water ballast until the waterline lay along the upper barrel band. The assumption was that the original *Turtle*, while transiting on the surface, would have had the deadlights (Bushnell referred to them as "three round doors") open to allow air to circulate. The freeboard would be necessary to keep waves from washing over the deck and entering the hull through these doors. In our case, we left the hatch open to facilitate communication with the test pilot.

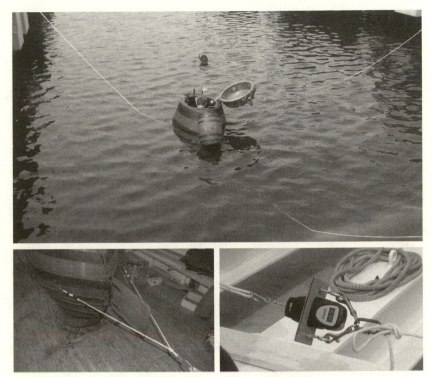

Figure 13.2a, top: The *Turtle*, with its waterline along the upper barrel band, in position for the static thrust testing. The line passing from the stern of the *Turtle* runs to the load cell, where the thrust is measured. (*Jerry Roberts*) Figure 13.2b, bottom left: A view of the bridle connecting the *Turtle* to the static thrust load line. (*Roy Manstan*) Figure 13.3, bottom right: The digital load scale used to measure propeller thrust. (*Jerry Roberts*)

The divers attached a rope yoke to lifting eyes on either side of the hull. The yoke was connected to a single line that extended from the *Turtle* horizontally aft; passed through a sheave; then led up through the water to a second sheave onshore where it then passed horizontally to the digital load cell. (See figures 13.2a, b, and 13.3.) The port and starboard tending lines were kept slack during the tests.

An important consideration was the selection of the line used to transmit the propeller thrust to a load cell. Considerable variation will occur in rpm when the propeller is turned by hand. This increase and decrease in rpm will cause a similar increase and decrease in thrust. A load line with zero elasticity will transmit the thrust directly to the load cell, and these variations in thrust will be evident. If there is excessive

Figure 13.4. This view of the *Turtle* and pilot Paul Mileski is from a single frame taken from the video documentation of the May 2008 testing at the Mystic Seaport Museum. (*Ken Beatrice*)

elasticity in the load line, it and the *Turtle* will act essentially as a spring and a mass in a viscous (water) medium. The damped oscillations will result in unreliable measurements at the load cell.

Initially, a 3/32˝ torque balanced stainless steel wire rope was considered for this application, but the bending resistance through the small-diameter sheaves was unacceptable. The elasticity (elongation under load) of 1/8˝ braided cotton line was subsequently tested, and found to be sufficiently low to satisfy our test requirements. The braided cotton line is also very flexible in bending, allowing it to pass through the sheaves with little resistance; being braided, it is also torque balanced such that when subjected to a load, it will not have a tendency to twist or spin at the terminations.

Communication between the pilot and test team was maintained via VHF radio. The pilot was instructed to turn the propeller within a specified range of rpm. Once the propeller was turning in the desired range, the pilot transmitted the tachometer readings to the test team. Figure 13.4 shows the pilot and *Turtle* in position to start the static thrust tests.

Using a combination of the hand crank and foot treadle, the test pilot operated the propeller for approximately one minute, endeavoring to remain at or close to the specified rpm. He was instructed to provide a tachometer reading once every few seconds. Between each reading the propeller made several rotations, the number depending on the rpm. It was difficult for the pilot to maintain a consistent torque while rotating the shaft with the treadle and hand crank. This resulted in variations in the rpm over a data cycle, and, therefore, thrust as measured at the load cell.

STATIC THRUST TEST RESULTS

Each time the pilot announced the tachometer reading, the test team recorded the rpm and resulting thrust. This procedure was repeated for multiple rpm ranges and for each of the four propeller designs. The data provided sets of rpm ranges and averages, along with the corresponding thrust, as shown in table 2.

TABLE 2. STATIC THRUST MEASUREMENTS

Propeller	RPM (Range)	Thrust (Range)	Thrust (Average)
(1)	35	0 lb. 5 oz – 0 lb. 7 oz.	0 lb. 6 oz.
(1)	41 – 44	1 lb. 3 oz. – 1 lb. 5 oz.	1 lb. 4 oz.
(1)	49 – 52	2 lb. 0 oz. – 3 lb. 0 oz.	2 lb. 8 oz.
(1)	60	3 lb. 11 oz. – 4 lb. 4 oz.	3 lb. 14 oz.
(1)	63	4 lb. 3 oz. – 4 lb. 10 oz.	4 lb. 7 oz.
(1)	75 – 76	5 lb. 12 oz. – 7 lb. 0 oz.	6 lb. 11 oz.
(1)	100 – 104	13 lb. 0 oz. – 14 lb. 0 oz.	13 lb. 8 oz.
(2)	33	4 lb. 5 oz. – 6 lb. 7 oz.	5 lb. 8 oz.
(2)	39	7 lb. 12 oz. – 9 lb. 1 oz.	8 lb. 10 oz.
(2)	41	10 lb. 0 oz. – 10 lb. 14 oz.	10 lb. 11 oz.
(3)	32	1 lb. 12 oz. – 1 lb. 12 oz.	1 lb. 12 oz.
(3)	38 – 41	3 lb. 2 oz. – 3 lb. 14 oz.	3 lb. 4 oz.
(3)	49	3 lb. 8 oz. – 5 lb. 9 oz.	4 lb. 4 oz.
(3)	52	5 lb. 7 oz. – 7 lb. 3 oz.	6 lb. 4 oz.
(3)	59 – 60	7 lb. 0 oz. – 12 lb. 5 oz.	9 lb. 4 oz.
(4)	31	2 lb. 3 oz. – 3 lb. 14 oz.	3 lb. 4 oz.
(4)	34 – 39	3 lb. 0 oz. – 4 lb. 5 oz.	3 lb. 15 oz.
(4)	50	5 lb. 15 oz. – 10 lb. 0 oz.	7 lb. 8 oz.
(4)	59	15 lb. 0 oz. (single data point)	
(5)*			

*Propeller 5 was not included in the static thrust tests.
Source: The *Turtle* Project

The above data is presented as a series of plots in figure 13.5.

The mass of water accelerated rearward by the propeller blades during the static thrust tests was very evident, particularly in the video documentation. Figures 13.6 and 13.7 are frames taken from the video of the wake created along the water line with propeller #1 rotating at ~ 60 rpm. Compare this waterline to figure 13.4 above with the propeller "at rest."

PILOT COMMENTS

Shortly after completing the static thrust tests, the pilot, Paul Mileski, recorded his observations. He makes particular note of the importance of

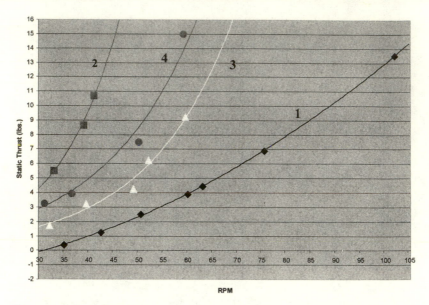

Figure 13.5. Static thrust vs. rpm for the four propellers tested. (*Turtle* Project)
Plot #1: Propeller (1), bronze triangular 25° blade angle
Plot #2: Propeller (2), aluminum triangular 45° blade angle
Plot #3: Propeller (3), stainless steel triangular 35° blade angle
Plot #4: Propeller (4), stainless steel elliptical (paddle-shaped) 35° blade angle

rotational momentum, or lack thereof, in his discussion of the effectiveness of each of the four propellers tested. He emphasized that unless there was significant momentum, the propeller had a tendency to stall when the treadle motion reached top and bottom dead center. The following are a transcription of Mileski's comments:

Regarding Propeller (1), bronze, triangular, 25° blade angle:

Can maintain ~ 40 to 50 rpm using treadle alone. Has sufficient mass to prevent stalling at stroke top/bottom at relatively modest effort. Use of hand crank and treadle enables smoother (more steady) rotation. I could not use the crank alone without great difficulty. RPM of 40–50 using crank and treadle together can be sustained.

Good combination of blade angle and mass because speed can be maintained at top and bottom of treadle stroke. My impression is that operation without the internal flywheel would be difficult or impossible due to stalling at top/bottom of stroke.

Figure 13.6, left. Video frame showing the propeller backwash across the *Turtle* hull with the bronze 25° propeller turning at ~ 60 rpm. Figure 13.7, right. Detail of the propeller backwash at ~ 60 rpm. (*Ken Beatrice*)

I felt that this propeller was the most efficient in terms of coupling my muscle power to the water because I could sustain the highest rpm operation with the same muscle force—hence greater horsepower.

Regarding Propeller (2), aluminum, triangular, 45° blade angle:

Very difficult to maintain rpm compared to (1)—very low inertia compared to the torque required to turn steadily—stalls unless handle is used together with foot treadle. My impression is that this propeller would likely be unusable in practice with the existing Turtle, even with the flywheel. No way without flywheel.[4]

Regarding Propeller (3), stainless steel, triangular, 35° blade angle:

Can maintain higher rpm than (2), but requires handle and treadle together to prevent occasional stalling. Significantly greater torque required than that for (1), but somewhat less than that for (2). The stainless blades are more massive than (2) and thus reduces stalling. Appears more useful than (2). Flywheel necessary (impression).

Regarding Propeller (4), stainless steel, elliptical (paddle shaped), 35° blade angle:

Overall noticeably less force required for a given rpm compared to (3) and (2). Somewhat lower moment of inertia (impression) compared to (3) but still greater than (2). Appeared less prone to stalling due to the ability to spin faster—offsetting the lower moment of inertia. Impression—flywheel necessary.

Impression—operation with treadle alone may be likely, but not as likely as with (1).

General comments by the authors:

1. Use of the hand crank alone to deliver torque with its small radius (2 1/2″) is difficult at low rpm. For these tests, it was used primarily to assist the treadle and particularly helpful during starts or when stalling occurred during low speed operation.

2. Operation without a flywheel seems really unlikely unless a) the required operating torque is very low,—i.e., small blade angle, and b) the blades are sufficiently massive to act as a flywheel.

3. Foot treadle seems most suitable for rpm ~ 60.

After completing the static thrust tests, Mileski quickly eliminated the 45° propeller from consideration. He was able to generate a high level of thrust at a low rpm compared to the other designs, but could not sustain the propeller rotation.

This allowed us to turn our attention to the 25° and 35° propellers. Mileski noted that the elliptical blades (propeller 4) required "noticeably less force for a given rpm than propeller (3)." The implication being that there is an increase in efficiency and reduced impact on the pilot with an elliptical, paddle-shaped propeller.

Mileski, however, was quite clear that he was able to sustain "the highest rpm operation with the same muscle force—hence greater horse-power" with the bronze 25° propeller. While some of the difference was accounted for in its higher mass and angular momentum, his overall impression was that compared to all of the other designs—this propeller was the most efficient.

Just as we did at Mystic Seaport, David Bushnell's pilots would have tested and trained with the vessel on the surface, and likely with the hatch open. David's brother Ezra, and possibly Phineas Pratt, would have expressed their concerns about the efficiency of this new propulsion device. Our tests showed how critical the blade angle is to the ability of the pilot to create and maintain thrust, and we came prepared to test a variety of designs. If Ezra had complained about the difficulty he had with rotating his brother's "pair of oars," David could have made adjustments to the blade angle, a task easily accomplished by a sympathetic blacksmith.

Chapter Fourteen

HULL RESISTANCE

MOVING A TWO-TON TURTLE

TEST 2: HULL RESISTANCE (TOWING SPEED)

In 1809, Robert Fulton experimented with "the resistance of bodies under water at a depth of 6 feet. In these calculations let it be understood that the weight or moving power runs thro' the same space in the same time as the body which is drawn through the water."[1] This quote is from Fulton's patent application for a steam-powered vessel, where he described the state of the art in experimental naval architecture during the late eighteenth and early nineteenth centuries. Fulton showed the results of tow tests of various shapes designed to separate the effects of skin friction along the length of a hull from the shape of a vessel's bow and stern.

As we have already mentioned, the original *Turtle* transited on the surface during extended operations. The pilot would need to submerge only when attempting to attach the mine to a ship's hull, or as a tactic used to evade a pursuing enemy. All *Turtle* propulsion testing has been associated with the vessel operating on the surface.

The *Turtle* can be classified as a submersible vessel, rather than a submarine, in that when transiting on the surface it, in Bushnell's words: "could swim so low on the surface of the water." The majority of the *Turtle* hull, as with a submersible vessel, is designed to remain afloat with little freeboard, submerging only when required. A submarine, on the other hand, is designed to operate primarily submerged.

The hydrodynamic forces acting against a submarine hull when underway submerged are termed "drag" forces. For vessels operating on the surface, the term "resistance" is used, and is the term we will refer to in this discussion.

The total resistance of a vessel moving through the water consists of frictional resistance of the water flowing past the hull, wave-making resistance created by the energy in producing a wake, and eddy resistance caused by flow passing appendages and abrupt changes in hull form. The contribution of each factor will depend on the hull design and its speed through the water. Because these three components act together and are often difficult to separate, we felt that for the *Turtle*, it would only be necessary to determine the resistance due to their combined effect.

According to Comstock (1967), "the [total] resistance of a ship at a given speed is the force required to tow the ship at that speed in smooth water."[2] As shown in the equation below, the resistance associated with these hydrodynamic forces is a function of the density of the fluid, the square of the towing velocity, a resistance coefficient, and the surface area of the hull. The contributions from wave making and eddies, however, will increase as the towing speed increases. The relation between vessel speed and total resistance becomes more complex and is reflected in the resistance coefficient.

$R_t = 1/2\rho V^2 C_t S$
Where:
R_t = Total resistance (towing load)
ρ = Mass density of the fluid (water)
V = Speed of the ship
C_t = Total resistance coefficient
S = Total wetted surface area

TEST PROCEDURES

The *Turtle* towing tests were designed to determine hull resistance at relatively low towing speeds where skin friction dominates. Four tows were conducted on the second day of testing and in conjunction with the transiting speed runs. The towing tests commenced at the end of the second through the fifth transit, the first being used for a "backing down" maneuvering test.

After each transit was completed, the support divers turned the *Turtle* to face back toward the transit starting point. The divers attached the same bridle and line used for the static thrust tests to a pair of forward lift eyes on the *Turtle*. The towline passed horizontally back to the lower

Figure 14.1, left. Old Saybrook High School student Will Tucker assisting with the tow tests to determine the hull resistance at various towing speeds. (*Ken Beatrice, from video frame*) Figure 14.2, right. Detail of the spring scale used to measure hull resistance. (*Roy Manstan*)

sheave, then up through the water to the topside sheave and to a spring scale held by a test team member. After taking up slack in the towline, the person holding the scale was instructed to slowly apply a load through the spring scale, and to continue pulling at a constant speed until the *Turtle* had returned to the transit start position.

A sliding marker on the scale retained the initial load necessary to overcome the inertia of the *Turtle* and to generate the water boundary layer that moves with the surface of the hull. Once in motion, the towline force dropped to that associated with maintaining the steady-state towing speed of the *Turtle*, i.e., R_t in the equation on page 164.

The towing speed was measured with a stopwatch recording the time the scale was pulled between two marks set fifty feet apart, essentially providing the average towing speed over that distance (figure 14.1). The test director recorded the start-up and the steady-state loads on the scale as it was being pulled. Figure 14.2 shows the spring scale with a recreation of typical startup (38 lb.) and towing loads (24 lb.).

TEST RESULTS

This procedure was repeated four times. The first two tows were accomplished by the same test team member. Two other individuals conducted the third and fourth tows. Table 3 shows the initial load, the observed range of towing loads over the fifty-foot timed track, and the average towing speed in feet per second (fps) and knots.

Note that the initial load of 38 lb. was the same on the 1st, 3rd, and 4th tows, while the 2nd tow test resulted in a 16% higher initial load at

TABLE 3. TOW TEST (HULL RESISTANCE)

Tow #	Initial Load (lb.)	Tow Load Range (lb.)	Tow Speed (f/s)	Tow Speed (knots)
(1)	38	25	1.58	0.937
(2)	44	28 – 30	1.75	1.039
(3)	38	22 – 25	1.48	0.874
(4)	38	10 – 15	1.13	0.672

Source: The *Turtle* Project

44 lb. This higher initial load may have been due to a more sudden initiation of the pull, i.e., a greater acceleration of the combined mass of the *Turtle* and the hydrodynamic mass of the boundary layer, while a more gradual startup occurred for the other three tow tests.

The towing load measured during tow #1 remained fairly consistent at 25 lbs. Tow #2 ranged from 28–30 lb.; tow #3 ranged from 22–25 lb.; and tow #4 ranged from 10–15 lb. The ranges in towing loads most likely result from variations in the walking speed as the scale was being pulled, and from the elasticity of the braided cotton towline.

The maximum towing speed was only slightly greater than one knot. From the video of tow #3, we could observe that a slight bow wave was forming (see figure 14.3). It is likely that the bow wave will increase at higher speeds, and in order to achieve speeds greater than one knot, the pilot would have had to provide significantly more horsepower.

When total resistance is primarily related to hull skin friction, the ratio of any two loads should be nearly equal to the ratio of their velocities squared:

$$R_a/R_b = (V_a/V_b)^2$$

This relationship between the *Turtle* towing loads and velocities is apparent when comparing the two higher speeds for tow #1 (towing load of 25 lb. at 0.937 knot) and tow #2 (average towing load of 29 lb. at 1.05 knot) to the lowest towing speed during tow #4 (average towing load of 12.5 lb. at 0.672 knot).

$$R_{\#1}/R_{\#4} = 2.0$$
$$(V_{\#1}/V_{\#4})^2 = 1.94$$
$$R_{\#2}/R_{\#4} = 2.32$$
$$(V_{\#2}/V_{\#4})^2 = 2.39$$

It can be concluded from these comparisons that skin friction is the primary contributor to total resistance at speeds up to one knot. With the video showing signs of a bow wave forming, we can speculate that wave-making resistance will begin to become a factor for the *Turtle* at speeds above one knot.

Figure 14.3. Video frame showing the wave generated with the hull moving (under tow) at about seven-eighths of a knot. (*Ken Beatrice*)

Figure 15.1. The *Turtle* underway during the transit speed tests as it approached the end of the finger piers that support the shiplift facility. (*Jerry Roberts*)

Chapter Fifteen

TRANSIT SPEED

A Tortoise or a Hare?

TEST 3: TRANSIT SPEED

Bushnell never makes any claim as to the speed at which his vessel was able to transit. Gale in his August 1775 letter to Benjamin Franklin asserts that "he can row it either Backward or forward Under water about 3 Miles an Hour." If Gale had actually witnessed a test of the *Turtle* and wasn't depending on hearsay, one can imagine him estimating that it had transited at three miles an hour, after watching the vessel submerge and then resurface a short time later, at a considerable distance.

The transit speed depends on the ability of the propulsion system to overcome the resistance of the water against the vessel's hull. Assuming that the power source (in this case the pilot) is capable of turning the propeller at a particular rpm, the vessel will transit in proportion to how effectively the pilot is able to generate and maintain thrust.

The series of measurements described below provide a small sample of transit speed vs. propeller rpm. From this information, it is possible to determine parameters that affect the overall capability of the *Turtle's* propulsion system, i.e., propeller "slip," wake generation, and a subjective assessment of the energy exerted by the pilot.

Based on the static thrust measurements and the pilot's comments, we eliminated the 45° propeller blade as a candidate design. Because of the similar performance between the stainless-steel triangular and elliptical blades, we decided to test only the triangular design. This would enable a better comparison between 35° and 25° blade angles. We also included propeller (5), the triangular 35° blade with the modified edges, allowing us to evaluate the propeller that was not included in the static thrust measurements.

Descriptions and illustrations of the five propeller designs referred to here were given in chapter 7.

TRANSIT TEST PROCEDURES

The *Turtle* was positioned on the shiplift and ballasted to float at the same draft as the static thrust tests. The transit speed, or speed of advance, was determined by measuring the time the *Turtle* passed between two marks. The total distance of the transit was approximately one hundred feet. The first fifty feet were allocated to accelerating the vessel and establishing a straight course parallel to the lift. A thirty-three foot range was designated for the timing of the vessel's speed. The remainder of the transit was used to stop and reposition the *Turtle* for the maneuvering and towing tests. The overall distance traveled was limited by the need to stay within the confines of the shiplift facility.

The test plan specified a total of five transit runs. The first three transits called for propeller (3), with the first transit at 35 rpm, the second at 45 rpm, and the third at 55 rpm. The fourth transit called for propeller (5) at 35 rpm. Propeller (1) was selected for the fifth and final run at 65 rpm.

The rpms specified on the test plan were target values. The pilot was instructed to attempt to operate at as close to the specified values as possible. Knowing that the pilot would be unable to maintain an exact rpm, he was also instructed to transmit the tachometer readings to the test director.

While the test director recorded the rpm readings, another member of the test team recorded the time the *Turtle* passed through the marks. The transit path of the *Turtle* was also recorded on video. Figure 15.1 shows the *Turtle* during one of the transits. The port and starboard tending lines were kept slack during the transits, and were only there as a precaution if the *Turtle* veered significantly from its path.

RESULTS

After reviewing the video, the transit path during run #2 was noted to have deviated significantly from a straight course, and the actual distance traveled could not be determined or estimated with any precision. Data was recorded for runs #1, #3, #4, and #5. The results are shown in table

TABLE 4. TRANSITING AND MANEUVERING TESTS

Transiting Dead Ahead

Run #	Propeller	RPM (range)	RPM (average)	Speed (ft/sec)	Speed (knots)
1	(3)	34 – 37	36.5	0.667	0.395
2*					
3	(3)	50 – 58	53.25	0.757	0.448
4	(5)	39 – 43	39.9	0.749	0.443
5	(1)	65 – 70	66.4	0.804	0.476

Backing Down

1	(3)	43 – 45	44	0.83	0.49

*Pilot veered off course; no data taken for run #2.
Source: The *Turtle* Project

4, and include the range and average rpm recorded during the transit, and the resulting speed over the thirty-three-foot timed track in feet per second (fps) and knots. Note that the pilot was able to maintain an average rpm close to that specified in the test plan.

The video also produced images of the wave generated at the bow (see figure 15.2 from run #4). As expected, at speeds of less than a half knot, the hull is only beginning to generate a diverging bow wave. There are no signs of transverse waves. This is an indication that the "total resistance" (chapter 14) of the *Turtle* transiting at these speeds is primarily due to hull friction with little contribution from the wake.

The evaluation of the effectiveness of the *Turtle* propellers requires an understanding of two parameters, "pitch" and "slip." These parameters were discussed in chapter 12 (see the section titled "Modern Propeller Design"). Chapter 12 also describes the difference between a modern constant pitch (and thus constant slip) propeller and Bushnell's flat propeller, where pitch and slip vary over the length of the blade.

Slip provides insight into the thrust obtained during a transit as well as a measure of the effort that must be sustained by the pilot. Maximum thrust occurs when slip is 100%, as in the static thrust testing. Once the vessel is underway, slip and thrust decrease with increasing speed. In order to maintain a specific rpm and speed, the pilot must sustain his input to the drive mechanism. The pilot's observations, while subjective, provide insight into how slip, thrust, and transit speed affected his sustainable level of effort.

Figure 15.2. Video frame showing the *Turtle* transiting at just under one-half knot. (*Ken Beatrice*)

Based on propeller pitch and the speed of advance of the *Turtle*, values of slip for the four successful runs (#1, #3, #4, and #5) are provided in table 5 at three locations along the length of the blade.

The following propeller evaluations are based on tables 4 and 5. As a reminder, propeller (3) has the 35° stainless blades, propeller (5) has the aluminum 35° blades with modified edges, and propeller (1) has the bronze 25° blades.

Comparing run #1 to run #3 (table 4), increasing the rotation of propeller (3) by 46% (from 36.5 rpm to 53.25 rpm) increased the *Turtle* speed by 28% (from 0.395 kts to 0.448 kts). The pilot, however, indicated that a great deal of exertion was necessary to achieve the higher rpm. The increased effort is not surprising, considering that the slip (table 5) also significantly increased.

Comparing run #3 to run #4, propeller (3) averaged 53.25 rpm to achieve 0.448 kts, while propeller (5) achieved almost the same hull speed (0.443 kts) at only 39.9 rpm. In addition, the slip is significantly greater for propeller (3) than propeller (5). These indicate that the triangular blade with the modified edges (propeller 5) is a more efficient design than the flat triangular blade.

Comparing run #1 to run #5 provided the most interesting results. The pilot indicated that the level of effort for run #1, where propeller (3) obtained nearly 0.40 kts at 36.5 rpm, was "about the same as" run #5, where propeller (1) achieved 0.48 kts at 66.4 rpm. This is a significant result in that the pilot was able to obtain a 20% increase in speed at nearly the same level of effort, borne out by the identical slip associated with these two runs. Ultimately, the best propeller design is one that provides the highest sustainable speed with the least effort.

PILOT COMMENTS

After reviewing the video documentation, test pilot Paul Mileski provided comments regarding each of the propellers tested. "Delivery of power to the stainless propeller [propeller (3)], even at the lower rpm, was dif-

TABLE 5. PROPELLER PITCH AND SLIP

(Measured at the Blade Centroid, at 70% of the Blade Length*, and at the Blade Tip)

Propeller (3) Stainless Steel Triangular Blades

Transit	Location On Blade	Blade Angle	Pitch[1] (inches)	(feet)	rps (n) (rev/sec)	Speed (ft/sec)	Slip[2](S) (%)
Run #1	Centroid	35°	36.3	3.025	0.608	0.667	64
Run #1	70%	35°	41.8	3.48	0.608	0.667	69
Run #1	Blade Tip	35°	59.4	4.95	0.608	0.667	78
(Run #2 Pilot veered off course and data was not recorded.)							
Run #3	Centroid	35°	36.3	3.025	0.887	0.757	72
Run #3	70%	35°	41.8	3.48	0.887	0.757	75
Run #3	Blade Tip	35°	59.4	4.95	0.887	0.757	83

Propeller (5) Aluminum Triangular Blades with Modified Edges

Transit	Location On Blade	Blade Angle	Pitch (inches)	(feet)	rps (n) (rev/sec)	Speed (ft/sec)	Slip (%)
Run #4	Centroid	35°	36.3	3.025	0.665	0.749	64
Run #4	70%	35°	41.8	3.48	0.665	0.749	68
Run #4	Blade Tip	35°	59.4	4.95	0.665	0.749	77

Propeller (1) Bronze Triangular Blades

Transit	Location On Blade	Blade Angle	Pitch (inches)	(feet)	rps (n) (rev/sec)	Speed (ft/sec)	Slip (%)
Run #5	Centroid	25°	24.2	2.02	1.107	0.804	64
Run #5	70%	25°	27.8	2.32	1.107	0.804	69
Run #5	Blade Tip	25°	39.6	3.3	1.107	0.804	78

*The pitch of modern propellers is often given for this location on the blade (Gerr, p. 25).
[1]See Chapter 12, equation on page 152.
[2]See Chapter 12, equation on page 154.
Source: The *Turtle* Project

ficult, often requiring two hands to get past top dead center. I had to concentrate on maintaining rpm rather than on course direction. When necessary, however, I could control the rudder."

The modified 35° propeller (propeller (5)) elicited some interesting comments from the test pilot. He noted that "acceleration was greatest with this propeller, that is, during the up/down strokes the propeller accelerated quickly and the *Turtle* likewise seemed to 'surge' forward. This propeller seemed to translate rotation into thrust in the most efficient manner compared to the other two propellers." He also noted that "it was significantly less fatiguing than the stainless prop, but about the same as the bronze prop."

Mileski's comments, plus the improved transit speed over the flat 35°
triangle, made this design an interesting concept. We feel, however, that
the modifications to the edges of this propeller more closely represent
the leading and trailing edges of modern propeller designs, and that it is
unlikely Bushnell would have incorporated this concept.

The final transit run was conducted with propeller (1). "I was a happy
camper [Ezra Lee may have used a different phrase] operating the
bronze propeller. I could use my left hand on the crank and my right
hand on the tiller." Mileski was able to sustain a higher speed and rpm
with the 25° design than with the 35° designs and noted that, as in the
static thrust runs, this propeller required the least overall effort. He
emphasized the significant impact that blade angle has on the *Turtle* and
pilot performance. Mileski, who is a long-distance cyclist, also commented that selecting the proper blade angle is analogous to selecting the best
gear ratio when peddling uphill grades.

Conclusions regarding Bushnell's propeller

Much of the technology that Bushnell incorporated—the pumps, the
depth gauge, the ventilation system—were adaptations of mechanical
devices in use, or at least had been described in the scientific and popular literature. The mode of propulsion, however, was truly unique. In
1785 Ezra Stiles, then president of Yale College and an early supporter
of Bushnell's submarine, published *The United States Elevated to Glory
and Honor* in which he provided a list of notable inventions. Included in
this list were Benjamin Franklin's "Electrical pointed Rods" and David
Bushnell's "Submarine Navigation by the Power of the Screw."[1]

The shape and construction of the replica propeller are based on cir-
cumstantial evidence and supposition as to what may have triggered
ideas in Bushnell's mind. Ezra Lee had provided the only clues to its
dimensions. We tested five configurations that fit within these design
constraints. If we had broadened the range of propeller size and shape,
we would not have been true to our goal of replicating form and function as best we could. We did not "reverse engineer" the *Turtle* replica in
an attempt to attain the three knots that Lee had claimed.

Because the effectiveness of a propeller is dependent on the ability to
sustain sufficient horsepower, feedback from the source of the power, i.e.,
the test pilot, was essential. This evaluation fell to test pilot Paul Mileski
who conducted the static thrust, transiting, and maneuvering tests. In

doing so, he had an opportunity to test and evaluate all five design configurations. The other two pilots, Dave Hart and Roy Manstan, only used the bronze 25° propeller.

Based on the preceding propulsion tests, our conclusions regarding Bushnell's propeller include the following.

1. The propeller was heavy; either a brass casting or an iron forging.

2. The blade faces were flat.

3. The blade angle was closer to 25° than the 35° recommended by Martin for windmill blades.

4. The blades may have been elliptical, i.e., more paddle or oar shaped.

5. Lee's memory may have underestimated the blade width, allowing a slightly broader blade with greater surface area.

The biggest hurdle that Bushnell faced was the human power that drove the *Turtle*. The use of steam, while not able to replace the pilot as a submarine power source, soon stimulated other inventors to consider the bladed propeller for surface vessels.

Colonel John Stevens produced his first propeller sometime between 1801 and 1804 for his experiments with steam propulsion. Figures 15.3 and 15.4 are from an article by Francis B. Stevens that appeared in the April 1893 issue of the *Stevens Indicator*. The vessel, a replica of the original boat Stevens used in New York Harbor, was produced for the 1892 Columbian Exhibition where his original steam engine and propeller were on display. The article describes the history of John Stevens's design and the testing of his bladed propeller, and its use during a patent dispute that began in 1844.[2] His blade was a flat triangle, narrow at the hub and very broad at the end. The hub was designed to accommodate either two or four blades. He devised a method to adjust the angle of the blade, and was able to conduct experiments to determine what the most efficient blade angle should be. His steam engine, initially designed to run a single propeller, was later configured to accommodate two counter-rotating propellers.[3] A more recent photograph of Stevens's propeller and steam engine can be found in *The National Watercraft Collection* by Howard Chapelle (1976).

Figure 15.3, left. The earliest propeller for which we have a photograph was used by Colonel John Stevens in his steam-driven vessel during the first few years of the nineteenth century. Figure 15.4, right. Detail of John Stevens's propeller. These photos were taken during the 1892 Columbian Exhibition. (*Stevens 1893, Stevens Institute of Technology*)

BUSHNELL'S USE OF A TREADLE AND FLYWHEEL

A treadle consists of a hinged platform connected by a rod to a crank on the rotating shaft. The foot-driven treadle moves up and down, pivoting on its hinge. This vertical motion is translated into rotary motion at the crank. The inherent difficulty with a treadle design is that the forcing function is only in the vertical direction. When the treadle is in the up or down position, the crank is at top or bottom dead center, and there needs to be a lateral force applied to rotate the shaft past these positions. This can come from one or a combination of three sources: (1) a handle on the crank that allows the operator to apply a lateral force by pushing and pulling; (2) a flywheel that once in motion applies its angular momentum to the spinning shaft; and (3) the object that the shaft is turning (i.e., the propeller) is in itself massive enough to apply its own rotational momentum.

The replica *Turtle* drive mechanism included a single treadle hinged in its center. One foot pressed down on the forward half forcing the connecting rod down. When the crank passed bottom dead center under the combined momentum of the flywheel and propeller, the pilot's other foot pressed down on the back half of the treadle, pushing the connecting rod up. When the angular momentum was insufficient, the pilot assisted the rotation with the hand crank.

Test pilot Paul Mileski felt that the 2 1/2-inch crank radius was insufficient and that increasing this dimension would enable the pilot to deliver and sustain more torque when using either or both the treadle and hand crank. Considering that Lee was able to maneuver the original *Turtle* successfully during an extended mission, it is likely that Bushnell's propulsion system also had a greater cranking radius.

Throughout both the static thrust and transit testing, Mileski continued to comment about the contribution of propeller weight to the overall power required to maintain the rpm. This parameter had not been considered when the testing was originally conceived, resulting in a wide range of propeller weights. Suffice it to say here that a more massive propeller adds to the total angular momentum of the rotating propulsion system. It was apparent to all of the pilots, however, that the added momentum from the flywheel was essential in providing a sustainable propulsion system.

The inclusion of a flywheel into our *Turtle* replica was first proposed by the project director, Fred Frese, and agreed to by the Navy coordinator and test director, Roy Manstan. The flywheel is a 17 1/2-inch diameter, fifty-pound weightlifting disc attached to the propeller shaft. The flywheel and the hand crank with its 2 1/2-inch radius can be seen in figures 12.5 and 12.6, chapter 12. Examples of flywheels from the eighteenth century were described in chapter 7, all of which would have been a familiar sight in Bushnell's world.

In general, the pilots and the authors feel that it is doubtful Bushnell produced a propeller large enough to preclude the need for an internal flywheel. On a modern engine-driven vessel where there is typically a massive flywheel and sufficient engine power to overcome the losses in the system, horsepower is delivered—and therefore the thrust needed to maintain ship speed—simply by increasing or decreasing the throttle. On the *Turtle*, however, horsepower is a function of muscle power.

Assessment of overall performance

The preceding propulsion tests were intended to determine the operational characteristics and limitations of the hull and propulsion system. We determined that with moderate effort, the pilot was able to obtain a speed of about one-half knot. The towing tests, designed to measure hull resistance, showed that up to one knot there was little sign of a bow wave. Under these conditions, where wake generation had not yet become a contributing factor to overall hull resistance, the thrust required to drive the hull through the water would be that required to overcome skin friction. It is therefore conceivable that the *Turtle* could have achieved one knot before the hull would begin to create a bow wave, at which point there is a significant increase in the thrust (and horsepower) that the pilot would need to produce.

Because the propulsion tests were designed to measure speed and not endurance, the test pilot, Paul Mileski, was only required to sustain his speed over a short distance. While, as a Navy diver at NUWC, his aerobic and muscular conditioning was appropriate for the physical strain associated with the tests, he was also about twenty years older than the original *Turtle* pilots during their missions in 1776. Thus when conducting the pilot endurance tests, a test pilot was selected from the pool of Navy divers who was nearly the same age as Bushnell's three pilots.

Chapter Sixteen

MANEUVERING AND SUBMERGING

ENGAGING THE TARGET UNDETECTED

TEST 4: MANEUVERING

The tests of the *Turtle* replica included four aspects of its maneuvering and submerging capabilities: backing down, S-turn (rudder control), submerging (vertical propeller), and the mine attachment (wood screw) mechanism. The first three were accomplished at the Seaport shiplift. The fourth test had to be postponed and was completed at a later date with the *Turtle* ashore.

BACKING DOWN TEST PROCEDURES

Maneuvering the *Turtle* was accomplished with a combination of the rudder and the propeller. In Bushnell's letter to Thomas Jefferson, he described that the propeller, "being turned one way, rowed the vessel forward, but being turned the other way rowed it backward." Bushnell also noted that "a rudder, hung to the hinder part of the vessel, commanded it with the greatest ease."

The *Turtle* maneuvering tests were run during the series of transit runs (see chapter 14). Because of the necessity to complete all transiting, towing, and maneuvering tests within the one-and-a–half-hour time window, we were limited to a single "backing down" run and a single "S-turn" maneuvering run.

The backing down run was conducted immediately after the completion of the first transit run. Note that propeller (3) had been installed for the initial transit runs and therefore used for the backing down test. The test pilot, Paul Mileski, was instructed to reverse the rotation of the propeller and back down along the same transit course within the confines of the shiplift facility. With the rudder set in the "dead astern" position, the pilot engaged the propeller. He was initially able to control the rudder sufficiently to back down along a straight track. After the first forty feet, however, the *Turtle* began to veer slightly off course. As the *Turtle* turned toward the shiplift walkway, we decided not to make any further course corrections and ended the backing down maneuver. The *Turtle* was then repositioned in preparation for the second transiting speed run.

The purpose of this test was to observe the ability to maintain steerage during a backing down maneuver. The pilot indicated that with one hand operating the crank, he was able to work the tiller with his other hand and could maintain some control over rudder position. The pilot also noted that the *Turtle* tended to pull to port when transiting forward or backing down. It is uncertain if this was caused by the propeller or by winds that were from starboard. Even with its low profile, the wind likely had some influence over the *Turtle*'s course. The pilot said that he was able to make course adjustments by keeping the rudder to starboard.

Because the primary interest in the backing down test was steerage, speed through the water and rpm were not recorded. However, after viewing the video and discussing the test with the pilot, it appeared that the *Turtle* backed down faster than when it transited forward. We then conducted a frame-by-frame review of the digital video to determine if we could estimate the speed and rpm. Each frame is provided with a time code used in video editing. The cyclic movement of the pilot's head corresponds to his hand and arm motions when driving the hand crank. By recording the time at the beginning and end of each cycle, we were able to determine the time interval between each complete rotation of the shaft, i.e., propeller rpm.

From the video review we were able to determine an accurate time for ten cycles, resulting in a range of 43 to 45 rpm, with an average of 44 rpm. The distance traveled along a straight track, as with the transit runs, was thirty-three feet. Again from the video, we determined that the time required to back down along this track was approximately forty seconds, giving an average speed of 0.83 fps. The above results have been included at the end of the transit run data presented in table 4 (chapter 15).

BOW VS. STERN-MOUNTED PROPELLER

Bushnell had to design and integrate multiple systems to create a vessel that could perform a complex underwater mission. Mounting the propeller in the bow to pull the *Turtle,* rather than pushing with a stern-mounted propeller, may only have been a matter of expediency. The stern had to accommodate the rudder and tiller. The mine was also mounted at the stern. The pilot navigated with a compass, and by facing forward viewing the horizon through the glass windows and open deadlights. It may have been simply easier to design and incorporate the propeller drive mechanism (treadle, crank, flywheel, and shaft) ahead of the pilot.

The implications of this are apparent when reviewing the backing down and the transit data on table 4. During transit #3, the *Turtle* moved forward at 0.757 fps with the propeller averaging 53.25 rpm. When backing down, however, the *Turtle* was able to transit at 0.83 fps while averaging only 44 rpm. The pilot was thus able to back down 10% faster while turning the propeller about 25% slower.

This has a major impact on the effort expended by the pilot maneuvering the *Turtle* for any prolonged mission. The reason for the difference between the *Turtle*'s bow-mounted propeller and one mounted in the stern becomes evident when viewing the static thrust video. (Refer to figures 13.6 and 13.7 for examples taken from the video.) When the *Turtle*'s propeller is turning to generate thrust, water is forced rearward across the hull. The effect is somewhat self-defeating. The propeller thrust is trying to pull the *Turtle* forward, yet the water flow (prop wash) is pushing against the hull. The effect is to reduce the net force driving the vessel forward. Because this effect is much more pronounced as rpm increases (i.e., proportional to prop wash velocity squared), we believe that the physical capabilities of the pilot would have been far exceeded well below the three knots claimed by Lee.

With Bushnell's bow-mounted propeller, the loss of efficiency of the overall propulsion system created even more work for the pilot. Ezra Lee would have had an easier time escaping from the British had he simply put the *Turtle* in reverse and backed up to Manhattan.

S-TURN MANEUVERING TEST PROCEDURES

While tight turns were not essential to the maneuvering of the *Turtle,* Lee mentioned that after being caught in the ebb tide he "hove about and rowed . . ." for two and a half hours in order to approach his target during slack tide. This implies that he had sufficient rudder control over the track of the *Turtle* to be able to maneuver to his specific target. As

Figure 16.1. The *Turtle* replica rudder set hard to port. (*Roy Manstan*)

we have mentioned, Bushnell was confident that his rudder "commanded it with the greatest ease."

Steerage of any vessel is governed by hull shape and by how responsive the hull is to rudder position. The hull shape selected for our *Turtle* replica was discussed in chapter 8. A round (barrel shape) vessel would respond to subtle changes in rudder position causing the vessel to quickly veer to port or starboard. A streamlined bow and stern will be less responsive to changes in the rudder and will more readily stay on course. The pilot will not have to concentrate on controlling the tiller, yet can still make gradual course changes as required.

The S-turn maneuver was run at the end of the final transit and tow test with the bronze propeller installed. With the *Turtle* repositioned at the start of the shiplift facility, the pilot was instructed to get underway with the rudder set at "dead ahead."

When the *Turtle* passed a designated location, the pilot was instructed to set the tiller "hard to starboard." The Turtle began its turn toward the west leg of the shiplift walkway and when it approached 90° to the original track, the pilot was instructed to set the tiller "dead ahead." As the *Turtle* responded and straightened its course, the pilot was directed to turn "hard to port." The *Turtle* then swung to port and began to transit across the original track heading toward the east leg of the shiplift walkway. At that point the *Turtle* had completed one half of the S-turn. As the *Turtle* closed in on the walkway, the pilot was directed to set the tiller "hard to starboard." The *Turtle* slowly responded in a gradual arc that brought it again across the original course, at which time, the S-turn maneuver was completed.

Note that when the pilot set the tiller "hard to port" or "hard to starboard," the corresponding rudder positions are 25° port (figure 16.1) and 25° starboard.

The entire transit took approximately one minute, fifty seconds. Based on a review of the video during the first half of the S-turn, the pilot was operating the propeller at an average of 75 rpm. Note that the *Turtle* was equipped with propeller (1), allowing the pilot to conduct the maneuver at the higher rpm. The S-turn covered a distance along the walkway of approximately sixty feet. Considering the "S" to consist of two U-turns, the *Turtle* was capable of performing a 180° maneuver over a distance of thirty feet in less than one minute.

OBSERVATIONS

The tests showed that the hull was sufficiently responsive to rudder position that the pilot could control his track while operating the propeller with his feet and one of his hands. While close-in maneuvering when alongside the target ship is critical, what the pilot relied on most would be the ability to operate the *Turtle* with, against, and across the prevailing current.

For small craft running on the surface, such as an eighteenth-century whaleboat, a double-ended hull shape is preferred. At both the stem and stern of the *Turtle*, the sides project at an angle of 45° from the centerline of the hull. In Bushnell's words: "The external shape of the submarine vessel bore some resemblance to two upper tortoise shells of equal size, joined together." He later added that the overall cross section "verged toward an ellipsis."

With water flowing parallel to an elliptical hull, there are pressure forces pushing against the forward sector of the hull. There are, however, counterbalancing forces associated with a pressure distribution caused by the water as it flows past mid hull and fills in along the aft sector. The net force acting against a double-ended hull due to the forward and aft pressure distribution is zero, or nearly so. In these conditions, as we have shown for the *Turtle*, the forces when transiting in still water are primarily associated with hull skin friction.

If the elliptical hull shifts to a shallow angle with respect to the flow, the force distribution begins to change. While skin friction acting parallel to the axis of the hull is still a component, the unequal port and starboard flow will result in an unequal pressure distribution acting perpendicular to the hull. The net force will tend to push the hull laterally. At shallow angles, the effect is analogous to lift on a wing where the pressure distribution on the underside of the wing is greater than along the top of the wing, resulting in a net upward force.

As the angle of the hull with respect to the current increases, the force component perpendicular to the axis also increases. When at 90° to the current, the force is all associated with water pressure acting perpendicular to the hull.

When a whaleboat, for example, crosses a river straight ahead toward a point on the opposite shore, the current perpendicular to the direction it travels will cause the boat to reach the shore well downstream of its target. By rowing upstream at an angle to the current, the lateral forces acting on the boat will push it sideways, as well as downstream. The path of the boat will then remain more perpendicular to the shoreline. The harder the occupants row, the closer they will come to their intended point on the opposite shore. The optimum transit angle is a function of hull shape and current speed. Using this strategy, a kayak, for example, requires little effort to cross perpendicular to a fast moving stream.

The hydrodynamics associated with the distribution of forces on a hull is complex and combines how the boundary layer, eddies, and turbulence all vary as a function of the angle the hull presents to the water flow. Certainly, the ability to move crosscurrent is more pronounced with a rowboat or kayak, where there is less of the hull below the surface than with the *Turtle*, where the majority of the hull is submerged. The purpose here, however, is to emphasize the conditions that would have affected the *Turtle*, whether our replica or the original, when maneuvering under real conditions.

TEST 5: SUBMERGED OPERATIONS

Bushnell described the process of submerging to Jefferson: "When the operator would descend, he placed his foot upon the top of a brass valve . . . through which the water entered at his pleasure; when he had admitted a sufficient quantity, he descended very gradually." Bushnell continued: "When the skillful operator had obtained an equilibrium [neutral buoyancy] he could row upward or downward, or continue at any particular depth, with an oar placed near the top of the vessel, formed upon the principal of the screw."

Neutral Buoyancy and Vertical Control

Submerging, as described by Bushnell, first required that the pilot obtain neutral buoyancy, and then engage the vertical propeller to drive the *Turtle* down to the desired depth. Once neutrally buoyant, i.e., the

Figure 16.2. The vertical propeller was mounted near the stem. The shaft passed through the hull and was turned by a hand crank just forward of the pilot. In the pitch-black interior of the *Turtle*, the crank was within easy reach when operated at night. (*Warner Lord*)

weight of the *Turtle* equals the weight of the displaced water, the *Turtle* will hover at that depth. A neutrally buoyant (or very slightly positively buoyant) condition is essential to provide the pilot with depth control.

The center of rotation of the replica vertical propeller is located at approximately the midheight of the hatch (figure 16.2), allowing the propeller to generate vertical thrust as soon as the hatch is awash. Many modern drawings and models of the *Turtle* incorrectly show the vertical propeller projecting above the hatch profile. In this position, the *Turtle* would have to become negatively buoyant in order to submerge the propeller, yet when negatively buoyant, the vessel will sink.

Submergence Test Procedures

The shiplift was lowered to a depth of about ten feet. This allowed sufficient room for the *Turtle* to submerge and hover before deballasting and returning to the surface. The pilot used the primary air supply consisting of a scuba tank mounted just behind him, connected via a long umbilical to his second-stage regulator (refer to figure 9.7). The original *Turtle* had to submerge to the keel depth of its target warship, which could be as much as twenty to twenty-five feet. For our purposes, it was only necessary to test the operation of the various mechanical controls that allowed the *Turtle* to submerge and resurface.

Figure 16.3, left. Prior to submerging, the support divers added small weights to the top of the hatch to test the vessel's sensitivity to obtaining neutral buoyancy. (*Bonnie Beatrice*) Figure 16.4, right. The *Turtle* successfully submerged and hovered for a short time before returning to the surface. The pilot maintained VHF communication to the surface through an antenna housed in the small buoy seen floating nearby. (*Ken Beatrice; from video frame*)

This test required a two-step process to obtain neutral buoyancy: (1) the pilot added water ballast until the hatch was approximately half submerged, and (2) the support divers then added small weights to the top of the hatch (figure 16.3) until it was just awash. This second step enabled us to measure how sensitive the *Turtle* was to incremental increases in ballast. The pilot could view the water rise and cover the windows along the side of the hatch. Communication via VHF radio was maintained throughout the submerged operations, using an antenna mounted in the white surface buoy seen in the photos above.[1]

After successfully obtaining neutral buoyancy, the plane of rotation of the vertical thrust propeller is four inches below water. After receiving instructions to begin rotation of the vertical propeller, the pilot was able to observe the *Turtle* moving downward. The tachometer used in the propulsion tests was not set to measure the rotation of the vertical propeller. The pilot, however, tried to maintain about one revolution per second. Prior tests of a full-scale model of this propeller determined that the static thrust would be approximately three to four pounds at 60 rpm.

After observing that the *Turtle* had submerged, the pilot stopped the rotation of the propeller. He communicated to the test personnel on the shiplift that the *Turtle* appeared to be hovering in position (figure 16.4). The support divers confirmed that the *Turtle* remained on station approximately eighteen inches below the surface and several inches above the shiplift platform. After a few minutes, the pilot began pumping out the ballast water and the *Turtle* slowly began its ascent. Once afloat on the surface, the shiplift was raised and brought the *Turtle* completely out of the water. The hatch was opened and the pilot emerged.

OBSERVATIONS

With the hatch awash, the test pilot could feel the resistance to rotation as he engaged the vertical propeller and submerged. Our tests confirmed that the vertical propeller as configured on the replica, provided sufficient thrust to drive the neutrally buoyant *Turtle* downward, where it could remain stationary at depth. We also confirmed that after hovering briefly, the *Turtle* pilot was able to resurface using the hand pumps to remove water ballast from the bilge.

The process of attaining neutral buoyancy is, and would also have been in 1776, a critical operation. Too much ballast and the *Turtle* would head to the bottom. If there were any significant leaks in the hull, maintaining neutral buoyancy would have been a daunting task. Working in the dark, Lee would have found it particularly difficult to keep watch on his depth gauge while trying to operate the other controls.

When Lee was heading back to Manhattan pursued by the British whaleboat, he maintained his course by keeping an eye on Manhattan through the open deadlight. In Lee's words: "I jogg'd on as fast as I could, and my compass being then of no use to me, I was obliged to rise up every few minutes to see that I sailed in the right direction, and for this purpose keeping the machine on the surface of the water, and the doors open." The only time the *Turtle* submerged during the several hours that Lee prosecuted his mission was when he was under the *Eagle*, possibly a total of thirty minutes.

MINE ATTACHMENT MECHANISM

The operational plan for the original *Turtle* was that the pilot submerge to the desired (keel) depth; row under the hull; pump out some of the water ballast to buoy up against the hull; turn the wood screw into the planking; disengage the wood screw; disengage the magazine; quickly move to a safe distance and return.

When speculating about Lee's inability to fix the wood screw into the hull of the *Eagle*, Bushnell stated that "had he moved a few inches, which he might have done, without rowing, I have no doubt but he would have

Figure 16.5, left. Remnants of copper sheathing found at an eighteenth-century wreck site. Figure 16.6, right. Detail of the copper nails found in association with the sheathing. (*Roy Manstan*)

found wood where he might have fixed the screw; or if the ship were sheathed with copper, he might have easily pierced it."

A contemporary writer, Samuel Richards, noted in his diary that "it [the *Turtle*] was sent one night, with a magazine of powder attached to it—under the command of a serjeant and 12 men—the party proceeded to the ship—having a pointed rod at top designed to be stuck into the ships bottom, but this point not taking effect—the tide which was strong—wafted the engine away from under the ship & the enterprise failed. The sergeant who had the command gave me a particular narrative of the proceeding, and said that he was of opinion that the projecting point struck the head of a bolt which prevented its success; but I judged it is as probable that the point was prevented from entering the ship by the copper sheathing."[2] It is now known that the *Eagle* was not copper sheathed at that time.[3]

We devised a test, however, to confirm Bushnell's confidence that the wood screw would have been able to penetrate copper. Ship hulls began to be sheathed in copper during the mid-eighteenth century, and it is not surprising that there was speculation that the *Eagle* was included. Copper sheathing was a very expensive addition. Hulls were typically planked with oak. Some included a sacrificial layer of a soft wood, with the sheathing added over this layer. Figures 16.5 and 16.6 show a sam-

Figure 16.7, left. A test panel with a sheet of thin copper was set on the *Turtle* hatch. The wood screw mechanism was pressed up against the copper and just as Bushnell had predicted, "if the ship were sheathed with copper, he might easily have pierced it." Figure 16.8, right. Detail of the screw after being threaded into the copper test panel and disengaged from the screw mechanism from inside the *Turtle* by the test pilot. (*Andrew Nilson*)

ple of hull sheathing and the copper nails used to secure it to the planking found off Nova Scotia in association with several cannons.

TEST PROCEDURES

A test platform was built that contained a 10 x 17-inch panel of copper backed with oak and a thin intermediate layer of pine. The test panel copper is made from flashing used in modern roofing construction and measures .025 inch thick. The thickness of the eighteenth-century hull sheathing varied from .023 inch to .028 inch.

The platform was designed to float above the *Turtle* such that the wood screw would be positioned just under the copper panel. Due to test personnel and facility scheduling limitations, there was insufficient time available to conduct this test at the Seaport. After the *Turtle* was returned to the high school, it was decided to recreate the test conditions as best as possible in a dry (rather than afloat) location. With the hatch closed, the test panel was lifted and placed on top of the hatch, such that the copper panel was centered above the wood screw (figure 16.7).

In Bushnell's description of the process of threading the wood screw, he noted that "by pushing the wood screw up against the ship, and turning it at the same time, it would enter the planks; driving would also

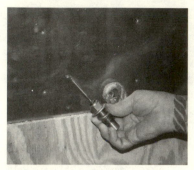

Figure 16.9. Detail of the wood screw and the hole where it had penetrated the copper. (*Roy Manstan*)

answer the same purpose." The use of the term "driving" implies that if using buoyancy and hand pressure to thread the screw were insufficient, driving the screw with a mallet would work.

With this in mind, the test pilot pushed the screw up against the test panel. After being unable to simply thread the screw into the copper sheathing, two light taps with a mallet drove the point of the screw through the copper. Using the hand crank, the pilot then easily threaded the screw past the copper sheathing, the pine layer, and well into the oak. He then continued with the process of disengaging the wood screw from the attachment mechanism (figures 16.8 and 16.9).

OBSERVATIONS

The tests confirmed Bushnell's claim that "if the ship were sheathed with copper, he might easily have pierced it." The wood screw threaded into the copper test panel with little effort. The pilot was able to readily disconnect the attachment mechanism leaving the screw threaded into the panel.

The mechanism that Bushnell designed would have to be reliable, and the point of the screw maintained sufficiently sharp. There were sources of high-quality steel available to artisans like Isaac Doolittle or Phineas Pratt who worked with Bushnell, and who had the skills and machinery necessary to produce a high-quality and durable wood screw.

There has been some speculation that when Lee pushed up against the hull with Bushnell's screw, this would simply have caused the *Turtle* to move in the opposite (downward) direction. In order to counter the downward reaction, the *Turtle* pilot would need to increase buoyancy by pumping out some of the water ballast. Lee described his attempt:

I then shut down all the doors, sunk down and came under the bottom of the ship. Up with the screw against the bottom but found that it would not enter. I pulled along to try another place, but deviated a little one side and immediately rose with great velocity and came above the surface . . . .

It is apparent that Lee pumped out a quantity of his water ballast to provide the upward force needed for pushing the wood screw up against the *Eagle*'s hull. When he "deviated a little [to] one side" the added buoyancy caused the *Turtle* to rush to the surface.

Bushnell mentioned that Lee "struck, as he supposes, a bar of iron, which passes from the rudder hinges, and is spiked under the ships quarter." While certainly a possibility, for Lee to have engaged the hull at this location, would have been a very unlucky occurrence considering that the iron covered only a very small percentage of the ship's bottom. Why the wood screw failed to pierce the hull will always remain a mystery requiring, of course, a bit of our own speculation.

Figure 17.1. The *Turtle* pilot had immediate access to the secondary air supply through a scuba regulator mounted inside the hatch. (*Roy Manstan*)

Chapter Seventeen

PILOT ENDURANCE

A MISSION NOT FOR THE FAINT OF HEART

TEST 6: PILOT ENDURANCE

From the beginning of David Bushnell's enterprise, his brother Ezra had assisted with the design and construction of the *Turtle*. He had also volunteered to be the pilot during combat operations against the British, and in anticipation of that, had thoroughly trained in the mechanical operation of the vessel. By midsummer 1776, the *Turtle* was ready to deploy against the British fleet assembling in New York Harbor, and David and Ezra waited in anticipation. Then, as with so many of the Continental troops, Ezra contracted a debilitating sickness. Unless a suitable replacement could be found, the mission would be scrapped and this most important opportunity lost.

Undaunted by this setback, Bushnell took his issue to the newly promoted Brigadier General Samuel Parsons of Lyme, Connecticut, who, as a colonel, had been commanding officer of the 10th Continental Regiment serving in New York at the time. Parsons selected three volunteers, one of whom later commented that he was to "learn the ways & mystery of this new machine and to make a trial of it." Bushnell then proceeded with the *Turtle* into various harbors along the Connecticut coast, evaluating each of these individuals. From this small group, Bushnell selected Sergeant Ezra Lee, "who appeared more expert than the rest," and devoted the training to this individual. By the end of August, they were operating in the Connecticut River. Word reached them of the British landing on Long Island and the retreat of Washington's army to Manhattan. The urgency to return to New York and make an attempt against the British fleet was obvious.

The flaw in Bushnell's original decision to have only one trained operator manifests itself on the night of September 6, 1776, when Lee embarks from Manhattan on a dangerous and complex mission. In Lee's words: "We set off from the City, the Whale boats towed me as nigh the ships as they dared to go, and then cast me off. I soon found that it was too early in the tide, as it carried me down by the ships. I however hove about, and rowed for 5 glasses [2 1/2 hours], by the ships' bells, before the tide slacked so that I could get alongside the man of war, which lay above the transports."

Lee was unsuccessful at completing his attack against the British. While he was a man of courage and mentally up to the task, the need to operate the *Turtle* for two and a half hours waiting for slack water would certainly have taxed his physical stamina. Also contributing to Lee's difficulties was the brief period of training in the operation of the complex mechanical systems that comprised Bushnell's "submarine vessel." Ezra Bushnell had mastered the controls and understood the subtleties of its operation; Lee had less than two weeks at the helm.

Other than their ages, there are no references to the physical characteristics of either Ezra Bushnell, who was thirty years old, or Ezra Lee, who was twenty-seven. What is certain, however, is that they were both of an adventurous character and dedicated to the Patriot cause. Ezra Bushnell had been a sergeant in the 3rd Company of the 7th Connecticut Regiment in 1775, serving directly under Captain Nathan Hale.[1] By 1776, Ezra's military service was dedicated to helping his brother, David, complete the *Turtle*.

Ezra Lee served as a sergeant in then Colonel Parsons's 10th Continental Regiment. The 10th had seen service during the Siege of Boston and, by the beginning of the summer of 1776, had joined Washington in New York.[2] It was early August when Parsons sent Lee as a volunteer to replace Ezra Bushnell. The 10th was engaged in the Battle of Long Island while Lee trained in the *Turtle*.

By the end of 1776, Congress had reorganized the Continental army to consist of eighty-eight infantry line regiments. Each state was required to provide a quota of regiments with enlistments that were to last until the end of the war. Connecticut's quota was eight. Lee served with the 1st Connecticut Line throughout the war, first as an ensign and later promoted to lieutenant.

Considering their military service and their willingness to venture out in the *Turtle*, it is obvious that both Lee and Bushnell were willing to risk their lives in what was a mission unprecedented in human history.

Test pilot

Remembering Lee's remarks that "the time for a trial must be when it is slack water, & calm, as it is unmanageable in a swell or a strong tide," it would have been obvious to both David and Ezra Bushnell and to Ezra Lee that operating the propeller was a strenuous undertaking and to make headway in any appreciable current would be nearly impossible. Being physically able to transit extended distances with the *Turtle* would have been essential, and any lack of this ability would have disqualified either of them.

There was no intention of subjecting a modern day version of Ezra Lee to two and a half hours behind the crank of the replica *Turtle*. The objective was to gain insight into the relationship between pilot and propeller over a moderate time period. The test pilot's observations as to the effort expended in comparison to other physical and aerobic activities would help understand what the endurance requirements would have been for pilots of the original *Turtle*.

The selection criteria for the endurance run test pilot was that he represent a best fit to the age and presumed physical conditioning of both Lee and Bushnell. The size and height of the pilot, and in particular his sitting position relative to the drive mechanism, will impact the muscles required to sustain the load he must apply to the treadle and crank. These factors are discussed in the Ergonomics section (chapter 18).

For reasons that have been noted previously, all *Turtle* Project pilots were drawn from a pool of past or present members of the Engineering and Diving Support Unit (EDSU) at the Naval Undersea Warfare Center. The person selected for the endurance testing was David Hart who, at age twenty-nine and an active member of the EDSU, maintains the rigorous physical standards associated with his Navy diver qualifications.

Test procedures

Bushnell included three deadlights in his hatch that could be opened to allow fresh air to circulate within the *Turtle* during transits. According to Lee, "three round doors were cut in the head (each 3 inches diameter) to let in fresh air until you wished to Sink, and then they were shut down & fastened." With the assumption that these "doors" were open while en route to the British fleet, the three deadlights on our hatch were also open during the endurance run.

When set into position for the endurance run, the *Turtle* was constrained from forward motion, as was the case for the static thrust measurements. This creates a 100% slip condition at the propeller with maximum thrust (and maximum exertion) associated with whatever rpm is used. When transiting at the same rpm, however, the slip (and thus thrust) will be less than 100% (see table 5). The previous test pilot, Paul Mileski, had used the same propeller (bronze with a 25° blade angle) for both the transiting and static thrust tests. He noted that the effort when transiting at an average of 66 rpm was similar to what he experienced during the static thrust tests at 50 to 55 rpm. The endurance testing was to be limited to a maximum of 45 rpm, estimated to be equivalent to transiting at approximately 60 rpm.

The test pilot was instructed that he could terminate the testing at any time. With the hatch closed, he would be breathing air entering through the three open deadlights. While air circulation through these openings was considered sufficient, the pilot had access to the scuba equipment inside the *Turtle*. A secondary source was available through an emergency system that supplied air from a scuba tank on the pier through a hose to a quick-disconnect fitting in the hatch (see figures 17.1, 17.2, and 17.3).

The test pilot was monitored with a Med Choice Fingertip Pulse Oximeter, Model MD300C1. The oximeter provides heart rate in beats per minute (bpm) and percent oxygen saturation ($HbO2$) of the blood. In brief, the oximeter measures the overall oxygen carrying capacity of the blood flowing past the sensor. One hemoglobin molecule is capable of carrying up to four oxygen molecules. As the blood circulates, some molecules are at full (100%) capacity, while others may only carry three oxygen molecules (75% capacity), resulting in a net saturation level measured by the oximeter sensor.

The pilot positioned himself inside the *Turtle*, test operated the propeller, and checked the tachometer and VHF communications. The pilot then tested the internal scuba air supply and the secondary surface supplied air system. He placed the pulse oximeter on his finger and checked its operation. Once the pilot and test director were satisfied that all sys-

Figure 17.2, left. Scuba tank used as the secondary air supply and a portion of the one-hundred-foot length of hose that would provide additional air in an emergency. Figure 17.3, right. The secondary air supply hose was connected to a quick disconnect fitting installed in the *Turtle* hatch. (*Roy Manstan*)

tems were operating satisfactorily, the hatch was closed and the shiplift lowered to its prescribed depth.

The safety swimmers were positioned at two of the deadlight openings. One swimmer was to view and maintain verbal contact with the pilot. The other held a small video camera that was connected to a topside monitor, allowing test personnel to also view the pilot. The test director was able to communicate directly to the pilot via VHF radio, as well as through the two safety swimmers. The swimmers were ready to open the hatch on command from the test pilot or the test director.

The pilot was instructed to provide readings of heart rate and HbO2 levels from the oximeter and propeller rpm from the tachometer at one-minute intervals. Throughout the test, safety divers visually monitored the pilot and would frequently inquire if he was okay to continue.

The criteria established by the test director for stopping the test would be if the HbO2 level dropped below 95% and/or the heart rate exceeded 130 bpm and remained at these levels for more than two minutes (two data cycles). We fully understood that these were conservative limits and that a heart rate approaching 140 bpm would have been acceptable. The goal was to subject the test pilot to moderate levels of exertion that could be sustained for an extended period of time.

Test results

The time-varying relationship between propeller rpm and heart rate are plotted on figure 17.4. Blood oxygen levels remained reasonably constant over the duration of the test and were not plotted.

Testing commenced at 2:15 p.m. Baseline "at rest" measurements of heart rate and blood oxygen levels were recorded while the pilot acclimated to conditions inside the *Turtle*. During the acclimation period, the pilot's blood oxygen level (HbO2) remained constant at 99%, while his heart rate ranged from 66 bpm to 76 bpm, averaging 70 bpm.

At 2:28 p.m., the pilot was instructed to operate the propeller at 45 rpm. Within the first minute, at 45 rpm, his heart rate increased from 76 bpm to 111 bpm. During this initial run of five minutes, his heart rate averaged 108 bpm. His HbO2 level dropped to 98% and remained at that level.

At 2:33 p.m., the pilot was instructed to reduce rotation to 35 rpm and maintain that rate for thirty minutes. His HbO2 level remained essentially constant at 98% throughout the entire 35 rpm run. His heart rate, however, fluctuated during the first ten minutes between a high of 113 bpm and a low of 93 bpm, averaging 100.6 bpm. This fluctuation in heart rate diminished, resulting in an average of 90.5 bpm over the next twenty minutes.

Near the end of the 35 rpm run, the test director inquired as to the pilot's condition and whether he would be willing to continue. With a response in the affirmative, at 3:03 p.m., the pilot was requested to increase propeller rotation to 45 rpm and continue for an additional ten minutes. Within one minute, the pilot's heart rate had increased to 116 bpm, and averaged 104 bpm for the duration of the ten minute run. His HbO2 level varied between 96% and 98%, averaging 96.8%.

At 3:13 p.m., the pilot was instructed to cease operating the propeller but to continue providing data from the oximeter. The test director continued to monitor him, and at 3:25 p.m., the hatch was opened and the pilot exited.

Pilot Comments

At the end of the day, the pilot was asked to document his observations regarding his experience during the endurance test. The following comments were provided to the test director and are included here as a supplement to the data. One could certainly imagine similar commentary coming from Ezra Bushnell in 1775, after an exhausting session in the Connecticut River with his brother's submarine.

Regarding propeller operation:

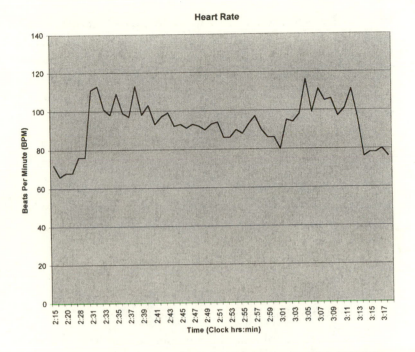

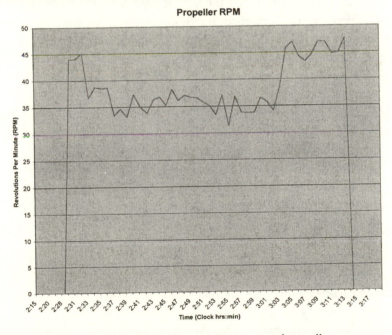

Figure 17.4. Plots of the relationship between heart rate and propeller RPM as a function of time during the forty-five-minute endurance test. (*The Turtle Project*)

I felt like I was not able to get a good rhythm with either hand cranking or foot cranking alone. I experienced difficulties rotating through top and bottom dead center. I settled on a combination of hand crank and foot power by pulling up on the hand crank for half of the rotation and pushing down on the treadle with my feet for the other half. This allowed me to establish a good rhythm without overexerting.

Regarding effort expended:

At 35 and 45 rpm, I was able to find a consistent breathing pattern, much like when you go for a jog (35 rpm) or a run (45 rpm), and muscle fatigue was such that I would have been able to continue for a long time at 35 rpm.

Regarding rudder/tiller:

It's difficult to determine how easy it would have been for me to find that rhythm while having to steer the Turtle.[3]

Paul Mileski offered additional comparisons during a posttest review of the static thrust video with the test director. It should be remembered that Mileski, while in excellent physical condition, is about twenty years older than the endurance pilot, Dave Hart, (but twelve years younger than the other test pilot).

When operating at 35 to 40 rpm, Mileski felt that the energy expended was "similar to a brisk walk along a moderate uphill grade; sustainable but noticeably more difficult when at 40 rpm." When running the static thrust test in the 50 rpm range, Mileski commented that "it was much like an uphill jog that, without an occasional break, I would have become fatigued." Pilots Mileski and Manstan agreed that operating at very high rpm is comparable to other activities that result in heavy breathing followed by a lengthy period of exhaustion.

ANALYSIS

In our evaluation of the physiological effects of operating the *Turtle*, we referred to the work of Per-Olof Åstrand and Kaare Rodahl (1986), who published their studies of the human response to physical exertion in their *Textbook of Work Physiology: Physiological Bases of Exercise*. While much of their work was with high intensity, short duration exercise, they did cover the effects of the prolonged exertion that more closely relates to the operation of the *Turtle*.

Our measurements of specific physiological parameters were limited to heart rate (beats per minute or bpm) and oxygen saturation (HbO2). We did not subject the pilot to a procedure where he would be required to operate under anaerobic conditions. The oximeter data, however, does provide insight into the ability of the pilot, and by supposition Ezra Lee, to sustain the rotation of the propeller over an extended mission. The pilots' observations, while subjective, were also essential to the overall evaluation of what Lee may have experienced.

Åstrand and Rodahl's measurements were conducted under very controlled conditions where parameters affecting each test subject's performance could be measured with a high degree of accuracy. They subjected the individuals to three categories of exertion: "submaximal," "maximal," and "supermaximal." From the data, they determined levels of effort leading to "exhaustion" and levels of effort that could be sustained in a "steady state" for a prolonged duration.[4]

The oximeter data from the *Turtle* tests does not allow a complete analysis of the physiological effects on the test pilot. There are, however, analogies that can be drawn between our testing and Åstrand and Rodahl's work; in particular their analysis of test subjects operating on a treadmill or stationary bicycle. The parameters in their studies that are relevant to the analysis of the *Turtle* data are oxygen uptake (volume of oxygen extracted from the inhaled air in liters/min) and heart rate.

STEADY STATE (SUBMAXIMAL EFFORT) AND EXHAUSTION (MAXIMAL EFFORT)

During vigorous exercise, respiration rate and heart rate increase in order to provide the necessary oxygen uptake to compensate for the increased metabolism; i.e., as cardiac output increases due to demand, oxygen uptake also increases. Åstrand and Rodahl recognized that age, sex, physical size, and genetic variations, as well as training, influence an individual's oxygen uptake and hence physical performance.[5]

According to Åstrand and Rodahl, for a person in good physical condition (as with our test pilot and Ezra Lee), heart rate and cardiac output approach steady state during the first five minutes of moderate exertion, such as walking or running, and can remain at about the same level for an hour. For their experiments, the "level of effort" at the treadmill or bicycle is measured in watts. Steady-state conditions were obtained after five minutes at levels ranging from 50 watts to 250 watts. When the subjects operated between 250 and 300 watts, there was no additional oxygen uptake, indicating that their test subject was operating under anaerobic conditions.[6]

Åstrand and Rodahl provide a list of ranges of heart rates for the "average individual, twenty to thirty years of age" corresponding to levels of "prolonged physical work." Their heart rate ranges include: up to 90 bpm for "light work"; 90—110 bpm for "moderate work"; 110—130 bpm for "heavy work"; 130—150 bpm for "very heavy work"; and 150—170 bpm for "extremely heavy work." Comparing this to the *Turtle* endurance test data (figure 16.4), our pilot would rate in the "moderate" category during most of the forty-five-minute test, exceeding 110 bpm only occasionally when run at 45 rpm.[7]

Oxygen uptake for Åstrand and Rodahl's test subjects during moderate to heavy prolonged exercise was about 50% of their maximum, or about 2.5 liters/min. for the athletic test subjects, and 1.5 liters/min. for individuals of average physical conditioning. Beyond an hour, however, this level could not be sustained.[8]

Measurements of oxygen uptake taken on their cycle ergometer was compared to that from other activities, such as cross-country skiing, swimming, running, heavy and moderate manual labor, climbing stairs, and others (but did not include driving a wooden submarine). For example, running at 10 mph and competitive cross-country skiing is roughly equivalent to 300 watts on the ergometer with an oxygen uptake of 4 liters/min.; running at 7 mph and exceptionally heavy manual labor corresponds to 200 watts and 2.8 liters/min.; while walking at 4.5 mph and gardening corresponds to 100 watts and 1.5 liters/min.[9]

While entirely subjective, the endurance test pilot related his level of effort while operating the propeller at 35 rpm as being equivalent to "jogging," while at 45 rpm, it was closer to "running." Considering that his HbO2 levels never dropped below 96% and his heart rate remained below 120 bpm during the 45 rpm intervals, we believe that his overall effort working the hand crank and foot treadle did not exceed 200 watts, well within the range for aerobic exercise. It is reasonable to conclude that the test pilot, and Ezra Lee as well, could have significantly increased the propeller rpm (and thus speed), while remaining within aerobic levels.

Åstrand and Rodahl measured oxygen uptake under heavy working loads to determine the time at which the subjects would be compelled to stop due to exhaustion. After a ten minute interval under steady-state conditions, the test subjects were instructed to increase their exercise rate to levels between 300 watts and 450 watts. The subjects could only sustain 450 watts for two minutes, and 300 watts for less than six minutes.[10] A similar situation may have occurred during our static thrust measurements when test pilot Mileski showed signs of exhaustion after only a few minutes operating propeller (2) with the 45° blade angle, whereas he was readily able to sustain the rotation of propeller (1) with the 25° blade angle (see chapter 13).

Intervals and "micropauses"

Further testing by Åstrand and Rodahl showed that providing ten-minute intervals and alternating between cycling and running, the test subjects could maintain their effort for several hours.[11] An extended rest, however, may not have been a good strategy for Lee. If he struggled against a one-knot current for an hour, a ten-minute rest would have carried the *Turtle* more than three hundred yards downstream.

Åstrand and Rodahl also studied the effect of what they termed "micropauses," or very brief periods of rest between short duration exertions. For athletes in training, these intermittent exercise routines allow longer overall training periods by reducing the potential for oxygen deficit and lactate buildup.[12] Short periods of maximum exertion with micropauses would not, however, be appropriate for operating the *Turtle*. Once in motion, it should remain in motion, yet Lee needed some strategy for maximizing his oxygen uptake without leading to anaerobic exertion and exhaustion.

Ezra Lee ... 1776

For Lee to complete his mission, he had to maneuver the *Turtle* against the prevailing current in New York Harbor and remain within striking distance of his target. In addition, once alongside the *Eagle*, he had to have enough reserve stamina to submerge, position himself, and attach the mine with only thirty minutes of air available within the *Turtle*.

For the intrepid Lee, undoubtedly the Revolutionary War equivalent of today's "lean, mean, fightin' machine," the mission required more physical endurance than would be expected of the average colonial farmer–soldier. Lee had already volunteered to man a fireship against the British fleet, and was a likely candidate to take the helm of Bushnell's submarine.

When Brigadier General Parsons approached him about the *Turtle* mission, Lee agreed, "but first returned with the machine down Sound, and on our way practiced with it in several harbours—we returned as far as SayBrook with Mr. Bushnell . . ."

During the brief opportunity Lee had to train along the Connecticut coast, he likely was able to establish an efficient propeller cranking rhythm (as did our test pilots) for transiting and maneuvering under basic operating conditions. When in the lower Connecticut River at Saybrook, however, he would have been subjected to river and tidal currents that would be at least as significant as what he would soon encounter in New York. He would have experienced how the *Turtle* handled in a heavy current and must have understood what level of exertion he would have to endure. Lee's strategy for sustaining this effort probably included some combination of shifting between muscle groups, plus allowing occasional brief pauses.

We feel that it is a reasonable assumption that Lee was a physically strong individual, used to hard labor. This, however, is not the only criterion for evaluating his adaptability to the operation of the *Turtle*. Lee's aerobic endurance may very well have been up to the task. However, if he was crammed into a vessel where there were restrictions on his ability to operate the mechanical controls, his body simply may not have been able to match his aerobic conditioning. To accommodate a twenty-first-century equivalent of Lee, the modern military strives to fully understand the ergonomics of the mechanical environment in which the soldier or sailor works.

Chapter Eighteen

PILOT ERGONOMICS
By Design or an Afterthought

TEST 7: ERGONOMICS

Ergonomics, also referred to as "human engineering," has become a significant consideration during the design of modern military technology. The position of the operator relative to the mechanical controls, whether in a tank, a jet, or a SEAL Delivery Vehicle (SDV, the modern equivalent of the *Turtle*) is critical because the success of the weapon depends on the interaction between the person and the machine, particularly during the stress of combat.

It was going to be David's brother Ezra who would take the helm of his submarine. We assume, therefore, that the original *Turtle* was built to accommodate Ezra's size and height, and his position relative to all of the onboard mechanical systems. Bushnell noted that "particular attention was given to bring every part, necessary for performing the operations, both within and without the vessel, before the operator, and as conveniently as could be devised; so that every thing might be found in the dark." Although human engineering was not in David Bushnell's curriculum at Yale, ensuring that "everything might be found in the dark" is evidence that he considered this in his design. It was important, therefore, that we also monitored how our three test pilots interacted with the *Turtle*'s operational systems throughout all of the sea trials and testing.

Bushnell's description continued: "A firm piece of wood was framed, parallel to the conjugate diameter, to prevent the sides from yielding to the great pressure of the incumbent water . . . This piece of wood was also a seat for the operator." The propeller mechanism "was directly before the operator, who sat upright . . ." The tiller "was within the vessel, at the operator's right hand . . ."

Bushnell also provided the pilot with "small glass windows in the crown [the brass hatch], for looking through" as well as "three round doors, one directly in front, and one on each side . . ." The cockpit arrangement would likely have been designed such that his brother could see through the open deadlights when in a sitting position. Gale, who had witnessed Bushnell's early tests where Ezra was most likely at the helm, commented that the hatch "receives the person's head as he sits on a seat."

TURTLE PILOTS EZRA BUSHNELL AND EZRA LEE

We believe that Ezra Lee may have been shorter than Ezra Bushnell. When Lee described his retreat back to the city, he noted that while being pursued by the British, "my compass being of no use to me, I was obliged to rise up every few minutes to see that I sailed in the right direction, and for this purpose keeping the machine on the surface of the water, and the doors open." From this comment, we surmise that his eye level was lower than the openings in the hatch. Every time Lee raised himself to look out through the doors, he would have had to release the hand crank. The treadle would have been difficult to continue operating while in a standing or crouched position. There are, however, no records of the physical characteristics of either Ezra Bushnell or Ezra Lee.

Any modern consideration of how David Bushnell organized the locations of the internal mechanical systems would have to be estimated based on the size of the average eighteenth-century male. The curator of the Revolutionary War uniform collection at the National Museum of American History estimates the average height of the colonial soldier to be five feet six inches.[1] To gain more insight into this, we pursued another source.

When a soldier enlisted, the recruiter recorded his age, height, and occasionally other characteristics, such as hair and eye color, skin tone, scars, verbal accents, and his civilian profession. In a note published in the *Connecticut Courant and Hartford Weekly Intelligencer* for March 17, 1777, recruiting officers of Colonel Swift's Regiment were instructed to "make a return to the Commanding Officer of Said Regiment of the men they have here inlisted, their Names, Age, Complexion, colour of Hair, size, and Place of Address."[2] Unfortunately, none of this information concerning Lee and Bushnell has survived.

> LYME, 30th JUNE, 1777.
> THIS is to notify all the Men in the Town of New-London that are detached to fill up the Continental Army, that they must march by Monday the 7th of Ju'y, or they will be deemed Deserters, and treated accordingly.   EZRA LEE, Ens.
> N. B. If they want Blankets or Arms they must apply to the Select-Men, who are directed to furnish all such Men with those Articles, as the other Soldiers are.

> DESERTED from Capt. Beardslee's company in Col. Swift's regiment, one Wm. Bostwick, an Englishman, 24 years of age, 5 feet 9 inches high, light complexion, black hair, light coloured eyes, had on a round brim'd hat with a red and blue cockade, a pepper-and-salt coat, a red and white flowered vest, good leather breeches, gray stockings, flowered silver shoe buckels. Whoever takes up said deserter secures him so that he may come to justice, shall receive five dollars reward and all necessary charges paid, by   PHINEAS BEARDSLEY, Capt.
> New-Fairfield, April 24, 1777

Figure 18.1, top. Four months after his mission in the *Turtle*, Sergeant Ezra Lee became Ensign Lee of the 1st Connecticut Line. Lee placed this not too subtle ad instructing the men who were being detached to the Continental army to immediately muster at New London. (*Connecticut Gazette, July 4, 1777*) Figure 18.2, bottom. A soldier who neglected to meet the obligations of his enlistment was quickly labeled a deserter. This ad provides a description of one such individual and offers a reward for his capture. (*Connecticut Courant, May 5, 1777*).

If not serving in the Continental army, members of Connecticut militia companies were still required by law to muster for training "in some or either of the Months of *March, April, May, September, October, or November.*"[3] During the campaigns of 1777, soldiers were drawn from these state militia units to fill the dwindling ranks of the Continental army. After his *Turtle* mission, Ezra Lee continued serving in the 10th Continental Regiment and later under Colonel Huntington of the 1st Connecticut Line.[4] He had been promoted to ensign (and then to lieutenant) and was responsible for assembling the troops who had enlisted in his regiment. He placed an ad (figure 18.1) in the July 4, 1777, issue of the *Connecticut Gazette; and the Universal Intelligencer.*

As Lee distinctly pointed out, when a soldier or new recruit failed to show when called, they were considered deserters. Ads for their capture and return were published in local newspapers and included the name and all of the personal information obtained by the recruiters. The data used in our analysis was obtained from the *Connecticut Courant and Hartford Weekly Intelligencer,* and includes the entire year for 1777. (This newspaper has been reproduced on microfilm and is available at the Connecticut State Library's Web site, http://www.cslib.org.) The quality of the majority of microfilm copies was, in general, reasonable enough to read, and a sample size of 110 individuals was obtained. A few original copies of the newspaper were also viewed; a sample appearing in the May 5, 1777, edition is shown in Figure 18.2.

Many of the names appeared multiple times, indicating that it took awhile before they were captured or, remembering that desertion was potentially a capital crime, returned on their own. Care was taken to ensure that these individuals were included only once in the sample. The individuals were primarily from Hartford area regiments, although a few soldiers from Massachusetts were included.

The data is presented in the bar chart, figure 18.3. Heights of the soldiers are provided to the nearest inch. When collecting the data, however, some ads provided a range of heights for the individual. For example, there were six ads where the soldier was described as being "5 feet 6 or 7 inches." Because our bar chart data is arranged in whole number increments, we selected the higher of the two heights. When calculating the average height, however, a value of 5′6 1/2″ was used for these six samples. For two individuals whose heights were given as "5 feet 8 or 10 inches", a value of 5′9″ was used in the analysis. Another ad provided a height of 5′7 1/2″. This height was rounded to 5′8″. The height of one individual was listed at "4 feet 4 inches". He is not shown on the bar chart, but was included in the calculation of average height.

Based on the 110 samples, the average height was determined to be 5′7.65″, slightly taller than that indicated in the Smithsonian uniform collection. The bar chart shows that twenty-four individuals (21.8%) were 5′6″. The number dips to sixteen (14.5%) for those in the 5′7″ and 5′8″ categories, and to fourteen individuals (12.7%) in the 5′9″ category. The number increases to twenty-one individuals (19.1%) listed at a height of 5′10″. Finally, at both ends of the sample, there were a total of twelve individuals (10.9%) at heights less than 5′6″, and seven individuals (6.4%) taller than 5′10″.[5]

Of the 110 total samples, there were 97 individuals where age was given and legible on the microfilm. The average age of this sample was 27.3 years, ranging from age 17 to 50.

Turtle test pilots

The position of the controls and seat on the replica *Turtle* were located in proportion to the size and height of the first test pilot, Roy Manstan.

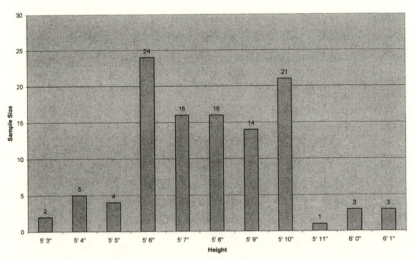

Figure 18.3. Bar chart showing the distribution of the heights of Connecticut soldiers from the Revolutionary War. The data was obtained from newspaper ads from 1777 describing soldiers who were listed as deserters. (*The Turtle Project*)

The seat position allowed him to view through the hatch windows, to have easy access to all of the *Turtle*'s mechanical systems, and to operate the treadle and crank from a sitting position.

All three of the *Turtle* test pilots are approximately six feet tall. When operating the treadle in the sitting position, an angle of about 90° is created between the pilot's thighs and upper body. Any downward push on the treadle is generated by muscles in the upper thigh at its junction with the hip. If the legs were at a steeper angle, more of the pilot's weight could be incorporated into the push, relieving some of the dependence on the muscles. After completing two days of testing, *Turtle* pilot Paul Mileski commented that in relation to his height, the seat was quite low, making it difficult to efficiently operate the treadle.

This same issue was noted by Manstan during the initial launch and operational trials in the fall of 2007. He also noted that with the 90° sitting position, he had to depend entirely on the muscles at his hip—thigh junction and had difficulty sustaining the driving force against the treadle. (Note, however, that this test pilot was more than twice Lee's age.) A test was later performed in the replica *Turtle* where the same pilot raised his sitting position and operated the treadle. While only a subjective observation, he indicated that an angle between his upper body and thighs closer to 120° would have been preferable.

Had the replica seat been higher and/or the treadle lower, the pilot's knees would have been below the level of his hips resulting in an angle greater than 90º. This would also have provided room for a greater cranking radius, as Mileski had suggested (chapter 15), for increasing the power provided by the pilot. Had the replica seat been set higher in the cockpit, however, the heads of all three test pilots would have hit the top of the hatch.

The testing made it apparent that the position of the pilot relative to the treadle is critical to sustaining sufficient power to the propeller during extended operations. The "typical" continental soldier, at just under 5´8″, would certainly have fit inside the original *Turtle* (and the replica) better than the three current test pilots, and his leg position may have been much better adapted to efficient operation of Bushnell's treadle.

ANALYSIS

Transiting and maneuvering the *Turtle* places physiological and psychological demands on the pilot. In the previous section, we looked at the intensity of the work required to operate the *Turtle* in relation to the pilot's heart rate and oxygen uptake. If exertion remains below a level where the muscles are functioning under aerobic conditions, the person can sustain the effort. As time and the level of effort increase and the oxygen uptake does not satisfy the body's needs, there is a greater likelihood that the individual's exertion will become anaerobic, leading to exhaustion.

The ability of an individual to function over an extended period of time is also related to muscle fatigue. While the overall body may be receiving a sufficient supply of oxygen, there may be specific muscles that are not. This may be due to a sitting or crouched position restricting blood flow to the very muscles performing the work, and the person may be compelled to stop before he has reached the exhaustion discussed in the endurance section.

The pilots, including Ezra Lee, operated their vessels in a sitting position while turning the propeller, using a combination of arm and leg muscles. The application of the loads through the treadle and hand crank are essentially cyclic in nature. First, the right foot presses down on the forward edge of the treadle to rotate the shaft clockwise through one-half cycle. The other foot then presses down on the back of the treadle to complete the rotation of the shaft. The motion essentially alternates the muscle contractions from the right leg to the left. The pilot has the option to employ the hand crank exclusively, or to supplement the treadle by pulling and pushing the hand crank mechanism through top or bottom dead center, where the treadle has no contribution to rotation.

The amount of time that Bushnell's pilots had training for the missions was important not only for the ability to gain familiarity with the controls, but also to adapt the specific muscles that would be required to turn the propeller. In their *Textbook of Work Physiology*, Åstrand and Rodahl (1986) discuss how training of muscles subjected to repetitive cyclic motions can, to some extent, adapt the muscles to potentially anaerobic conditions.[6] We assume that Ezra Bushnell had trained for many hours from the summer of 1775 until the following summer in preparation for his mission. He would have had the time to adapt the specific muscles that had to be employed while sitting. As with any athletic training, repetitive exercise would also have affected the pumping efficiency of his heart, resulting in a slower heart rate to supply the oxygen demand. Ezra Lee, on the other hand, only had a matter of days in late August 1776 to prepare his body.

Each test related to *Turtle* performance required the test pilot to maintain a consistent propeller rotation (rpm). All of the pilots found a particular rhythm that combined both the treadle and hand crank. Occasionally, the initial test pilot, Roy Manstan, used the hand crank alone in order to briefly relieve his leg muscles.

Åstrand and Rodahl also describe the effect of prolonged strenuous exercise on the skeletal muscle groups (i.e., arms and legs) that *Turtle* pilots (past and present) relied on. The muscles used to maintain the propeller cranking rhythm experience cyclic contractions. The muscles of the upper thigh and hip are the most active in the operation of the treadle. The arms and shoulders control the hand crank. Using, for example, a propeller operating at 60 rpm, the pilot's muscles extend and contract once per second.

Åstrand and Rodahl compared the ability to sustain rhythmic contractions to an individual's maximum isometric (static) load capability for fingers, hands, arms, and legs. Their measurements ranged from 90% of the maximum isometric load at 5 contractions per minute to 60% at 30 contractions per minute. Their conclusion was that rhythmic contractions "can probably be performed for long periods of time only if the developed strength does not exceed 10%–20% of the maximal isometric strength."[7]

We noticed that during the static thrust tests, propeller (2) with a 45° blade angle created the greatest effort on the part of the pilot. At the maximum rotation obtained of 41 rpm, the pilot was not able to sustain the cranking function more than one minute. When operating propeller (1), with a 25° blade angle, at more than 60 rpm, this same pilot was confident that he could have sustained the rotation for a significant length of time. In other words, the pilot was likely employing his cranking muscles below 20% of their maximal isometric strength.

The primary physiological factors affecting muscle fatigue are maintaining a sufficient supply of oxygen and minimizing lactic acid buildup. Blood flow to the muscles is restricted during contractions affecting the exchange of metabolic products. The pilot's sitting position, where his thighs are approximately 90° to his hips, also constricts blood flow to the active muscles. This constriction would be relieved to some extent if the thighs were situated at a greater angle, as mentioned earlier in this discussion.

Åstrand and Rodahl emphasize that there are also psychological factors that affect the response of an individual when the muscles would otherwise take control, compelling the person to cease his activity. At the end of their "Neuromuscular Function" chapter, Åstrand and Rodahl make a comment appropriate to Ezra Lee's situation: "In healthy subjects, young as well as old, untrained and well-trained, the problem during maximal effort is, as a rule, that the muscles eventually stiffen and refuse to obey the subject's will. However, it has been repeatedly stressed that people usually perform better if motivation is maximized."[8]

One can only imagine Lee's motivation as the *Turtle* passed Governors Island en route to Manhattan when, in Lee's words: "I was abreast of the fort on the island 3 or 400 men got upon the parapet to observe me—at length a number came down to the shore, shoved off a 12 oar'd barge with 5 or 6 sitters and pull'd for me." No doubt with muskets at the ready.

PART FOUR

SPECULATION

The *Turtle* replica produced by Handshouse Studio and students from Massachusetts College of Art is launched with a horse-drawn cart into the North River in Massachusetts. (*Cary Wolinsky*)

It was only a matter of circumstances that the *Turtle* was unable to complete its mission. George Washington, commenting about the failed attack in his letter to Thomas Jefferson, noted that "one accident or another always intervened. I then thought, and still think, that it was an effort of genius." Bushnell was certain, as were his supporters, that in his submarine vessel, he had conceived an effective and potentially dangerous weapon.

His reluctance to include dimensional details in his letter to Thomas Jefferson is a limiting factor in creating a replica. The term reluctance is used here because Bushnell was quite clear in the cover letter he included when forwarding a copy of the Jefferson letter to Yale College president Ezra Stiles. In the Stiles letter he notes, "I could wish that what I have written should not come to the knowledge of the public." Bushnell would have been concerned that too many details in the wrong hands would encourage the pursuit of submarine warfare. He was right in some ways. Even now, over 230 years later, we are still laboring to recreate his ideas.

We have attempted to give due diligence to replicating the form and function of the original submarine based on our understanding of eighteenth-century engineering, and our interpretation of the contemporary written accounts. The subsequent investigation of the operational characteristics of our *Turtle* has led us to the following speculation about the practical limitations of Bushnell's vessel, and what his intrepid pilots would have experienced. We have included our understanding of where Bushnell built and tested his submarine, the vessels that may have accompanied him, and the pilots that he trusted with the operation of his vessel. The final chapters take a close look at the events that occurred during Lee's attempted attack on HMS *Eagle*.

Chapter Nineteen

THE TURTLE IN THE CONNECTICUT RIVER

A CONVENIENT LOCATION

POVERTY ISLAND

"If you seek the oldest inhabitant of Saybrook, and ask him to point out its locality, he will say, with boyhood's fondness for olden play-grounds in his time: 'Ah yes! It is *Poverty* Island that you mean. It used to be there, but spring freshets and beating storms have washed it away.'"[1]

Beginning in 1899 with Everett Tomlinson's narrative *David Bushnell and his American Turtle*, there has been speculation that Poverty Island was where David and Ezra Bushnell stashed their submarine.[2] Tomlinson claimed (probably based on his conversations with that "oldest inhabitant of Saybrook") that the brothers had purchased a fisherman's seine and reel on the island and soon built a shed where they could store and test their secret weapon and train Ezra in its operation.

Regardless of the source of Tomlinson's information, his account continued to influence writers throughout the twentieth century. Reference to Poverty Island appeared in 1963 in Frederick Wagner's *Submarine Fighter of the American Revolution*, then in Frank Anderson's commentary that accompanied his 1966 reprint of Abbot's book. In 1976, Marion Hepburn Grant included Poverty Island in *The Infernal Machines of Saybrook's David Bushnell*, written for the Bicentennial Committee of Old Saybrook, Connecticut. The most recent account to include Poverty Island was in 2004 in Arthur S. Lefkowitz's *Bushnell's Submarine, The Best Kept Secret of the American Revolution*.

Because of the role that Poverty Island purportedly played, we have included a discussion of this location in our version of the *Turtle* story.

One of the *Turtle* Project's volunteers, Ken Beatrice, spent a great deal of time with the investigation and credit is due to his efforts on this subject. His inputs and insights are included in the following speculation regarding Poverty Island and Bushnell's operations in the Connecticut River.

Wagner recognized the uncertainty as to where Bushnell conducted his trials, and states in his notes to chapters 3 and 4: "The exact site where Bushnell built the Turtle is not stated in any contemporary records; tradition says it was either on Sill's Point or on Poverty Island, and evidence points more strongly toward Poverty Island."

Wagner doesn't indicate what evidence he uses for this, but in his book maintains that:

> he needed a secluded spot for construction. Poverty Island, out in the Connecticut River, seemed a suitable site. Long since eroded away by the river, the island was then not far [actually about three and a half miles downriver] from Sill's Point, near the Saybrook ferry. It seemed unlikely that any British ships or British soldiers would come snooping around, but to prevent idle speculation, he built a shed to conceal the work that would be going on. Then he gave out the story that he had become a fisherman and that the shed held the reel on which he wound his seine.[3]

The one accurate statement regarding Poverty Island from Tomlinson's "oldest inhabitant of Saybrook" is that "storms have washed it away." At the mouth of the Connecticut River is a large salt marsh shown on modern navigation charts as Great Island. What remained of Poverty Island became incorporated into Great Island and is now shown as Poverty Point. Charts from the nineteenth century show the process, with a distinct island still shown in 1815, but turning into a small peninsula by 1838.[4]

Surprisingly, Captain Abner Parker's 1771 "Chart of the Saybrook Barr" does not indicate any small island in that area.[5] River sedimentation over the millennia had created a shallow sandbar extending more than a mile out from the mouth of the river. Parker's chart concentrated on defining the four natural channels through the sandbar that enabled access to and from the river. Poverty Island was probably considered of little interest.

An 1853 chart shows a narrow strip of marshland called Poverty Island Beach, although it was not an island.[6] This same chart shows very shallow water, all less than five feet, surrounding Great Island and Poverty Island Beach. The closest distance to the channel is a quarter of a mile from the tip of the beach. Ken Beatrice kayaked around the southern end of Great Island for the *Turtle* Project. His soundings, made with a lead line, confirm that these same shallow depths still exist where Poverty Island was once located.

Bushnell ventures into Long Island Sound

David Bushnell was raised in the Society of West Saybrook (also referred to as the Pachaug District), a parish in the town of Saybrook. In 1810, this parish was incorporated to form the current town of Westbrook. Bushnell had sold his half of the family farm to his brother Ezra in order to secure the funds to attend Yale. It is likely that much of the initial construction of his submarine occurred at the farm, but as it neared completion, a more suitable location was needed close to water.

From the beginning of the *Turtle* Project, we discussed what Bushnell would have required for conducting any testing or training with his submarine. In Bushnell's words: "In the first essays with the sub-marine vessel, I took care to prove its strength to sustain the great pressure of the incumbent water, when sunk deep, before I trusted any person to descend much below the surface: and I never suffered any person to go under water, without having a strong piece of rigging made fast to it, until I found him well acquainted with the operations necessary for his safety."

In Benjamin Gale's August 1775 letter to Benjamin Franklin, Gale described: "The Experiments that has Yet been Made are as follows—— In the Most Private Manner he Convey'd it on Board a Sloop In the Night and Went out into the Sound, He then sunk under Water, where he Continued about 45 Minutes without any Inconveniency as to Breathing." It is surprising that Bushnell ventured out into Long Island Sound after dark. Night operations to avoid British curiosity and intervention would have been nearly impossible considering the difficulty navigating, even during daylight, the narrow channels at the mouth of the river.

Nonetheless, Bushnell had access to a sloop and sought out water where he could test the *Turtle* at depths at least as great as the keel of a British warship, and may have occasionally risked taking his submarine out beyond the mouth of the river. During the summer of 1775, however, the British were patrolling the waters around Saybrook making Long

Island Sound a dangerous place to carry on any extensive training activities. From the July 9, 1775, entry in the diary of Samuel Tulley: "Sabbath—9th Wind Southerly & hot PM Were alarmed in time of Service by firing of Cannon from a ship in ye Sound which took Several Vessels; schooner upon ye Bar Was Siez'd but released as also several Vessels."[7]

THE SILL HOUSE

A more suitable location where the Bushnell brothers could safely put the final touches to their submarine and conduct their trials was most likely further up the Connecticut River at the home of Richard Sill, a veteran of the French and Indian War, on what is now called Ayer's Point. The homeowner was the uncle of a Lieutenant Richard Sill of Lyme, Connecticut. The younger Sill was a classmate of Bushnell, also graduating from Yale in 1775. Sill and Bushnell, along with Nathan Hale, were members of Yale's Linonian Society.[8]

The earliest mention of a relationship between Bushnell and the Sill family is found in a genealogy published in 1859. "During the revolutionary war in 1775, David Bushnell became an inmate in Captain [Richard] Sill's family, occupying for a long time a room called the blue bed room, about 8 by 10 feet square, where he contrived and perfected that wonderful piece of mechanism called the torpedo, for the destruction of British ships infecting our coasts."[9]

Gilman Gates in his 1935 book *Saybrook at the Mouth of the Connecticut River*, states that "David Bushnell originated the idea of submarine warfare and built the first submarine boat, the 'American Turtle.' The boat was built in the Ferry District in the present town of Old Saybrook, probably at Carbine Point, also called 'Ayer's Point' and 'Sill's Point.'"[10]

Gates likely relied on local tradition when suggesting that the *Turtle* had been built at Ayer's Point. When we speculated on this location as well, we considered several factors. Bushnell needed access to water in a secluded location. He also needed access to local artisans, vessels, and resources. Rumors of his designs would certainly begin to spread, and he had to conduct his operations out of public view and in a neighborhood with known patriotic sympathies. He also had to live close enough to his "experiment" that he and his brother could readily commute (for lack of a better word) to his submarine workshop.

The Sill homestead was an ideal location, satisfying all of these requirements (see figure 19.1). The earliest part of the structure was built in 1690, followed in 1750 by a larger section facing the river; a final addition was built in 1802.[11] Lieutenant Sill served in the 10th Continental Regiment during the Siege of Boston in 1775. There was no doubt about the Sill family's commitment to the American Revolution, and there was plenty of room for the *Turtle* at their Ayer's Point home. Because Sergeant Ezra Lee served under Lieutenant Sill until volunteering to join Bushnell in August 1776, we can speculate that through this association, Lee was already aware of the *Turtle* when Colonel Parsons approached him.

Figure 19.1. View of the Sill House from the Connecticut River, where we believe the original *Turtle* was completed and launched in 1775. (*Roy Manstan*)

One of Bushnell's principal sources of mechanical support was Phineas Pratt whose shop was in Essex, known as the Potapaug district of Saybrook at that time. Essex was also the home of the Hayden Shipyard where the Connecticut warship *Oliver Cromwell* was built in 1776. There was no lack of patriotic fervor in Bushnell's neighborhood.

So, how was the commute? One of the authors, Roy Manstan, grew up less than a mile from the Bushnell homestead and has traveled the local winding roads for decades. These modern roads follow the same path, with only minor straightening of the twists and turns, since the area was settled in the late seventeenth and early eighteenth centuries.

Before the advent of dynamite, the geology of Connecticut was never conducive to creating roads in a straight line. Nonetheless, on a leisurely morning, Bushnell could have walked to Phineas Pratt's home in less than an hour and a half [a leisurely pace being twenty minutes per mile].[12] Another twenty-minute stroll and he could have been at the shipyard discussing access to a sloop with Uriah Hayden. And all three of them could have sat together pondering the whole operation, gathering inspiration over a pint or two at Hayden's Tavern adjacent to the shipyard.[13]

Figure 19.2. Aerial view near the mouth of the Connecticut River where the *Turtle* was launched and tested during the summer of 1775. 1. Connecticut River looking southeast. 2. Essex waterfront at the Connecticut River Museum and the location of Uriah Hayden's shipyard and tavern in 1775. 3. Ayer's Point and the Richard Sill House. 4. Lord Cove. 5. Goose Island. 6. Calves Island. 7. Poverty Point on Great Island at the mouth of the Connecticut River. 8. Long Island Sound. (*Bruce and Janna Greenhalgh*)

If Bushnell was ambitious one morning, he could also have hiked to the Sill house on Ayer's point in about two-and-a-half hours to polish the *Turtle*'s brass hatch, stopping again on his way home at Hayden's Tavern. Certainly, the Bushnell brothers would have used the family buggy to do all this traveling. The point, however, is that to create a "machine" as complex and unique as was his submarine vessel, Bushnell had to consider the accessibility of human and material resources, and needed to find local solutions to logistical issues.

THE *TURTLE* IS LAUNCHED

We believe that the Sill house was the home of the completed *Turtle* and where it was first launched—or maybe set afloat would be a better term. Access to water was practically in the Sill's front yard. A narrow road is all that separates the house from the Connecticut River. The river along Ayer's Point is relatively shallow, however, and not conducive to testing a submarine. There was plenty of water to float the *Turtle*, but it was not deep enough for submerging. In his letter to Ben Franklin, Benjamin

Figure 19.3. View of the passage to Lord Cove behind Goose Island where we believe David Bushnell brought the *Turtle* when training his brother, Ezra, in 1775. (*Roy Manstan*)

Gale was quite specific that "the Whole Machine may be Transported in a cart." The photo of the Handshouse *Turtle* replica being launched with a horse-drawn cart (figure on part 4 intro page) provides an accurate portrayal of how Bushnell may have launched his submarine into the shallows in front of the Sill house.

Figure 19.2 is an aerial view of the lower Connecticut River looking south. Long Island Sound can be seen in the distance. Poverty Island was at the very mouth of the river. Near the foreground is Ayer's Point where the Sill house is located. Directly across the river are Calves Island and Goose Island. Calves Island to the south is a tree-covered sandy hill separated by a narrow strip of water from the mainland cliffs that overlook the river. Goose Island to the north is a low marshland that rises just high enough to create Lord Cove behind it. As with Calves Island, there is a narrow strip of water between Goose Island and the mainland where Lord Cove empties into the Connecticut River.

Armed with his kayak and lead line, Ken Beatrice surveyed the depths in the waterways behind both islands. His results were very promising, and the three of us returned there to photograph the area. Figure 19.3 is a view looking north into the passage that runs between Goose Island and the mainland. Lord Cove is very shallow. However, the passage seen in the photograph is surprisingly deep. This strip of water is about six hundred yards long and one hundred yards wide with depths ranging from fifteen to twenty feet. The outflow current from Lord Cove is minimal, and only influenced slightly by tidal changes in the river. The channel behind Calves Island, while as deep as that behind Goose Island, is directly open to flow entering from and exiting into the Connecticut River. These observations allowed us to speculate that the variety of conditions in these locations would have been ideal to train Ezra Bushnell in all of the *Turtle*'s mission requirements.

With the ever-changing shoals along the Connecticut River, however, the question arises as to whether Calves Island existed in 1775. A hand-drawn sketch of the mouth of the Connecticut River by Ezra Stiles (later president of Yale College) shows the distinctive bend in the river at Ayer's Point.[14] Stiles drew an elongated island directly across from Ayer's Point showing, as exists today, a narrow channel between the island and the Lyme shoreline.

Bushnell could have readily brought the *Turtle* in tow across the half-mile-wide river to the seclusion afforded him behind Goose Island. There he would have sufficient room to maneuver and submerge the *Turtle*. The tall marsh grass on the island and the mainland cliffs would provide protection from curious and unwanted eyes. The submarine could be easily salvaged if lost, and if problems occurred with any of his many mechanical contrivances, he could return to the Sill house for repairs. His mechanic, Phineas Pratt, was just up the road.

Bushnell described his training process: "I made him descend and continue at particular depths, without rising or sinking, row by the compass, approach a vessel, go under her, and fix the *wood-screw* . . . into her bottom, &c. until I thought him sufficiently expert to put my design into execution."

Training and testing would have initially occurred during daylight hours, moving into the night only after the pilot had gained a high level of proficiency with the controls. To help the pilot while training under a ship with the wood screw mechanism, the hatch contained small windows along the side and top. This allowed the pilot "to see objects under water" [Gale to Franklin, August 7, 1775] and to "look upward" [Gale to Dean, November 9, 1775]. Lee noted that on a clear day, the windows allowed him "to read in three Fathoms of water," although it's doubtful that he carried a newspaper with him. Daylight training, however, also had its dangers. The cliffs overlooking Goose Island would provide a vantage point for someone keeping watch over the river—to provide the Bushnell brothers ample warning of any approaching vessels while their training proceeded.

Even today, sailboats nearly forty feet in length anchor in this waterway. The brothers could have performed multiple simulated attacks on a vessel provided to them by a local shipyard. Bushnell understood the need for repetitive training for a successful military mission. "I found, agreeable to my expectations, that it required many trials to make a person of common ingenuity, a skillful operator; the first I employed, was very ingenious, and made himself master of the business." Bushnell was, of course, referring to his brother.

After multiple trials behind Goose Island, the brothers could have shifted to the less placid water behind Calves Island. Still secluded from view, Ezra would have an opportunity to experience the operation of the *Turtle* in river currents more representative of reality, and it is here where they could have practiced crosscurrent transits.

It is conceivable that Bushnell also brought Lee to Goose and Calves Islands (not Poverty Island). Lee noted that they "returned with the machine down Sound and on our way practised with it in several harbors—we returned as far back as SayBrook with Mr. Bushnell, where some little alterations were made in it." Bushnell had to provide Lee with a crash course and would have wanted to train in familiar waters. He apparently also needed to work on the *Turtle,* and there would be no better place than back at Ayer's Point.

The importance of a suitable location for Bushnell to run the early tests of his submarine and to train his pilots cannot be overstated. Repetitive training is essential for the success of any military mission that requires a complex set of operations and maneuvers. His brother Ezra had ample opportunity to gain that experience, and it is likely that Phineas Pratt, who helped with the construction of the *Turtle* and was an accomplished mechanic, would also have had time at the controls.

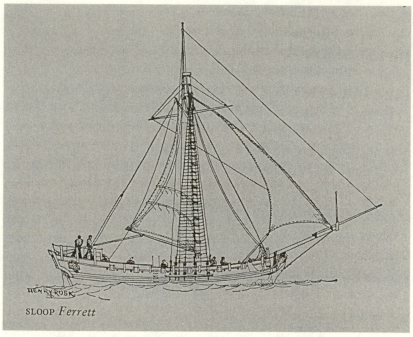

SLOOP *Ferrett*

Figure 20.1. With a length of over sixty feet and a twenty-foot beam, a vessel the size of the armed sloop *Ferrett* would have been more than sufficient to support the launch and recovery of the nearly two-ton *Turtle*. (*Chapelle 1935*)

Chapter Twenty

THE TURTLE DID NOT OPERATE ALONE

BUSHNELL'S NAVY

VESSELS TO SUPPORT BUSHNELL

By the time the *Turtle* was ready to launch and test, the logistical issues associated with handling and transporting the vessel had been accounted for. When Benjamin Gale wrote to his friend Ben Franklin in 1775 describing the submarine, he made specific reference to Bushnell having the use of a sloop when experimenting with the vessel. Later in the letter Gale added that "the Whole Machine may be Transported in a Cart." With the onset of winter, the *Turtle* was likely stored at the Sill house. In early spring when the ice began to break up in the Connecticut River, the *Turtle* would have been set afloat to ensure that the wooden hull could swell and seal the caulked seams. It would soon be time for David and Ezra Bushnell to prepare for their mission. There is no record of when the brothers ventured out of the river and sailed for New York City, but it is likely that they had arrived by midsummer.

Having established a relationship with the owner and crew of the sloop mentioned by Gale in 1775, we believe the same vessel was used during the summer of 1776, when the Bushnell brothers were at New York waiting for an opportunity to send out the *Turtle* against the British fleet. After Ezra Bushnell became sick, and Ezra Lee and two others volunteered to take his place, Bushnell then proceeded along the Connecticut coast with the *Turtle* and his volunteer pilots. Lee noted that they "returned with the machine down Sound and on our way practiced with it in several harbors," finally arriving in Saybrook. We have to assume that Bushnell was still depending on his sloop. After hearing of the British attack on Long Island, Lee then noted that they all "went back as far as New Rochelle and had it carted over by land to the North River."

When Bushnell's team of adventurers arrived at New Rochelle, the British were well established in New York Harbor. The fear of capture precluded transport down the East River, and it was decided to release the sloop and carry the *Turtle* over land to the North (or Hudson) River. From there it could be brought downriver to the relative safety of Fort George and the Battery at the southern tip of Manhattan. When word of the situation reached Washington, Brigadier General Parsons passed a letter to Major General William Heath requesting that he provide a vessel to support Bushnell:

> Sir.—As the machine designed to attempt blowing up the enemy's ships is to be transferred from the East to the North River, where a small vessel will be wanted to receive it, I wish you would order one for the purpose. As all things are now ready to make the experiment, I wish it may not be delayed. Though the event is uncertain, the experiment under our present circumstances is certainly worth trying.[1]

Any vessel of a size capable of supporting Bushnell's mission would have been a valuable resource and may have been what Washington referred to when writing to Thomas Jefferson in 1785 that "he [Bushnell] wanted nothing that I could furnish to insure the success of it."[2] Heath, who commanded the American troops at King's Bridge near the Hudson, was able to supply the vessel and safely transport the *Turtle* with its anxious crew to the southern tip of Manhattan.

Lee's night attack on the *Eagle* was unsuccessful. Several days later, Lee made another attempt, also unsuccessful, against a frigate anchored in the Hudson River. Bushnell then turned to another pilot, his mechanic Phineas Pratt, who made one more try that also failed. According to Lee, "Soon after [this third attempt by Pratt] the Frigate came up the river, drove our Crane galley on shore and sunk our Sloop, from which we escaped to the shore."

BUSHNELL'S SLOOP

Based on these accounts, Bushnell had access to a sloop at least as early as the summer of 1775, when he was experimenting in the Connecticut

River and Long Island Sound. Apparently a vessel of this type was sufficient to handle the logistics of transporting, deploying, and retrieving the *Turtle*. There is no mention of the name of the sloop he used while in Connecticut. There is also no indication of what vessel (or vessels) Heath supplied to Bushnell, or whether it was the same sloop that Lee claimed carried the *Turtle* when sunk by a British frigate in the Hudson River.

A sloop such as Bushnell would have sought to support the *Turtle* would have been a fast-sailing, seaworthy merchant vessel, capable of conducting extended trading excursions along the American coast, possibly ranging as far afield as Bermuda and the West Indies in the rum trade. A sloop of this size may have been sixty feet long with a nearly twenty-foot beam. It would have had a deep cargo hold that would easily contain the *Turtle* and the several "magazines" or mines. There would also have been room to carry Bushnell and his team, as well as the captain and crew of the sloop.

Because of the inherent danger that merchant vessels faced in their coastal trading, many carried some limited armament, typically swivel guns and possibly a small caliber cannon or two. If purchased and adopted into the colonial "navy," a merchant sloop might be modified to carry as many as ten or twelve small cannons, along with the swivels that could be quickly moved from one position to another in an engagement.

Howard I. Chapelle, in his *History of American Sailing Ships* (1935), provides a lengthy description of the early eighteenth-century sloop *Ferrett*, along with an illustration by Henry Rusk, shown in figure 20.1.[3] The *Ferrett* was a larger than typical colonial sloop, certainly seaworthy and with sufficient carrying capacity to represent a vessel capable of supporting the *Turtle*. The drawing shows the gunports, the intermediate "swivel stocks," and the "sweep ports" on either side of the gunports to accommodate oars, or "sweeps," used to maneuver in calm air. A vessel built along the lines of the *Ferrett* would have served Bushnell well.

THE *CRANE* GALLEY

Lee's mention of the *Crane*, however, presents an interesting insight into the ships that Washington had at his disposal. When General Parsons requested that General Heath provide a vessel to support Bushnell's "experiments," Heath may have assigned both the *Crane* and the unnamed sloop. Considering there were several galleys operating in the

Hudson at the time, Lee's specific mention of the *Crane* leads us to believe that Bushnell's sloop and the *Crane* were working together, or were at least in close proximity, during the *Turtle's* mission.

A "galley" was a specific type of naval vessel designed to be rowed as well as sailed. In *The History of American Sailing Ships*, Howard I. Chapelle describes a similar ship, the Continental galley *Washington*, built in 1776 on Lake Champlain (captured by the British on October 13, 1776). Its dimensions were 72′4″ on deck, 19′7″ beam, and 6′2″ depth in hold. It carried eleven cannons and eight swivels. Chapelle's illustration of the *Washington* by George C. Wales is reproduced here (see figure 20.2). Another galley named *Washington* was used on the Hudson River during 1776, but there are no known details.[4]

With rowing capability and a relatively flat bottom, a galley could be maneuvered in shallow water, and when winds and currents were not favorable, thus providing an additional offensive option during a naval engagement. Because Lee made a specific reference to the *Crane*, the ship has an interesting and potentially relevant history. The *Crane* was built in the town of East Haddam along the Connecticut River, only about ten miles north of Saybrook.

The following discussion illustrates the interest by Connecticut's Council of Safety in establishing a maritime force to counter, or at least hinder and annoy, the British naval threat along the coast. Early in January 1776, the Council began consideration of the construction of four galleys, and by the end of the month, had established specifications for their construction. A week later, on February 5 (see chapter 3), Bushnell approached the Council about his "machine contrived to blow [up] ships."

At a meeting of the Governor and Council of Safety at Lebanon, Friday 5th January, 1776:

> Motion to conclude where the 4 row-gallies lately ordered by the Assembly under the direction of said committee shall be built, and on consideration resolved, that one of them be built at Norwich, and one at East Haddam, and tho't proper not now to determine as to the other two.[5]

Again on January 29:

> Cap. Lester came in from viewing the row-gallies at Philadelphia and also Mr. Winslow from the same view at Providence and

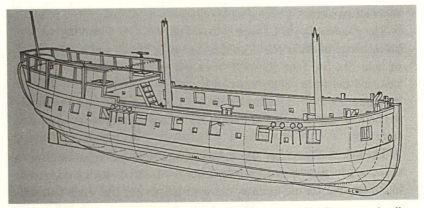

Figure 20.2. While smaller than Connecticut's galley *Crane*, the Continental galley *Washington* was built along similar lines. Having been mentioned by Ezra Lee, we believe that the *Crane*, if not a participant, was at least a witness to the *Turtle* missions in the Hudson River. (*Chapelle 1935*)

reported the construction &c. and shew'd plans, and were consulting about them &c. forenoon.

P.M. The said builders came in, having consulted a plan of building two of said row-gallies, and considered the same; and agreed that two to be built, one by said Lester, and one by Winslow, shall be 60 feet keel, 18 feet beam, and 5 feet hold, four inches dead rising.[6]

In February a slight change: "On further consideration voted, that the dead rising for the row-gallies ordered to be built by Capts. Lester and Winslow be about 7 inches instead of four as before agreed."[7]

In April, the Council of Safety voted: "Mr. Job Winslow for three hundred pounds, for the purpose of building the row-galley at East Haddam. *An order was drawn April 2d, 1776, and delivered to Mr. Job Winslow for £300,*" a partial payment as records show an additional £400 to Winslow in June. On July 5, Winslow received the balance of £313 6*s*. 10*d.*, making the total cost of the vessel just over £1,013.[8] Bushnell, on the other hand, had received £60.

Council of Safety, June 1, 1776: "The letters and bonds to the captains of the row-gallies being to be forwarded &c. and names not being given to two of them, is concluded to call Cap. Stanton's, built at Norwich, the Shark; and Cap. Tinker's, at East Haddam, the Crane." At last the galley has a name.[9]

It was barely four months between the time the plans were reviewed and when the galleys were launched and ready to sail. Jahiel Tinker had been appointed captain of the *Crane* in February, and David Brooks of Haddam appointed lieutenant in April. On June 20, Captain Tinker was ordered to "proceed directly to New London," where he was to "receive two of the nine pounders of the continental cannon."[10]

On the 10th of July, the Council of Safety authorized:

> that Capt. Tinker of the galley Crane take two three-pounders at the old fort at N. London for said galley, to be delivered him by the keeper thereof; that he receive eight swivel guns, as soon as they can be obtained; that he receive ten muskets, to be delivered him by Col. Williams he procured, (viz. of Jona. Goodwin;) that he receive of Mr. N. Shaw junr such powder, ball and military stores as he, said Shaw, shall judge necessary; that he receive from time to time a supply of salted beef, pork, bread, flour, sufficient for the support of the officers, sailors and marines on board said galley . . .[11]

Six days later, on July 16, 1776, the Council of Safety recorded that "His Excellency Gen. Washington having requested the loan of the row-gallies belonging to this Colony, to assist in the defense of New York now attacked by the enemy: Resolved and voted, that row-gallies Whiting and Crane proceed immediately to New York, and there be under the command of his Excellency Gen. Washington until further orders. *Orders sent.*" After providing the above resolutions of the Council of Safety, Hoadley (1890) added the following quote: "They participated in the attack on the Phoenix and Rose in the Hudson, off Tarrytown, Saturday August 3d. The Crane had one man wounded; the Whiting one killed, three wounded. *Courant*, Aug. 12 No. 603."[12]

On October 9, 1776, the *Crane* was run aground and captured as recorded in the journal of HMS *Tartar*, Captain Cornthwaite Ommanney:

> October 1778 Sailing up the North River
> Wednesday 9 Mode and Cloudy wr at 7 AM weighed and came to Sail
> In Co with his Majesty's Ships *Phoenix Roebuck Tryal* Schooner and two Tenders at 8 do Five Batterys on the

York and two on the Jersey Shore began to fire on us
Likewise hove a number of Shells with a Continual Fire-
ing till 1/2 past 9 after Hulling on [us] several times,
wounded our Masts and cutting a great deal of the Rig-
ing and Sails a Shott went thro the Mizen Mast and
afterwards killed a Midshipmn the Splinters of the Masts
wounded the Captn Lieut of Marines and Pilot after passing
the Batterys the Enemy began to Fire Small Arms from the
Woods, which they continued for several Miles up the River
At 10 gave chace to the Enemys Galleys, at 11 drove on shore
several of their Merchmn and the *Independence* Galley at
Noon drove on Shore the *Crane* Galley and continued chace
Light Breezes and Cloudy, found that the rest of the Galleys
Had got out of Gun Shott, by the Assistance of their Oars
At 1/2 pt 1 came too abreast of the *Crane* Galley sent Lieut
And some Men and took possession of her.[13]

Two other Connecticut galleys, the *Whiting* and *Shark*, were also
eventually taken by the British in the Hudson River. The naval careers of
these three vessels were brief, lasting less than three months, but all
served with distinction, losing several killed and wounded during August
against HMS *Rose* and *Phoenix* in the Hudson River. The three captains
continued their maritime careers as privateers during the Revolution.
Through Lee's comments, the *Crane* is the only vessel with a known
association to the *Turtle* story. If the *Crane* or any of the other captured
galleys had been supporting the *Turtle* operations, we can only imagine
what the logbooks for these vessels may have contained.

Figure 21.1. Portrait of Ezra Lee. (*Bishop 1916*)

Chapter Twenty-One

BUSHNELL'S BIGGEST MISTAKE

Too Few Turtle Pilots

The pilots . . . how many?

Ezra Lee began his 1815 letter to David Humphreys with an explanation of his selection as the *Turtle* pilot.

> In the Summer of 1776, he [David Bushnell] went to New York with it to try the Asia man of war; his brother being acquainted with the working of the machine, was to try the first experiment with it, but having spent until the middle of August, he gave out in consequence of indisposition. Mr Bushnell then came to General Parsons (of Lyme) to get some one to go and learn the ways & mystery of this new machine and to make a trial of it.

Lee, however, was not alone. He continued his account: "General Parsons sent for me, & two others, who had given in our names to go in a fireship if wanted, to see if we would undertake the enterprise . . ." There were two other volunteers with Lee, and each undoubtedly took turns in the *Turtle*. Bushnell was unequivocal about having tried more than one of these: "After various attempts to find an operator to my wish, I sent one who appeared more expert than the rest, from New-York, to a 50 gun ship [the *Eagle* actually carried sixty-four guns] lying not far from Governor's Island." As is discussed below, we feel that Bushnell's mechanic, Phineas Pratt, also piloted the *Turtle* during one of the three combat missions. As the submarine designer, it is likely that David Bushnell himself also spent time at the helm while testing and troubleshooting the vessel with his brother in 1775. From this, we surmise that at least six individuals had some degree of experience with the operation of the *Turtle*.

DAVID BUSHNELL

He had, however, planned to depend entirely on his brother Ezra to perform all of the missions and certainly most of the testing and training (along with perhaps Phineas Pratt). The one credible account that indicates the likelihood that David Bushnell may not have been physically up to the task of operating the *Turtle* comes from his longtime friend David Humphreys: "The Inventor, whose constitution was too feeble to permit him to perform the labour of rowing the Turtle, had taught his brother to manage it with perfect dexterity . . ."[1] Humphreys, who writes with honest praise for Bushnell's "wonderful machine," would have been equally honest about his assessment of Bushnell's capacity to operate the *Turtle*. While Bushnell's physical strength made him an unlikely candidate to conduct a combat mission in his brother's absence, his curiosity and the inherent desire of an engineer to optimize his invention would have made it nearly impossible to stay on the sidelines when experimenting with his vessel in 1775.

EZRA BUSHNELL

During 1775, Ezra Bushnell is listed as a sergeant in the 3rd Company (under Captain Nathan Hale), in Colonel Charles Webb's 7th Regiment. This regiment was raised in July 1775, and participated in the Siege of Boston. In January 1776, Colonel Webb's regiment was reorganized as a unit in the Continental army. In April, the 7th Regiment was ordered to New York where it served until the end of the year.[2] It is possible that Ezra Bushnell had spent time around Boston with Nathan Hale's 3rd Company. It is also likely that Hale was very familiar with the *Turtle*, having been a close friend of David Bushnell at Yale. As company commander, Hale had the authority to furlough his sergeant, Ezra Bushnell, to assist with putting the finishing touches on his older brother's submarine and participate on what everyone understood to be a dangerous clandestine mission.

We don't know what "indisposition" put Ezra Bushnell out of action. David only noted that his brother, who had "made himself master of the business . . . was taken sick in the campaign of 1776, at New-York, before he had an opportunity to make use of his skill, and never recovered his

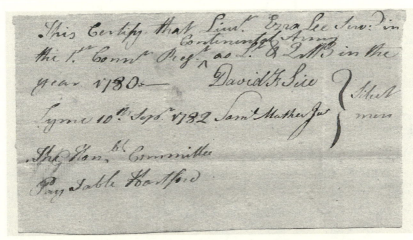

Figure 21.2. This document in Ezra Lee's hand certified his service as a lieutenant and quartermaster in the "1st Connt Regt, Continenntal Army" in 1780, and was used when Lee requested back pay from the Connecticut Pay Table in 1782. One of the signers was David F. Sill. Lee had served under Captain, later Lieutenant Colonel, Sill throughout the war. David Sill was the nephew of Richard Sill of Ayer's Point where Bushnell put the finishing touches on the *Turtle*. (*Roy Manstan*)

health sufficiently, afterwards." It is likely that Ezra, as with so many among the American forces, was overcome by "camp fever." According to David McCullough, in his book *1776*, "'Camp fever' or 'putrid fever' were terms used for the highly infectious deadly scourges of dysentery, typhus, and typhoid fever, the causes of which were unknown or only partially understood."[3] Malaria also took its toll on both the British and Continental troops.

It may have been camp fever that overcame Ezra Bushnell that August, but there is no doubt that he and Ezra Lee were in excellent condition when they initially undertook their adventures in the *Turtle*. David Bushnell and all of his supporters understood the part played by the pilot, and the physical and psychological demands that would be placed on them. George Washington, in a letter to Thomas Jefferson, commented that

> when it was to operate against an enemy, it was no easy matter to
> get a person hardy enough to encounter the variety of dangers to
> which he would be exposed; first, from the novelty; secondly,
> from the difficulty of conducting the machine, and governing it

under water; and thirdly, from the consequent uncertainty of hitting the object devoted to destruction, without rising frequently above water for fresh observations, which when near the vessel, would expose the adventurer to discovery and to almost certain death.[4]

EZRA LEE

Washington recognized the uniqueness of the mission, and the difficulty in finding volunteers who would have the skills and mental tenacity needed. One brute force tactic used against an enemy navy was to sail fireships into a fleet at anchor. The individuals who volunteered for these missions would have little margin for escape. The options would be swimming to shore, or having a small boat alongside that could be used once the momentum of the fireship would take it into the anchored enemy vessel. Ezra Lee had volunteered to run a fireship into the British fleet, which is likely why he was selected as a candidate to replace Ezra Bushnell.

The Continental army also employed scouts and specially trained units such as the Connecticut Rangers, commanded by Lieutenant Colonel Thomas Knowleton, that would be called on to accomplish the most dangerous missions. Knowleton's Rangers were formed immediately after the retreat from Long Island as "a small body of select troops . . . for special service along the lines."[5] Nathan Hale was one of three captains who volunteered for this "special service," one specific mission being to stay behind as a spy after Washington left New York City to the British.

These units performed a similar function as today's Army Rangers and other special warfare teams, such as Navy SEALs. Ezra Lee is probably the one individual serving in the Revolutionary War that could be classified as a SEAL, an acronym for SEa Air Land. Lee's obituary included the following observation: "Died at Lyme, Connecticut, on the 29th ult., Capt. EZRA LEE, aged 72, a revolutionary officer. It is not a little remarkable, that this officer is the only man of whom it can be said, that he fought the enemy upon land, upon water, and under the water."[6] Certainly Lee qualifies having served on land and sea, and one can only speculate that he would have volunteered to operate a hot-air balloon had that technology been available to Washington. (The first manned

Figure 21.3, left. The Old Lyme, Connecticut, grave site of Ezra Lee. The grave-stone inscription "He was a Revolutionary Officer. And esteemed by Washington," no doubt refers to Lee's experience with the *Turtle* and the attempt to sink the *Eagle*. (*Daniel Manstan*) Figure 21.4, right. The grave site of Phineas Pratt is locat-ed in the River View Cemetery in Essex, Connecticut. This marker is set behind Pratt's original gravestone, now barely legible. (*Roy Manstan*)

ascent in a hot air balloon occurred in France in November 1783 . . . too late for Lafayette to have brought one with him).

PHINEAS PRATT

This book has concentrated on the two primary pilots, Ezra Bushnell and Ezra Lee. When David Bushnell wrote to Thomas Jefferson, he described the events of September 6, 1776, but also mentioned the sub-sequent operations in October. "Afterwards, there were two attempts made in Hudson's river, above the city, but they effected nothing. One of them was by the aforementioned person [Lee]." Bushnell did not include any information about who piloted the second attempt. There is evidence that the pilot was Phineas Pratt. Pratt, along with Isaac Doolittle, had been involved with building the critical mechanical components for Bushnell. He was certainly familiar with the *Turtle*, and would have had many opportunities to test its operation while in the Connecticut River. Pratt was twenty-nine years old.

Bushnell would have understood that having a skilled mechanic on-site increased the likelihood that if a technical problem occurred it could be resolved, allowing the mission to continue. It would have been likely that Pratt accompanied Bushnell in August when Lee was training in the Connecticut River. He may very well have also been with them on Manhattan in September, and on the Hudson River in October.

There are two nineteenth-century accounts of Pratt's involvement with Bushnell: one, a Pratt family genealogy published in 1864 by Reverend F. W. Chapman; the second a letter written by Pratt's son, also Phineas, in 1870. The following is included to give proper credit to another patriot and submarine pioneer.

F. W. Chapman noted that after Lee's attempt:

> Mr. Pratt to whose history these notes are attached, and the con-structor of the Turtle, made another trial of its powers, descend-ing in it himself. A cloudy night was selected, but coming with-in a few feet of the vessel which lay in the East [actually the Hudson] River, by the sudden breaking of the moon-light through a cloud, he was discovered by the watch on deck. The watch calling out Who's there? Who's there? Mr. Pratt immedi-ately descended and came up about a half mile distant. The enemy gave chase and fired but before they could reach him the Turtle was securely packed in the hold of a vessel in waiting, and being sunk by the firing of the British, no trace of the Turtle was discovered by them. It was afterward transported to Say-Brook, where some portion yet remains as a relic of those times.[7]

Chapman added an intriguing comment suggesting the later use of some of the *Turtle* components: "Mr. Pratt being a skillful mechanic afterward used the metallic portions of the torpedo in the construction of clocks, some of which in addition to the usual qualities would indicate the day of the month, age, and appearance of the moon &c. One of said clocks is now in good repair and has been running more than 50 years."[8] In the nineteenth century, particularly during the Civil War when Chapman's book was published, the word "torpedo" was associated with mine warfare. If this is the case, Pratt may have been salvaging spare clockwork timing mechanisms. There had been, however, references to the *Turtle* itself as being a torpedo (e.g., Thatcher 1823, 75), so

Chapman may have been indicating that Pratt had adapted the *Turtle's* brass hatch, for example, as the backing for the town clock.

An article titled "Torpedoes in Warfare" appeared in the March and April 1870 issues of the *Boston Journal of Chemistry*. The author had written a brief history of this technology, emphasizing the advancements that had been employed during the Civil War. Included was a brief account of Bushnell's exploits with the *Turtle*. Phineas Pratt junior had read the article and, apparently interested in having the Journal print a more thorough account, wrote a letter describing the family tradition that his father was one of the *Turtle* pilots. Pratt also reiterates Chapman's claim that the senior Pratt had disassembled components from the "torpedo" and used them in the town clock. The following are excerpts from Pratt's letter:

> Seventy years ago, in A. D. 1800, I saw the metal part of that torpedo worked up into a town clock in my father's shop. I was then about sixteen years of age. I then could hear the whole history and view the dissected parts . . . My father, who afterwards bore the name of Dea. [Deacon] Phineas Pratt who was then in the American army, having in youth learned the ship carpenter business, and afterwards the brass clock making and gold and silver business, was qualified to undertake the work. He was requested to join and by leave went and joined them. The first trial the man whose station it was went to the ship, went under and could not cause the instrument made to enter wood to enter the bottom of the ship. He dropped down under the stern to arise and saw the men looking over. He started back, but the tide was so strong against him that he cast the magazine adrift and made for the City. The magazine soon went off and made a very loud report, throwing pieces high in the air. The men on board the ship instantly cut their cable and made off as quick as wind and sails would carry them. Gen. Washington felt highly pleased as he said what he wanted was accomplished without destroying life. They were ever afterwards very cautious how they came too near with their ships both in the first and last war. A second trial was made by my father, who went off under water to a war vessel which lay on the river. It was a cloudy night; he came up close to the vessel; at that instant the full moon burst through a cloud; he saw the

men on board looking over and they hailed him. He went down and came off to their own vessel that lay near the shore, put the torpedo on board and made for the shore. The British thought it not a comfortable place to lie and moved up the river, fixed on the vessel that had the torpedo in, and sunk her, came in a boat to look and went off. Mr. Bushnell nor any of the Company, having spent all the time and expense they could have, never received any compensation from that day and after gave it up.

My father died in 1812 not having received any pension though he served his country during the war.

After including a crudely penned sketch of the *Turtle* as he remembered it (figure 21.5), Pratt ends his letter with "Excuse an old man in his 87th year."[9]

Before he died, the elder Phineas Pratt had become an accomplished inventor, having produced a machine for manufacturing ivory combs. The business was established nearby in a portion of the town of Essex that later became known as Ivoryton, in reference to the significant ivory trade that continued in the town for many years.

As his son noted, Pratt never received a pension for his service. Joseph Plumb Martin, in his memoir, echoed the treatment of many of the soldiers who fought in the war:

When those who engaged to serve during the war enlisted, they were promised a hundred acres of land, each, which was to be in their own or adjoining states. When the country had drained the last drop of service it could screw out of the poor soldiers, they were turned adrift like old worn-out horses, and nothing said about the land to pasture them upon. Congress did, indeed, appropriate lands under the denomination of 'Soldier's lands,' in Ohio state, or some state, or a future state, but no care was taken that the soldiers should get them. No agents were appointed to see that the poor fellows ever got possession of their lands; no one ever took the least care about it, except a pack of speculators, who were driving about the country like so many evil spirits, endeavoring to pluck the last feather from the soldiers. The soldiers were ignorant of the ways and means to obtain their bounty lands, and there was no one appointed to inform them. The truth was, none cared for them; the country was served, and faithfully served, and

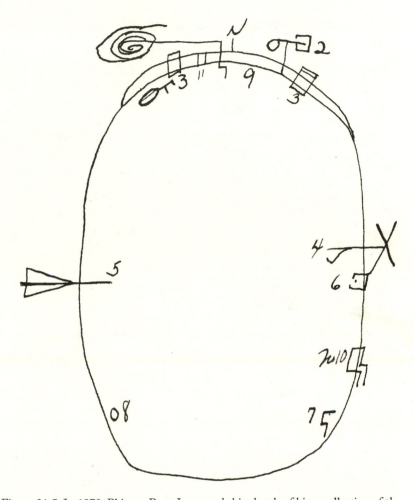

Figure 21.5. In 1870, Phineas Pratt Jr. created this sketch of his recollection of the *Turtle* claiming that as a teenager in his father's shop in 1800, he "could hear the whole history and view the dissected part." He identifies with numbers (1) "The magazine with the powder in it and clock work in it (to strike fire at a given time)." (2, 3) "Instruments in the cap that would open when at the surface and shut when the torpedo sunk (wood in one end)." (4) "Instrument to go ahead or back in water formed like a windmill." (5) "Helm." (6) "Mariner's compass with rotten wood called bonfire to give light." (7) "Force pump." (8) "Place to let in water and go down. The vessel was balanced with lead so that when at the surface it would just show the cap above water." (9) "Cap." (10) "Instrument to tell how deep they were under water. Some attempts were made the last war, but not like original which made the British very cautious how they lay too near with their ships. They swept their ships bottom every half hour with a rope." (*Connecticut Historical Society, Hartford, Connecticut*)

that was all that was deemed necessary. It was, soldiers, look to yourself; we want no more of you.[10]

Ezra Lee was no more successful. Frustrated with the lack of compensation, Lee submitted a petition for a stipend from Connecticut for his Revolutionary War service:

And I do further swear, that my occupation, when abler was Seamanship, but now it is nothing, by reason of age and infirmities; That ever since I went in Bushnell's Submarine Machine, under the British Ships in New York Bay, and being so long confined, partly immersed in water, taken inside, to conduct it under water, I have been much troubled with Rheumatics and Cramp in my legs; which, together with my sufferings at Valley Forge in the Winter of 1778 broke my constitution very much, and that I am the Same Sergeant Ezra Lee, named in the Annexed Extract from Silliman's journal of Sciences . . .[11]

Lee's petition was submitted on June 26, 1820. He died four months later.

Chapter Twenty-two

SEPTEMBER 6, 1776

THE TURTLE SETS OUT ON ITS MISSION

Anxiety among members of the Continental Congress grew as the British fleet converged on New York during the summer of 1776. Knowledge of what Charles Griswold later referred to as an "engine of devastation" brought Connecticut's delegate, William Williams, to send an urgent inquiry to Washington's Commissary General, Colonel Joseph Trumbull (son of Connecticut Governor Jonathan Trumbull): "Where is Bushnell? Why don't he attempt something when will or can be a more proper Time than is or has been etc I was knowing to his coming etc and that you was acquainted with the Plan etc."[1]

AUGUST

It was the middle of August when Ezra Lee and at least two other volunteers joined Bushnell. They soon began training along the Connecticut coast, ending in Saybrook where Bushnell made a few modifications to the *Turtle*, probably at the Sill house with Phineas Pratt. (See chapter 19.)

The events that brought the *Turtle* into its moment in American history begin at the end of the month. The British had landed on Long Island and in rapid succession, cornered the American army on Brooklyn Heights. On the night of August 29, the surviving troops escaped across the East River to Manhattan. Washington and his generals regrouped, assessed their losses, and began to debate their next move. Word was immediately passed to the Continental Congress and quickly spread across the colonies.

According to Lee, they had been training for "8 or 10 days" when news arrived that the British had "got possession of Long Island & Governor's Island." It was obvious to Bushnell and everyone involved that there could be no more delays, no more training, no more opportunity to modify or fine-tune any of the *Turtle*'s mechanical systems. It is likely that by August 30, the *Turtle* was onboard the vessel supporting their operations, and heading toward New York. What Benjamin Gale had described to Ben Franklin as a "new Machine for the Destruction of Enemy Ships" was about to embark on its first combat mission.

September

September 1: Bushnell, Lee, and likely Phineas Pratt were moving down Long Island Sound with the *Turtle*. By the time they arrived at the western end of the sound, they would have learned that the British were in control of the East River. Bushnell knew it would be nearly impossible to bring their sloop carrying the *Turtle* downriver to the southern tip of Manhattan. From Lee's account, the *Turtle* was off-loaded at New Rochelle and carried overland to the Hudson River, also known as the North River.

September 2: Back in New York City, Washington consulted with his general staff. His concern was that the British navy would soon control the two rivers, and would land their troops above the city, trapping the American army. Washington organized his remaining troops, numbering about twenty thousand, into three divisions. One under General Spencer was to march to Harlem and prevent the British from landing their troops. Another division, under General Heath, was to occupy Kings Bridge and the surrounding countryside. The division under General Putnam, including Parsons's brigade, was to remain in the city.

September 3: Congress passed a resolution specifying that brigades of troops from North Carolina and Rhode Island immediately march to the aid of Washington, adding a request for assistance from all of the colonies north of Virginia. The intent was not to retain possession of New York, but to provide Washington with additional troop strength to contend with whatever transpired.

September 4: The morning saw the British frigate *Rose* anchored in the East River across from Kips Bay. Two frigates, the *Repulse* and the *Pearl*, entered the Hudson River while more transports were moving into

Figure 22.1, top. This colonial powder horn is only about eight inches long but includes many famous locations along the Hudson and Mohawk Rivers as far north as "Onyda" Lake and Lake Champlain. The "map horn" includes Fort "Ticondroga," Fort Edward, Fort Stanwix, Fort Hunter, Saratoga, and many other locations associated with the French and Indian War and the Revolution. Figure 22.2, bottom. This mosaic of photos taken around the circumference of the "map horn" shows New York City with the ubiquitous windmill and the many churches that dominated the skyline. This would have been Admiral Howe's view from the deck of HMS *Eagle* in September 1776 as he considered what action he would take against the rebellious colonies. David Bushnell, sitting with his *Turtle* under the protection of the formidable Fort George, would have been pondering the *Eagle*'s fate. (*Daniel Manstan*)

the East River. It was becoming apparent that the British were preparing for a major offensive against the American forces on Manhattan.

September 5: There was disagreement among the general officers as to whether the city could or should be defended. General Greene, who favored an immediate withdrawal, requested that Washington convene a council to consider this option. By this time, Bushnell had transported the *Turtle* overland to the Hudson River. When word of their status had passed to General Parsons, he immediately requested General Heath to provide Bushnell a vessel for use in support of the *Turtle* (see chapter 20 for Parsons's letter). We can assume that Heath accommodated Parsons's request, allowing Bushnell to transport the *Turtle* to the protection of the Battery and prepare for its first mission. Yet even this would be a perilous transit, with British frigates already anchored in the Hudson.

SEPTEMBER 6, 1776

The consensus among Washington's generals was that the army should remain and defend their positions on Manhattan. Washington, however, realized that he would soon have to retreat from the city, but that would not happen before Bushnell's machine would be sent into the midst of the British fleet. The David and Goliath mission would send the tiny submarine into what David McCullough described as a "British armada [that] numbered nearly four hundred ships large and small, seventy-three warships, including eight ships of the line, each mounting 50 guns or more."[2] We can imagine that while David Bushnell sat with Ezra Lee and the *Turtle* onboard his sloop pondering the future, Admiral Howe (figure 22.4) onboard HMS *Eagle* was doing the same.

It has always been assumed, and we perpetuate that assumption, that with Lee at the controls, the *Turtle* went into action on the night of September 6, returning the following morning. This would have been a tight schedule considering that Parsons's request to Heath was sent the previous day. Many modern authors, including those publishing around the time of the Bicentennial, have used this date.[3] Bushnell, however, makes no reference to the timing of the event, nor does his friend David Humphreys, who published his account in 1788. The date was never recorded, but with Washington already planning his evacuation of the city, there would be little time to delay the operation. (In the next chapter, we look at evidence regarding the validity of the September 6 date.)

As described in part three of this book, we had investigated the operational capabilities of a replica of Bushnell's *Turtle* and the physical limitations of the pilot. The testing was conducted under optimal conditions. Speculating about the successes and failures of the original *Turtle*, however, requires insight into the environmental conditions that Lee experienced during what may have been a six or seven-hour mission.

Figure 22.3."Plan of the City of New York in North America," a map published in 1776 by Bernard Ratzer showing the location at the Battery (lower west side of Manhattan Island) where the *Turtle* set out on its mission against HMS *Eagle* anchored off Bedlow's Island, also known as Kennedy's Island, (far left) on September 6, 1776. (*New York Public Library*)

Lee was not exaggerating when he noted that "the times for a trial must be when it is slack water, & calm, as it is unmanagable in a Swell or a Strong tide." With this in mind, we have attempted to reconstruct the tides and currents in New York Harbor on the night of September 6/7. We have also looked at evidence of the weather during that time, as recorded in the journals and logs of British vessels on-site, including that of Lee's target, the British flagship HMS *Eagle*.

WEATHER

The following journal entries from five ships moored in and around the harbor give brief mention to the weather.[4] Abbreviations for "weather" include "Wr" and "Wear." The journals and logs also abbreviate "Ditto" with "Do," such that "Do Wr" indicates that the weather was as reported on the previous entry, or previous day. Extracts regarding weather are provided here:

> Journal of HMS *Eagle*, Captain Henry Duncan:
> Septemr 1776 Moored off of Bedlows Island
> Friday 6. AM . . . Fresh Breezes & Cloudy Wr with rain at times

> Journal of the Fireship *Strombolo*, Captain Charles Phipps:
> Septemr 1776 Moored off Red Hook in New York River
> Thursday 5th Fresh Breezes and Cloudy AM . . .
> Do Wear (PM) . . .
> Friday 6. Do Wr at 5 PM . . .

> Journal of HMS *Chatham*, Captain John Raynor:
> September 1776   Moor'd off Staten Island
> Friday 6th   Fresh Breezes with Rain at 3 PM . . . .

> Journal of HMS *Asia*, Captain George Vandeput:
> September 1776 Moor'd in the No. River at New York
> Saturday 7 Do [Moderate & hazy] with Rain at times.

> Journal of HMS *Eagle*, Captain Henry Duncan:
> Septembr 1776 Moored off of Bedlows Island
> Saturday 7th . . . . The first part [of the day] fresh Breezes and hazy the Middle light airs & do, latter fresh Breezes & cloudy with rain . .

> Master's Log of HMS *Roebuck*:
> Septr 1776   Governors Island ENE 1/2 of a mile
> Saturday 7th . . . . Fresh gales & Rain . . . . PM . . .

From these accounts, we surmise that the weather during the day on the 5th and 6th, was breezy and cloudy with occasional rain. Under these conditions, the *Turtle* could very well have been transported down the Hudson River to the Battery. By nightfall, as the temperature cooled, the breezes may have settled down, leaving the surface of the water relatively flat. Light rain would not have interfered with Lee as he transited in the *Turtle*. According to the *Eagle*, the morning of the 7th saw the breezes return. If we consider the *Roebuck* log entry, the weather had deteriorated by that afternoon.

Figure 22.4. Admiral Sir Richard "Black Dick" Howe. (*The History of the Civil War in America, 1780*)

Lee's description of the attack included a brief mention of the moon and his ability to see men standing on the deck of the *Eagle*. He also noted that it was almost daybreak when he suspended the attack and headed back to the city. Insight into the timing of his mission can be found at the U.S. Naval Observatory, Astronomical Applications Department Web site, http://aa.usno.navy.mil. Through this Web site, we obtained the "Sun and Moon data . . . for New York (longitude 73.9° west, latitude 40.7° north)" throughout the month of September 1776.

TIDES AND CURRENTS

The weather may have been acceptable for the mission, but the tides and currents would present the greatest deterrent to a successful operation. The Tide and Tidal Current Predictions Department of the National Oceanographic and Atmospheric Administration (NOAA) provided us with the "Tide Predictions (High and Low Waters)" at The Battery, New York, also for September 1776. The timing of the high and low tides, however, is only part of the picture. The magnitude of the currents that Lee experienced during his attack is not as easy to predict.

Because tidal volume and flow rates are related to bathymetry, it would be of interest to compare depth soundings made in the eighteenth century to soundings made when the modern tidal current charts were published. We referred to three charts of New York City and the harbor published at the beginning of the Revolutionary War. The chart produced by John Montressor (surveyed ca. 1766 and published 1775, (Van Erman 1990, 76)), the Bernard Ratzer chart (surveyed 1766–1767, and published 1776, (New York Public Library)), and the J. F. W. De Barres map (drawn 1776 and published 1777, (Dept. of the Navy, 1972, map 7)) all provided soundings in fathoms in the area where the *Turtle* operated. These depths were then compared to Coast and Geodetic Survey charts produced in 1961 (Dodson 1961, charts 541 and 745) and the recent NOAA chart of New York Harbor (chart no. 12327, 94th ed., Jan. 13, 2001).

Soundings from the Ratzer and De Barres charts show depths in the main channel between New York City and Staten Island, generally ranging from 9 to 14 fathoms (54 to 84 feet). The Montressor chart, however, indicates much shallower depths ranging from 5 to 8 fathoms (30 to 48 feet). The 1961 and 2001 charts indicate depths in this same area as about 38 to a maximum of 63 feet. This seems to place the bathymetry of the modern charts at a midrange between the three eighteenth-century charts. Considering the influence that the Hudson River has on tidal currents in the harbor, it may also be significant that the depths shown on the Ratzer chart in the channel near the mouth of the river, are nearly identical to those from the 2001 NOAA chart.

In general, we believe that the hourly variations in the tidal current of New York Harbor that existed in 1776 would approximate that found on modern charts. Differences would more likely occur along the shoreline, which has drastically changed, while the primary current profiles within the harbor channel would be similar. When speculating about the *Turtle* and Ezra Lee, we referred to the detailed *Tidal Current Charts, New York Harbor*, published in 1956 from Coast and Geodetic Survey data.

While it may take a leap of faith to accept these assumptions, our intent here is to propose a level of difficulty for the mission, and how Lee may have adjusted his operational tactics to overcome this.

Ezra Lee begins his attack on HMS *Eagle*

The *Turtle* mission lasted throughout the night, covering the portion of a tidal cycle that would provide two opportunities for Lee to engage the *Eagle* during slack water. The following scenario attempts to correlate the timing of the tides and currents with Lee's account.

September 6
Sunset at 6:19 p.m.
Twilight ended at 6:46 p.m.
It was dark, as the moon would not rise until after midnight.
Low Tide at the Battery, 9:57 p.m.

Lee described his transit to the *Eagle* in five sentences, giving only minimal insight into what he encountered. He begins: "We set off from the City—the whale boats towed me as nigh the ships as they dared to go, and then cast me off." There is no contemporary record of the time when Lee was towed out from the Battery, only that it occurred at night. In 1820, Charles Griswold (see appendix E) published an account claiming that it began at 11:00 p.m., possibly based on an interview with Lee, made at least forty years after the event.

Once set adrift, Lee encountered his first problem. "I soon found that it was too early in the tide, as it carried me down by the Ships." According to the charts, at low tide the current flowing south past the Battery and Governors Island would have been nearly two knots. This would certainly have made the job of the whaleboats easier, but the southerly flow that carried Lee "down by the ships" would last more than two hours before turning to an incoming flow. Lee had his compass set to reach his target, and likely maintained visual contact through the open deadlight. If he didn't find a suitable transiting strategy, he would speed past the *Eagle*, past Staten Island, through the Narrows and out to sea. These conditions also lend credibility to a start time of 11:00 p.m. If it had been after midnight, it is likely that the *Turtle*, after being set adrift, would have been caught in the incoming current and carried back toward Manhattan.

HMS *Eagle* was anchored off Bedlow's Island where the Statue of Liberty now stands. On charts and maps of the area drawn during the Revolutionary War, however, military cartographers occasionally referred to this small piece of land as Kennedy's Island. The island was named for Isaac Bedlow who settled there in the seventeenth century. After his death in 1672, the island passed through several hands until owned by Capt. Archibald Kennedy, who had commanded the British naval station in New York Harbor. It was called Kennedy Island until 1759, when he sold it to the colony of New York, and it was renamed Bedlow's Island.

To the east of Bedlow's Island is the main channel that runs from Manhattan and through the Narrows. To the west is a shallow bay, what was then called Oyster Banks, along the present Jersey City, New Jersey shoreline. When Lee realized that he would drift past the *Eagle*, he had to maneuver into a safe haven and wait for the tide to slack. His best bet would have been to move across the current and onto the Banks, where he could stay within the lee of Bedlow's Island.

The water on Oyster Banks was very shallow. The early charts do not provide soundings, but the Montressor chart shows, just to the south of Bedlow's Island, "an Island of Rock not seen at High Water." The charts of the harbor were drawn primarily to assist the deep draft warships with navigation through the channel and for finding appropriate anchorage. We know from the ship's log that the *Eagle* was anchored off Bedlow's Island, but near enough for Lee to have found refuge on Oyster Banks while remaining relatively close to his target.

> September 7
> Phase of the Moon is a waning crescent, 23% illuminated.
> Moonrise at 12:25 a.m.
> Morning twilight begins at 5:01 a.m.
> Sunrise at 5:29 a.m.
> High Tide at the Battery, 3:49 a.m.

From our testing, we believe that Lee could manage about one-half to one knot for an extended transit in still water, and a bit more than one knot for short bursts. Once caught in the tide, he knew that trying to head directly upstream would be fruitless. He could have gained some control over his tidal drift by rowing hard at an angle to the current and let the downstream flow push him laterally toward the Banks. This would have allowed him some maneuvering control, albeit downstream, by changing speed and direction as he rowed.

That Lee performed extensive maneuvers in response to being caught in the current is evident in his description: "I however hove about and rowed for 5 glasses, by the ships' bells, before the tide slacked so that I could get alongside the man of war . . ."

Without knowing the time that Lee started his mission, it is difficult to correlate the timing of his maneuvering to the speed and direction of the current during the transit. We speculate, however, that after expending a tremendous amount of energy, he was able to bring the *Turtle* into the protection of Bedlow's Island. Here he could rest in relative safety. The current in the vicinity of the *Eagle* would have subsided by 12:30

a.m., but in all likelihood, Lee had become exhausted and simply needed to rest. He would have to wait several hours for the next opportunity.

High tide at the Battery was at 3:49 a.m., but the current in the vicinity of Bedlow's Island would not become manageable until about thirty to forty-five minutes later. Shortly after 4:00 a.m., having rested sufficiently, Lee ventured out into the diminishing current to engage the *Eagle*: "The moon was about 2 hours high, and the daylight about one—when I rowed under the stern of the ship. could see the men on deck, and hear them talk." Keeping in mind that Lee's account was written nearly forty years after the event, his details may be a bit vague. That Lee remembers seeing the moon indicates the sky had cleared, and he would have had some natural illumination to be able to see men onboard the ship. The moonlight would, however, have made him vulnerable to detection and anxious to submerge.

UNDER HMS *EAGLE*

By 4:30 a.m., the current would have subsided sufficiently to allow him to maneuver into position at *Eagle's* stern. He had to act fast, as he knew that daylight would soon arrive. As with most of his descriptions, Lee is brief and gives only a summary of the event. What he felt and thought during the thirty minutes he had under the *Eagle* can only be imagined. While alongside the ship, he had been close enough to hear the British seamen talking. He was tired, and aware that in a few minutes he would be submerging in total darkness. "I then shut down all the doors, sunk down, and came under the bottom of the ship . . ."

When he "shut down the doors," he had committed to making the attack; all of his physical senses would be heightened. He reached out and could feel the handle on the mechanism for turning the wood screw. He grasped the vertical propeller crank and gave it a practice turn. Within the blackness of the *Turtle's* hull, he could see a faint glow from the foxfire on his compass and depth gauge. He tried to remember all of the instructions Bushnell had given him while in the Connecticut River, and thought through what each step would be once he slipped under the *Eagle's* hull. The adrenaline was kicking in.

After taking a few deep breaths to relax, he pressed down on the foot-operated valve and could hear water rushing in through the opening. He could feel a slight pressure in his ears as the water filled the bilge, compressing the air inside the *Turtle* and forcing it out through the ventilation pipes. He knew that if he added too much, he would sink. He had practiced letting in just enough to increase his depth a few inches at a time, but that was during daylight when he could see the water rising past the hatch windows. Now he would have to depend on the foxfire-covered cork in the depth gauge. But he found that the rise of the cork was nearly imperceptible. Bushnell's depth gauge was just not accurate enough to discern such small increases in depth.

Lee then remembered from his training that when he rotated the vertical propeller, he could feel when the blades were beneath the surface. A few turns and he could tell that the propeller was spinning freely; the hatch was not awash and he had to let in more water. He pressed down on the fill valve. Water rushed in, air escaped through the pipes. The *Turtle* settled a few inches deeper. He spun the vertical propeller again and finally could feel resistance; the vessel was now neutrally buoyant and Lee could attempt to seek the bottom of the *Eagle*'s hull.

As Lee descended, he had his compass and depth gauge to watch his orientation under water, but in the total darkness, there was no use looking through the small windows to see where he was with respect to the *Eagle*'s hull. Moving into position would require maintaining physical contact with the hull as he descended. Benjamin Gale, in his November 9, 1775, letter to Silas Deane, briefly described the process: "When he comes under the side of the Ship, he rubs down the side until he comes to the keel."

Whatever technique Lee used, it would all be by feel. He may have descended a few feet, then transited until he contacted the hull, then repeated the process until his depth gauge told him he was close to keel depth. If a slight current shifted him away from the hull, it would only be by chance that he would be able to regain contact. In the darkness, the ship would be invisible; for Lee to find his target, the *Turtle* would have to bump against the *Eagle*'s hull. Yet in reading Lee's account, we have to assume that in spite of the difficulties he must have encountered submerging and keeping in physical contact with the *Eagle*, he did manage to maneuver under its hull.

Once in position, he immediately ran into a problem: ". . . up with the screw against the bottom but found that it would not enter." Lee could not secure the wood screw to the hull. He had to think quickly. There was only a half-hour of air in the *Turtle*, but he was determined to try again:

I pulled along to try another place, but deviated a little to one side, and immediately rose with great velocity and came above the surface 2 or 3 feet between the ship and the daylight—then sunk again like a porpoise. I hove partly about to try again, but on further thought I gave out, knowing that as soon as it was light the ship's boats would be rowing in all directions . . .

It was probably twilight, 5:00 a.m., when Lee surfaced. He had depleted the air inside the *Turtle*, and would have opened the three deadlights in the hatch. He undoubtedly rested and caught his breath while on the surface alongside the *Eagle*, pondering his next move.

Lee had spent several hours operating the *Turtle*, yet avoided detection by the hundreds of British vessels moored in the harbor. He had rested and made a final run to his target, trying to complete the mission before daybreak. He was able to submerge and maneuver under the *Eagle*'s hull, but had to return to the surface. Less than ten days earlier, the American army, including Lee's regiment, had suffered severe losses during the Battle of Long Island. This was his chance to return the favor. Lee's frustration at being unable to secure one hundred and fifty pounds of gunpowder to the vulnerable underbelly of the *Eagle* must have been unimaginable. Now he had to explain the failure to Bushnell.

Figure 23.1. The caption under Cyril Field's original illustration reads "Bushnell's navigator abandons his attempt upon H.M.S. *Eagle*." (*Field 1908*)

Chapter Twenty-three

THE NEXT DAY

WHY THE TURTLE MISSION FAILED

Lee does not discuss in his written account why the wood screw would not enter the hull. Immediately after the event, however, Lee provided his speculation to Bushnell who later related Lee's comment to Jefferson. "He went under the ship, and attempted to fix the wooden screw into her bottom, but struck, as he supposes, a bar of iron, which passes from the rudder hinge, and is spiked under the ship's quarter." The slope of the hull of a British warship along the stern quarter, and in particular near the attachment of the rudder, is too steep for Lee to have stationed the *Turtle* and attempted to engage the wood screw. We believe that it was highly unlikely that Lee could have slid down the side of the hull to the precise depth that would allow him to move to the underside of the keel, exactly in the area where the rudder hinge was located. We have already dismissed the possibility that the *Eagle*'s hull was sheathed with copper (see chapter 16). We believe there were factors other than simply having the misfortune to strike "a bar of iron" on a hull almost entirely devoid of iron. Lee may have had to contend with issues associated with controlling buoyancy while submerging; with contacting the hull in a flat area amidships; and with the process of operating and engaging the wood screw.

When we ballasted and submerged our replica, the test pilot, author Roy Manstan, performed each step that Lee was also required to perform. With our hatch secured, the pilot opened the valve to admit water into the bilge. He could feel and hear the effects of the water rushing through the valve. He also noticed that the rise in the float in our depth gauge was negligible and could not be depended on to tell him when the hatch was awash. The best method was to spin the vertical propeller. Once awash, the pilot could readily sense that the propeller blades were pushing against the water above, creating the thrust needed to submerge the vessel. Our tests allowed us to understand how critical this initial operation was for Lee. Once he left the surface, however, Lee encountered other obstacles that could potentially have caused the mission to fail.

Buoyancy issues

Beyond the inherent problems initiating the process of submerging, there is another environmental condition that Lee would likely have had to overcome when driving his submarine down and under the *Eagle*'s hull. New York Harbor experiences significant variations in salinity (and thus density) due to the contributing water masses from the ocean and rivers. The degree of mixing of these water masses is affected by seasonal and daily variations in tidal volume and by specific weather events. What results from these various factors are salinity and density profiles that increase from the surface to the bottom, the degree of which varies with time.

We now take another leap of faith and use modern salinity data to predict what Lee may have encountered. Our assumption (see the section titled "Tides and Currents," in chapter 22) is that the water masses and their salinities today are similar to what they were 230 years ago, and the mixing dynamics in September 1776, when Lee was in the *Turtle*, being a function of seasonal variations, will likewise be similar to September 2008. The Web site, http://www.stevens.edu/maritimeforecast, was developed by Stevens Institute of Technology as part of their New York Harbor Observing and Prediction System (NYHOPS). The Web site provides a regional map from which an area of interest can be selected, and will display color-coded real-time predictions of environmental data such as wind, current, tides, and salinity as a function of time. The site also archives previous data that can be reviewed for the past several years.

Lee's mission against the *Eagle* occurred about eight nights after the previous full moon. Reviewing September 2008 NYHOPS salinity data in the vicinity of Bedlow's Island for the same time period, we found that during a twenty-four-hour interval, the surface water salinity varied from about 19 to 23 ppt (part per thousand), while bottom salinity varied from 23 to 26 ppt. The hourly differences between the surface and the bottom ranged from 3 to 5 ppt. If we assume that the *Turtle* was some distance above the bottom while under the *Eagle*, we can speculate that he experienced a 2 to possibly a 4 ppt increase in salinity as he descended.

As we discussed in chapter 8 (see the section titled "Salinity and the Effect on Buoyancy"), an increase of 1 ppt results in a three pound increase in buoyancy of the *Turtle*. From our submergence tests (chapter 16), we determined that our replica vertical propeller, and probably Bushnell's as well, was capable of about three to four pounds of thrust, not enough to overcome as much as twelve pounds of increased buoyancy as he submerged to the keel. Was this an issue for Lee? We are confident that he experienced some degree of increased buoyancy during his descent. Bushnell, who understood salinity and buoyancy, may very well have anticipated the possibility, and forewarned Lee that he would have to compensate by adding water ballast as he descended.

POSITIONING THE *TURTLE* AMIDSHIPS

The *Eagle* was an Intrepid class vessel and what the navy referred to as a "Third-Class" warship carrying sixty-four guns. Without a well-defined representation of the *Eagle*'s hull, we have turned to a model of HMS *Agamemnon*. Although the *Agamemnon* was an Ardent class vessel, we believe that it provides a reasonable approximation of the general lines, particularly below the waterline, of British warships that Bushnell had targeted.[1]

It is difficult to envision the relative size of the *Turtle* to the massive hull of a British man-of-war such as the *Eagle*. We have added a similarly scaled profile of the *Turtle* alongside the illustration of *Agamemnon*. The *Turtle* is shown at the stern of the ship (where Lee had approached *Eagle* on the surface) and below the hull amidships, where it would have attempted to attach the mine (figure 23.2).

We presume that there was sufficient water beneath the *Eagle* for Lee to have maneuvered into position at the deepest portion of the hull. The little submarine, with its external ballast, was at least six feet tall. Add a few feet to maneuver and there would need to be about ten feet of clearance. The contemporary charts indicate that where the channel rides along the edge of the Oyster Banks shallows near Bedlow's Island, the water ranged from 5 to 7 fathoms (30 to 42 feet). A fully loaded warship

Figure 23.2. The *Turtle* is shown in relation to a British man-of-war similar to HMS *Eagle*. The *Turtle* approached from the stern and slid down the side of the hull in a attempt to locate a flat area amidships. 1. "When I rowed under the stern of the ship. . . could see men on the deck." (Ezra Lee: February 20, 1815) 2. "When he comes under the side of the Ship, he rubs down the side until he comes to the keel. . ." (Benjamin Gale: November 9, 1775) 3. "[I] came under the bottom of the ship, up with the screw against the bottom but found that it would not enter." (Ezra Lee: February 20, 1815) (*Paul Dobson*)

may have a draft of nearly twenty-three feet.[2] Add ten feet for the *Turtle* and it was probably good that Lee submerged under the *Eagle* during high tide.

Having carefully descended under the *Eagle*, the next step was to engage the wood screw. Lee would ideally want to position the *Turtle* in an area where the *Eagle*'s hull was the most horizontal, i.e., amidships. But to find a nearly flat hull contour where the curvature would support the *Turtle* when buoyant, Lee would have had to maneuver as much as forty feet forward along the underside of the ship. The *Agamemnon* hull illustrates the steep curvatures of a warship between the stern, where Lee first contacted his target, and the flat portion amidships.

ENGAGING THE WOOD SCREW

Once Lee had positioned the *Turtle*, he would have pumped out a sufficient amount of ballast to buoy up against the underside of the *Eagle* . . . or did he? The deballasting operation was essential for keeping the hatch in contact with the hull and allowing Lee to maintain position. The amount of added buoyancy to accomplish this task was critical. Bushnell's mine was "lighter than the water, that it might rise up" against the targeted hull. It would be likely that the mine was at least ten to twenty pounds positively buoyant in order to maintain its contact with the hull. Once the mine was released, the *Turtle* would now be carrying ten to twenty pounds less buoyancy. Lee had to plan to pump out enough water to keep himself in position under the *Eagle*'s hull and to counter the loss of buoyancy to avoid sinking, had he released the mine.

This buoyant force would also have countered the force Lee used when attempting to press the wood screw up against the hull, and it is at this point where he ran into a problem. If Lee had not pumped a sufficient amount of ballast water, "want of an adequate pressure, to enable the screw to get a hold upon the bottom" would, as suggested by Charles Griswold in his account of Lee's attack (appendix E), cause the nearly neutrally buoyant *Turtle* to have "rebounded from the ship's bottom." Without a visual reference, Lee would not be able to sense if the *Turtle* remained stationary or moved downward when he pushed the wood screw up against the *Eagle*'s hull.

If we assume that Lee had removed a sufficient amount of ballast to remain in contact with the *Eagle*, his task was then to thread the wood screw into the hull. In order to ensure sufficient holding power, the wood screw would have to enter at least an inch into the hull planking, or as much as two inches if the water-saturated wood had softened. Bushnell designed the wood screw mechanism to slide vertically six inches (this device is described and illustrated in chapter 11), allowing the mechanism to move upward as the pilot turns and threads the screw into the hull. The freedom to slide six inches was intended to accommodate the two-inch penetration into the wood, as well as some degree of curvature of the hull contour. As can be seen in figure 23.2, the slope of the hull below the waterline, particularly aft, is significant. It is conceivable that Lee positioned the *Turtle* at a location where, after sliding the mechanism upward, the tip either never touched the hull or did not thread deep enough to hold. Lee, working in total darkness, may have thought that the tip was rotating against some metal on the *Eagle*'s hull whereas, in reality, the mechanism had slid upward the full six inches and was simply turning against the socket, which supported it within the *Turtle* hull.

Another unknown factor that may have contributed to Lee's inability to engage the wood screw was the amount of marine fouling on the *Eagle*'s hull. Because we know that the ship did not have copper sheathing when deployed to America, its hull would certainly have been covered to some degree with barnacles and algae. This fouling may have restricted the *Turtle* from coming in close enough contact for the wood screw to thread into the hull.

THE *TURTLE* RETREATS TO MANHATTAN

After being unable to insert the wood screw, Lee tried to reposition the *Turtle*. He later reported that he "deviated a little to one side," apparently shifting to a location where the slope of the hull could not support the buoyancy of the *Turtle* and "immediately rose with great velocity." Remaining under the flat portion of the hull would have been critical.

Back on the surface alongside the *Eagle*, Lee had decided that with the sun about to rise, he would return to the Battery and hope for another opportunity. He did not want to risk capture.

> I thought the best generalship was to retreat,. . . So I jogg'd on as fast as I could, and my compass being then of no use to me, I was obliged to rise up every few minutes to see that I sailed in the right direction, and for this purpose keeping the machine on the surface of the water, and the doors open—I was much afraid of getting aground on the island as the Tide of the flood set on the north point.

If we assume that Lee left the *Eagle* shortly after 5:00 a.m. when twilight was beginning to illuminate the harbor, he would have caught a favorable tidal current of between one-half and one knot carrying him toward the Battery. It may have been about 6:00 a.m. when British troops on Governors Island spotted him. According to Lee:

> When I was abreast of the fort on the island 3 or 400 men got upon the parapet to observe me—at length a number came down to the shore, shoved off a 12 oar'd barge with 5 or 6 sitters and pulled for me—I eyed them, and when they had got within 50 or 60 yards of me, I let loose the magazine in hopes that if they should take me, they would likewise pick up the magazine, and then we should all be blown up together, but as kind Providence would have it, they took fright, and returned to the island to my infinite joy.

Lee escaped capture, but was still a long way from Manhattan. While being pursued by the British he had jettisoned his mine, automatically setting the clockwork in motion. Between 6:00 a.m. and 7:00 a.m., the tide would have slacked in the harbor and begun to ebb. He began to maneuver past the north end of Governors Island. "I then weathered the Island and our people seeing me, came off with a whale boat, and towed me in—the Magazine after getting a little past the Island, went off with a tremendous explosion throwing up large bodies of water to an immense height."

The explosion in the harbor proved the efficacy of Bushnell's timing device. The *Turtle* had not been blown to bits during the attack. Lee was able to release the mine when he needed to, engaging the clockwork timer that detonated the gunpowder after the set time had elapsed. This undoubtedly gave Bushnell at least some satisfaction, but must also have impressed both the British and Americans who witnessed the explosive spectacle. As General Israel Putnam is reported to have exclaimed while watching the event from the city: "God's curse 'em. That'll do it for 'em."3

SEPTEMBER 6, 1776 . . . OR NOT?

Determining a specific date is not a matter that will change the historic implications of what Bushnell and Lee accomplished. Our investigation into what the pilot experienced and how the mission was conducted, however, is directly related to the environmental conditions to which he was subjected; namely, weather, tides, and currents. For the *Turtle*, as with any military mission, timing is everything.

In his 1815 manuscript letter, Lee simply noted: "I will now endeavor to give you a short account of my voyage in this machine.—The first night after we got down to New York with it that was favourable . . . [we] set off from the City." The sentence is underlined to emphasize that Lee does not mention that the event occurred on the night he and the *Turtle* arrived, but on the first night that was favorable. It is possible that the night of September 6 was the best opportunity, but Lee was not specific.

In February 1820, after what we perceive was an interview with Lee, Charles Griswold wrote a lengthy letter to Professor Benjamin Silliman describing the *Turtle* and the attack on the *Eagle* (see appendix E). Griswold described the selection of Lee to assist Bushnell, their training in Long Island Sound, and the return of the *Turtle* to New York:

> They therefore had the machine conveyed by land from New Rochelle to the Hudson river, and afterward arrived with it at New York. The British fleet now lay to the north of Staten Island, with a large number of transports, and were the objects against which this new mode of warfare was destined to act; the first serene night was fixed upon for the execution of this perilous enterprise, and Sergeant Lee was to be the engineer. After the lapse of a few days, a favorable night arrived, and at 11 o'clock, a party embarked in two or three whale boats, with Bushnell's machine in tow.

Based on Parsons's request to Heath, the *Turtle* could not have been transported to the Battery before the night of September 5, and in all probability arrived during the day on the 6th. Based on Lee's own account, "the first night that was favorable" may or may not have been the 6th. Griswold's interview with Lee mentioned that "a lapse of a few days" passed before Lee ventured out. This could imply that the attack occurred as late as September 9.

SEPTEMBER 7?

If the attack began the night of the 7th, ending the morning of the 8th, the tides and currents would shift by about fifty minutes. The NOAA predictions for that time period put low tide at the Battery at 10:50 p.m. and high tide at 4:50 a.m. If, as Griswold claimed, Lee left the Battery at 11:00 p.m., he may have experienced strong outgoing currents on the order of two knots, when the whaleboats set out with the *Turtle* in tow.

The conditions associated with his transit to the *Eagle* and finding refuge in the lee of Bedlow's Island would remain essentially the same. Lee would still have ample time to rest and wait for the tide to subside before making his attack. Slack in the area around the *Eagle* on the morning of the 8th, however, is now at about 5:00 a.m. or 5:30 a.m. With the current weakening, Lee could have maneuvered quietly to the stern of the *Eagle* by 4:30 a.m., just as he would have done on the morning of the 7th.

He would have submerged, made his attempt to attach the mine, and surfaced all within the half hour of air available in the *Turtle*. Sunrise would have shifted only about one minute, so his decision to retreat to Manhattan because of the approach of daylight would have been the same. The current moving him toward Manhattan, however, would have been favorable for an additional hour, turning to ebb at Governors Island at about 7:00 a.m. rather than 6:00 a.m.

SEPTEMBER 8 OR LATER?

Not likely. An additional day would have shifted slack water at the *Eagle* to at least a half hour after sunrise. Lee would have to move out from Bedlow's Island during twilight, making it less likely (although still possible) that he could have approached the *Eagle* undetected. It would also be several days before the current would be slack at Bedlow's Island, at a time corresponding to just prior to daylight. Lee also mentioned that the moonlight was sufficient to enable him to see the British sailors. It was, however, approaching a new moon on the night of September 8/9, with only a sliver of the crescent visible. The evidence strongly suggests that the mission began the night of September 6, or possibly on the 7th.

THE HUDSON RIVER

The *Turtle* had proven itself as a functional submarine. Systems critical to the survival of the pilot enabled Lee to spend several hours at the controls, submerge, slip undetected beneath the British flagship HMS *Eagle*, and return to Whitehall under the guns of the Battery. The detonation of Bushnell's magazine near Governors Island demonstrated that a clockwork timing device was a practical solution for deploying an underwater weapon. The complexity of all of these systems, the difficult environmental conditions, and the brief time Lee had to train and develop contingency decision-making ability all conspired to cause the mission to fail.

Back at Whitehall, the *Turtle* was safely secured, either afloat and tied alongside its sloop, or stowed onboard waiting for another opportunity. No doubt Lee and Bushnell spent hours discussing the issues that had interfered with securing the mine to the *Eagle*'s hull, and were considering what changes in the tactical operations would be appropriate, should another opportunity present itself. By September 7, however, there were more urgent concerns that took the attention of the decision makers on Manhattan. The council that General Greene had requested was taking place. The decision of how and when to evacuate New York City was coalescing in Washington's mind. There would be little time to provide Bushnell with another chance to prove the worth of his submarine.

Another council was formed on September 12, where ten generals voted to withdraw against three who felt compelled to hold the city. We do not know what Bushnell's moves were during the week after Lee returned from the *Eagle*. As soon as it was evident that there would be no more opportunities for his *Turtle* in the harbor, it is likely that the sloop, with Bushnell's crew and the *Turtle* onboard, proceeded up the Hudson with the *Crane* and other vessels under Washington's command. It was none too soon. On September 13, according to a letter by Colonel Oliver Babcock, four British warships "kept up an incessant fire, assisted by the cannon at Governor's Island."[4] On the 14th, the British had landed at Kips Bay along the eastern shore of Manhattan, and the following day, the British flag was raised over Fort George.

Lee did not provide dates, but he summarized the situation: "Before we had another opportunity to try an experiment our army evacuated New York and we retreated up the North River as far as fort Lee—A frigate came up and anchored off Bloomingdale. I now made another attempt under a new plan."

Bushnell only noted:

Afterwards, there were two attempts made in Hudson's river, above the city, but they effected nothing. One of them was by the aforementioned person [Ezra Lee]. In going towards the ship, he lost sight of her, and went a great distance beyond her: when he at length found her, the tide ran so strong, that as he descended under water, for the ship's bottom—it swept him away. Soon after this, the enemy went up the river, and pursued the boat which had the sub-marine vessel on board—and sunk it with their shot.

Though I afterwards recovered the vessel, I found it impossible, at that time, to prosecute the design any farther.

Bushnell knew that it would be difficult to use the *Turtle* again. He had run into several operational issues that made it most uncertain that he could have gained the support of those who made the operations around New York possible:

I had been in a bad state of health, from the beginning of my undertaking, and was now very unwell; the situation of public affairs was such, that I despaired of obtaining the public attention, and the assistance necessary. I was unable to support myself, and the persons I must have employed, had I proceeded. Besides, I found it absolutely necessary, that the operators should acquire more skill in the management of the vessel, before I could expect success; which would have taken up some time, and made no small additional expense. I therefore gave over the pursuit for that time, and waited for a more favorable opportunity, which never arrived.

The two attempts in the Hudson River occurred in late September and early October. Had Bushnell or Lee or Pratt been captured when their sloop was sunk in the Hudson River on October 9, one can only speculate as to their ultimate fate. The treatment that many American soldiers and sailors experienced in the derelict warships used by the British to confine their prisoners was well-known. The conditions in these prison ships, as related by Samuel Young during depositions made on December 15, 1776, and by several other survivors on February 17, 1777, were reported in the June 6, 1777, issue of the *Connecticut Gazette*. The following are extracts from Young's deposition:

Samuel Young being solemnly sworn after the manner used in Scotland, by lifting up the right hand and interrogated, deposeth.

That he was taken prisoner at fort Washington by the English army on Saturday the 15th of November; . . . [that he and others] . . . were carried on board a ship, where about five hundred of them were confined below deck; that during their confinement they suffered greatly with cold, not being allowed fire; . . . that great numbers died in this confinement, sometimes three, sometimes four or more, men in a day, and one day nine died; and that they themselves are in a frail state of health, occasioned by this barbarous usage; and many of those who were released died upon the road before they reached home.

Such may have been the fate of Bushnell and his supporters had they been captured . . . or they may have joined Nathan Hale on the gallows, an event that happened barely two weeks after Lee returned from the *Eagle.*

EPILOGUE

"... the ingenuity of these people is singular in their secret modes of mischief."—Commodore John Symons to Rear Admiral Sir Peter Parker, August 15, 1777

The *Turtle* represented Bushnell's vision of a method to counter the overwhelming power of a British warship. Yet this very complex submarine was conceived only to perform the function of conveying a powerful weapon to its target. Bushnell, Lee, and the *Turtle* represent a military tactic known today as "asymmetric warfare." After the disappointments with the *Turtle* at New York, Bushnell set his submarine ideas aside and returned to his original interest by redesigning his mine for deployment as a "floating magazine."

In spite of the failure of the *Turtle*, Bushnell retained his credibility as an engineer, and convinced Governor Trumbull that an alternate approach to submarine warfare, not requiring a submarine vessel, had potential. On April 22, 1777, the Connecticut Council of Safety had assembled in New London, where they "viewed all the adjacent hills &c. about Fort Trumbull . . . and proceeded to the fort and viewed and examined the works there &c. and went on board the ship *Oliver Cromwell* and viewed that &c., and then pass'd over to the fort at Groton &c. and returned to N. London about 3 o'clock p.m."[1] The Council, having assembled along the New London waterfront, had gained a new appreciation for their coastal fortifications and the impressive *Oliver Cromwell* built within walking distance of Bushnell's home.

The following day, Bushnell brought a sample of his newly constructed underwater explosive device and presented it to the Council. "Mr. David Bushnell, with Col. Worthington, apply'd and exhibited a specimen of a new invention for annoying ships &c. &c., and on motion &c.

voted him an order on officers, agents, commissaries, to afford him assistance of men, boats, powder, lead &c. as he shall call &c. &c. *Delivered him at large.*"[2] Colonel William Worthington's association with Bushnell was more than incidental. We believe that Worthington was very familiar with this new device. In 1777, Phineas Pratt, who is the person most likely to have produced Bushnell's latest invention, was serving in the 7th Connecticut Regiment under Colonel Worthington.[3] Worthington, Bushnell, and Pratt, were all residents of Saybrook.

Armed with enthusiasm and resources from the Council of Safety, Bushnell was able to make a first attempt with his "new invention for annoying ships" (certainly an understatement) on August 13, 1777. This latest invention consisted of a flotation device, from which the magazine filled with gunpowder was suspended. A geared mechanism was included, whereby a toothed wheel would be turned as the device slid along the targeted hull. As the wheel turned, it was designed to trigger a firing mechanism that, as with the clockwork on the *Turtle* mines, would detonate the gunpowder contained within the magazine.

Bushnell's target that night was HMS *Cerberus,* at anchor near New London harbor. The floating mine, however, was spotted and retrieved by a few unfortunate sailors onboard a schooner anchored aft of *Cerberus*. After the event, Commodore Symons, commanding officer of the *Cerberus*, described the device and its destructive power in his official report to Rear Admiral Sir Peter Parker:

> The schooner had got hold of it (who had taken it for fishing line), gathered in near fifteen fathoms, which was buoyed up by little bits of sticks at stated distances, until he came to the end, at which was fastened a machine, which was too heavy for one man to haul up, being upwards of 100 cwt.; the other people of the boat turning out assisted him, got it upon deck, and were unfortunately examining it too curiously, when it went off like the sound of a gun, blew the boat to pieces, and set her in a flame, killed the three men that were in the stern; the fourth, who was standing forward, was blown into the water.[4]

Later in the war, the *Cerberus* was scuttled and set on fire by its crew while in Narragansett Bay to avoid capture by the French fleet that had descended on Newport in 1778. The resulting explosion sent bits of the *Cerberus* into the nearby countryside.[5]

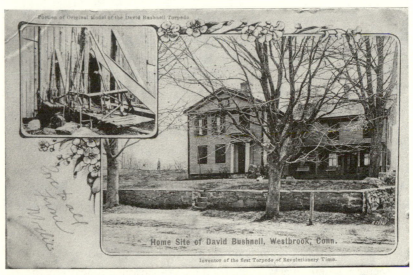

Early twentieth-century postcard showing the Bushnell homestead along Route 154 in what is now the town of Westbrook (the Pachaug District of Saybrook during the Revolution). The caption above the inset notes that it shows a "Portion of the Original Model of the David Bushnell Torpedo." (*Roy Manstan*)

Bushnell then turned to yet another concept of a floating mine. This was a simpler device that consisted of a gunpowder-filled keg detonated by a standard flintlock musket mechanism. A lever was attached to the top of the floating keg, such that when the keg came in contact with a ship's hull, the lever would rotate, triggering the flintlock. On a cold December night in 1777, Bushnell with help from local inhabitants sent several of these floating mines adrift in the Delaware River, upstream of the British fleet anchored near Philadelphia. Ice in the river slowed the progress of the mines, such that they did not appear in the vicinity of the warships until daylight. Two curious boys pulled one of them from the water, causing it to explode. Word quickly passed among the British, who immediately set to the task of ordering volleys of musket fire against Bushnell's devices. The whole affair was soon chronicled by Francis Hopkinson in his satirical poem "The Battle of the Kegs." Descriptions of Bushnell's design and use of floating mines can be found in Wagner (1963) and Roland (1978). Lefkowitz (2006) has included the entire text of Hopkinson's poem (also in Humphreys 1794; and Thatcher 1823).

Frustration seemed to follow Bushnell in all of his attempts against the British. He took a break from the war, returning to Yale in 1778 to continue his education, a decision that served him well later in life. After a year, he completed his masters degree and was probably pondering the future with some friends, when a British raiding party searching for American officers took Bushnell and several other civilians prisoners.

Here was a man who had designed, built, and made several attempts to destroy British warships and their crew using clandestine "infernal" forms of warfare. Now Bushnell was in British hands, and it was urgent that a way could be found to secure his release before his true identity became known. The reputation of British treatment of captured American soldiers, and prison ships in particular, was always on their mind. David Osgood provided a graphic description of the conditions in his December 11, 1783, "Thanksgiving Sermon."

> The prison ships of our enemies were filled with our unfortunate seamen, where they were exposed to suffocation and death by the stench and miseries of their confinement. . . . Through the whole course of the war, how were our feelings wounded, when we reflected on the cruelties inflicted upon multitudes of our people shut up in those floating dungeons of infection and disease; of misery and despair! The accounts we have had of their sufferings, were enough to harrow every bosom susceptible to the feelings of humanity. Whose blood is not chilled at the recollection of a son, a brother, or a friend, thrust into those gloomy regions; where 'no refreshing breezes entered', where poisonous exhalations were continually drawn in, and 'the tainted element was charged with pestilence and death'! How many thousands of our brethren have perished in the prisons of our enemy.[6]

Word of Bushnell's situation quickly reached Governor Trumbull who understood the significance, and potential consequence, of his capture. Trumbull immediately went into action, proposing a prisoner exchange before the British discovered Bushnell's identity. In May 1779:

> Upon a representation to this Assembly made by the Honble Major General Putnam, that sundry persons are confined in the several counties in this State, which persons he is desirous to exchange for an equal number of citizens of this and the neighbouring States who are prisoners to the British army: Resolved by this Assembly, that Daniel Smith, Thacher Seers, Ephraim

Treadwell, Ebenezer Leech, Arthur Newman, Eleazer Hall, George Crandell, Eli Lyon, and John Hendrick, be delivered into the hands or care of Brig. General Parsons, to be by him sent into New York in exchange for Samuel Webb of Stamford, David Bushnell of Saybrook, Simeon Coffin and Elihu Coffin, both of Martha's Vineyard, Samuel Silliman, Jesse Burr, Ebenezer Knapp, and Abraham Morehouse, all of Fairfield, and Lemuel Sherman of Stratford; and the several goalers and other persons who have the care and oversight of the above named British prisoners are hereby ordered and directed to take notice.[7]

With this prisoner exchange, David Bushnell may have literally "dodged the bullet" and was able to return to Connecticut. Bushnell was a civilian, but his precarious history of having conducted clandestine warfare against His Majesty's navy made him vulnerable to capture and treatment as a terrorist, a spy, or whatever charge that would likely result in hanging. The best alternative for Bushnell, who still desired to serve his country, would be to become a military officer.

Between May of 1778 and May of 1779, Congress passed several resolutions pertaining to the establishment of a Corps of Engineers. The requirement for service being that:

The commissioned officers to be skilled in the necessary branches of mathematics; the non-commissioned officers to write a good hand . . . . On March 11, 1779—Resolved, That the Engineers in the service of the United States shall be formed in a corps and styled the 'Corps of Engineers' . . . Resolved, That the Board of War be empowered and directed to form such regulations for the Corps of Engineers and companies of sappers and miners as they judge conducive to the public service . . . . On Feb. 7, 1780—Resolved, That the officers attached to the companies of sappers and miners be commissioned, and rank as follows. . . .

What followed was a list of officers and their enlistment dates, including "Mr. Bushnell, captain-lieutenant, August 2, 1779."[8] David Bushnell had turned again to serving his country.

By the summer of 1780, with insufficient recruits to the Corps of Engineers, also referred to as the Corps of Sappers and Miners, Washington had requested that one man from each regiment be assigned to the various companies of sappers and miners. As part of Connecticut's quota, Joseph Plumb Martin was selected to join David Bushnell's company and is listed as a corporal on January 1, 1781 (later promoted to Sergeant), remaining under Bushnell's command until discharged in 1783.[9] Martin related his experiences throughout the Revolution, including the more than two years he served under Bushnell, in a memoir published in 1830.[10]

Bushell was promoted to captain on June 8, 1781. Martin never discussed his relationship with Bushnell, nor his personality, only referring to him as "the captain" or "the old man." Martin does, however, relate situations where several soldiers, on one occasion in particular, planned a prank against their captain involving gunpowder with the potential for very hazardous consequences. The incident that occurred in September 1782 provides insight into how the rank and file soldier viewed Bushnell:

> One day, two or three of our young hotheads told me that they and some others of the men, whom they mentioned, were about to have some fun with 'the old man' as they generally called the captain. I inquired what their plans were, and they informed me that they had put some powder into a canteen and were going to give him a bit of a hoist. I asked them to let me see their apparatus before they put their project in execution. Accordingly, they soon showed me a wooden canteen with more, as I judged, than three pounds of gunpowder in it, with a stopper of touchwood for a fuse affixed to it, all, they said, in prime order. I told them they were crazy, that the powder they had in the canteen would 'hoist' him out of time, but they insisted on proceeding. It would only frighten him, they said, and that was all they wished to do—it would make him a little more complaisant. I told them that if they persisted in their determination and would not promise me on the spot to give up their scheme, I would that instant go to the captain and lay the whole affair before him. At length, after endeavoring without effect to obtain my consent to try a little under his berth, they concluded to give up the affair altogether, and thus, I verily believe, I saved the old man's life, although I do not think that they meant anything more than to frighten him. But the men hated him and did not much care what happened to him.[11]

If the rank and file weren't particularly enamored with their captain, there were many that understood and respected Bushnell's genius. The inventor had truly believed in the potential of his submarine, and others did as well. His supporters certainly included several of America's founding fathers. Benjamin Franklin was interested from the beginning, having written to Silas Deane during the summer of 1775: "I lament with you the Want of a naval Force. I hope the next Winter will be employ'd in forming one. When we are no longer fascinated with the idea of a speedy Reconciliation; we shall exert ourselves to some purpose. 'Till then Things will be done by Halves . . . I shall be curious to hear more Particulars of your new mechanical Genius."[12]

George Washington's comment that "it was an effort of genius," when solicited for information by Thomas Jefferson, indicates Washington's level of serious consideration and belief in Bushnell's vision. Long after the war was over, Jefferson continued to praise and defend Bushnell's achievements.

The story we have told of David Bushnell's incredible five-year journey with his "sub-marine vessel," from its concept while at Yale to the attack on HMS *Eagle*, ends in September 1776. For the *Turtle* Project and the Old Saybrook High School students, the journey began again in September 2003. What we tried to accomplish after 230 years was to uncover and understand as much as we could about the conditions that inspired and drove a small group of individuals led by Bushnell to create a truly unique vessel. This understanding could only be achieved by recreating this vessel, and subjecting it and a few willing volunteers to rigorous testing. The experiments we conducted at the Mystic Seaport Museum, two hundred years after Bushnell's "experiments made to prove the nature and use of a sub-marine vessel," put the finishing touches on gaining a better understanding of the revolutionary thought that led Bushnell and his collaborators to put their lives at risk in the pursuit of American independence. The success of Bushnell and his team has to be measured in terms of their persistence and creativity in a time of war, and not whether Ezra Lee failed in his attempt to sink the *Eagle*.

APPENDIX A

On October 13, 1787, David Bushnell responded to Thomas Jefferson's request for a description of the *Turtle*. Jefferson had initially asked George Washington for information, but Washington had deferred him to one of Bushnell's close friends: "I am sorry that I cannot give you full information respecting Bushnell's projects for the destruction of ships . . ." and later in the letter, "but Humphreys, if I mistake not, being one of his converts, will give you a more perfect account . . ." [The full text of Washington's letter to Jefferson is provided in appendix F.]

Humphreys interceded for Jefferson, and convinced Bushnell to pass on a detailed description of the *Turtle*. Bushnell sent the letter to Jefferson and a copy of the letter to Ezra Stiles, president of Yale College at the time. The following are Bushnell's cover letter that accompanied the copy to Stiles and his introductory letter to Jefferson from the *Naval Documents of the American Revolution* (Vol. 6, pp. 1500—1502). This is followed by a facsimile copy of the text of Bushnell's description of "The General Principals and Construction of a Sub-marine Vessel," published in the *Transactions of the American Philosophical Society* in 1799, when Thomas Jefferson was the Society's president.

DAVID BUSHNELL'S COVER LETTER TO EZRA STILES

Stamford October 16th 1787.

Sir

Induced by the desire you intimated in your Letter to me, of seeing what I should write to his Excellency Governor Jefferson our Ambassador at Paris, I have together with this, inclosed a Copy of what I have sent to his Excellency. The Original is forwarded by Colonel Humphreys, a Gentleman to whom I am much indebted, who wrote more than once upon the affair, and whose friendship, I have no doubt, I owe the attention of the Governor to the Subject, and his desire of information, agreeably to what you and Colonel Humphreys wrote long since.

I beg leave to thank you for your advice, and your kind offer to take the charge of forwarding my Letter to his Excellency. I could wish that what I have written should not come to the knowledge of the public, for the same reason, as I have written to the Governor, that I have ever wished to be silent upon the subject. Should what I have written to the Governor miscarry, I wish these might be ready to be forwarded to him, if I should be obliged to make use of them.

If you are desirous of any information which is not contained in this packet, I shall esteem it a favor, if you will give me the opportunity of satisfying you. Should you think proper to write to me or receive anything from His Excellency Governor Jefferson which respects me, I could wish they might be directed to the care of Major John Davenport in Stamford. I am Sir &c.

David Bushnell

DAVID BUSHNELL'S INTRODUCTORY LETTER TO THOMAS JEFFERSON:

Stamford, In Connecticut Octr. 13th 1787

Sir

In the latter part of the year 1785, I received a Letter from Colonel David Humphreys, and soon after, another from Doctor Ezra Stiles, President off Yale College in Connecticut, informing me, that your excellency desired an account of my Submarine Vessel, and the Experiments which I had made.

At the time I received those Letters, I was seized with a severe illness, which disabled me from writing, & though I attempted it several times, obliged me to desist. Ever since I recovered my health, my situation has been such, that until this time, it has not been in my power to write to your Excellency, upon the subject.

I shall think myself happy if this, arriving thus late, meet with your Excellency's acceptance, and give you the information you desired; and shall only regret, that I had it not in my power to write, as soon as I received the communications of those Gentlemen.

Doctor Stiles, in his Letter to me, transcribed from yours the following, 'If he thought proper to communicate it, I would engage never to disclose it, unless I could find an opportunity of doing it for his Benefit.' In answer to this declaration, I shall submit the disclosure of it entirely to your Excellency, to do as you shall think proper; & beg leave to return you my sincere thanks for your generous intentions.

I have ever carefully concealed my Principles & Experiments, as much as the nature of the subject allowed, from all but my chosen Friends, being persuaded that it was the most prudent course, whether the event should prove fortunate or otherwise, although by the concealment I never fostered any great expectations of profit, or even of a compensation for my time & expenses; the loss of which has been exceedingly detrimental to me.

With this your Excellency will receive a sketch of the general principals and construction of the Submarine Vessel blended together, as they occur at this time, with many of the Minutiae. I should gladly exhibit everything with the utmost minuteness, but apprehend I have not been sufficiently clear in what I have written, and have a doubt whether I could explain the whole intelligibly, without drawings, which I cannot easily execute or obtain. But should this not be sufficient, & you should wish to have a more minute description of the whole, or of any particular part not sufficiently explained here, I shall be happy to receive your Excellency's commands, and shall obey them, as soon as they come to hand, without any reserve.

Induced by the desire this should not fall into improper hands, I could wish, if it were not too great a favour, to hear that this finds a safe conveyance to your Excellency.

In the mean time, with the most respectful sentiments, I am &c.

David Bushnell.

P.S. Should your Excellency think proper to inform me of the safe arrival of this packet, I could wish such information might be directed to the care of Doctor Stiles.

His Excellency Thomas Jefferson Esquire.

BUSHNELL'S LETTER FROM THE *TRANSACTIONS*.

(303)

No. XXXVII.

General Principles and Construction of a Sub-marine Vessel, communicated by D. Bushnell of Connecticut, the inventor, in a letter of October, 1787, to THOMAS JEFFERSON *then Minister Plenipotentiary of the United States at Paris.*

Read June 8, 1798. THE external shape of the sub-marine vessel bore some resemblance to two upper tortoise shells of equal size, joined together; the place of entrance into the vessel being represented by the opening made by the swell of the shells, at the head of the animal. The inside was capable of containing the operator, and air, sufficient to support him thirty minutes without receiving fresh air. At the bottom opposite to the entrance was fixed a quantity of lead for ballast. At one edge which was directly before the operator, who sat upright, was an oar for rowing forward or backward. At the other edge, was a rudder for steering. An aperture, at the bottom, with its valve, was designed to admit water, for the purpose of descending; and two brass forcing-pumps served to eject the water within, when necessary for ascending. At the top, there was likewise an oar, for ascending or descending, or continuing at any particular depth—A water-gauge or barometer, determined the depth of descent, a compass directed the course, and a ventilator within, supplied the vessel with fresh air, when on the surface.

The entrance into the vessel was elliptical, and so small as barely to admit a person. This entrance was surrounded with a broad elliptical iron band, the lower edge of which was let into the wood of which the body of the vessel was made, in such a manner, as to give its utmost support to the body of the vessel against the pressure of the water. Above the upper edge of this iron band, there was a brass crown, or cover, resembling a hat with its crown and brim, which

which shut water tight upon the iron band: the crown was hung to the iron band with hinges so as to turn over side-wise, when opened. To make it perfectly secure when shut, it might be screwed down upon the band by the operator, or by a person without.

There were in the brass crown, three round doors, one directly in front, and one on each side, large enough to put the hand through—when open they admitted fresh air; their shutters were ground perfectly tight into their places with emery, hung with hinges and secured in their places when shut. There were likewise several small glass windows in the crown, for looking through, and for admitting light in the day time, with covers to secure them. There were two air pipes in the crown. A ventilator within drew fresh air through one of the air pipes, and discharged it into the lower part of the vessel; the fresh air introduced by the ventilator, expelled the impure light air through the other air pipe. Both air pipes were so constructed, that they shut themselves whenever the water rose near their tops, so that no water could enter through them, and opened themselves immediately after they rose above the water.

The vessel was chiefly ballasted with lead fixed to its bottom; when this was not sufficient, a quantity was placed within, more or less, according to the weight of the operator; its ballast made it so stiff, that there was no danger of oversetting. The vessel with all its appendages, and the operator, was of sufficient weight to settle it very low in the water. About two hundred pounds of the lead, at the bottom, for ballast, would be let down forty or fifty feet below the vessel; this enabled the operator to rise instantly to the surface of the water, in case of accident.

When the operator would descend, he placed his foot upon the top of a brass valve, depressing it, by which he opened a large aperture in the bottom of the vessel, through which the water entered at his pleasure; when he had ad-

mitted

SUB-MARINE VESSEL.

mitted a fufficient quantity, he defcended very gradually; if he admitted too much, he ejected as much as was neceffary to obtain an equilibrium, by the two brafs forcing pumps, which were placed at each hand. Whenever the veffel leaked, or he would afcend to the furface, he alfo made ufe of thefe forcing pumps. When the fkilful operator had obtained an equilibrium, he could row upward, or downward, or continue at any particular depth, with an oar, placed near the top of the veffel, formed upon the principle of the fcrew, the axis of the oar entering the veffel; by turning the oar one way he raifed the veffel, by turning it the other way he depreffed it.

A glafs tube eighteen inches long, and one inch in diameter, ftanding upright, its upper end clofed, and its lower end, which was open, fcrewed into a brafs pipe, through which the external water had a paffage into the glafs tube, ferved as a water-gauge or barometer. There was a piece of cork with phofphorus on it, put into the water-gauge. When the veffel defcended the water rofe in the water-gauge, condenfing the air within, and bearing the cork, with its phofphorus, on its furface. By the light of the phofphorus, the afcent of the water in the gauge was rendered vifible, and the depth of the veffel under water afcertained by a graduated line.

An oar, formed upon the principle of the fcrew, was fixed in the forepart of the veffel; its axis entered the veffel, and being turned one way, rowed the veffel forward, but being turned the other way rowed it backward; it was made to be turned by the hand or foot.

A rudder, hung to the hinder part of the veffel, commanded it with the greateft eafe. The rudder was made very elaftic, and might be ufed for rowing forward. Its tiller was within the veffel, at the operator's right hand, fixed, at a right angle, on an iron rod, which paffed through the fide of the veffel; the rod had a crank on its

S f   outfide

outfide end, which commanded the rudder, by means of a rod extending from the end of the crank to a kind of tiller, fixed upon the left hand of the rudder. Raifing and depreffing the firft mentioned tiller turned the rudder as the cafe required.

A compafs marked with phofphorus directed the courfe, both above and under the water; and a line and lead founded the depth when neceffary.

The internal fhape of the veffel, in every poffible fection of it, verged towards an ellipfis, as near as the defign would allow, but every horizontal fection, although elliptical, yet as near to a circle, as could be admitted. The body of the veffel was made exceedingly ftrong; and to ftrengthen it as much as poffible, a firm piece of wood was framed, parallel to the conjugate diameter, to prevent the fides from yielding to the great preffure of the incumbent water, in a deep immerfion. This piece of wood was alfo a feat for the operator.

Every opening was well fecured. The pumps had two fets of valves. The aperture at the bottom, for admitting water, was covered with a plate, perforated full of holes to receive the water, and prevent any thing from choaking the paffage, or ftopping the valve from fhutting. The brafs valve might likewife be forced into its place with a fcrew, if neceffary. The air pipes had a kind of hollow fphere, fixed round the top of each, to fecure the air-pipe valves from injury: thefe hollow fpheres were perforated full of holes for the paffage of the air through the pipes: within the air-pipes were fhutters to fecure them, fhould any accident happen to the pipes, or the valves on their tops.

Wherever the external apparatus paffed through the body of the veffel, the joints were round, and formed by brafs pipes, which were driven into the wood of the veffel, the holes through the pipes were very exactly made, and the iron rods, which paffed through them, were turned in

a lathe to fit them; the joints were alſo kept full of oil, to
prevent ruſt and leaking.   Particular attention was given to
bring every part, neceſſary for performing the operations,
both within and without the veſſel, before the operator,
and as conveniently as could be deviſed; ſo that every thing
might be found in the dark, except the water-gauge and
the compaſs, which were viſible by the light of the phoſ-
phorus, and nothing required the operator to turn to the
right hand, or to the left, to perform any thing neceſſary.

No. 2.

*Deſcription of a magazine and its appendages, deſigned to be
conveyed by the ſub-marine veſſel to the bottom of a ſhip.*

In the forepart of the brim of the crown of the ſub-marine veſſel, was a ſocket, and an iron tube, paſſing through
the ſocket; the tube ſtood upright, and could ſlide up and
down in the ſocket, ſix inches: at the top of the tube, was
a wood-ſcrew (A) fixed by means of a rod, which paſſed
through the tube, and ſcrewed the wood-ſcrew faſt upon
the top of the tube: by puſhing the wood-ſcrew up againſt
the bottom of a ſhip, and turning it at the ſame time, it
would enter the planks; driving would alſo anſwer the
ſame purpoſe; when the wood-ſcrew was firmly fixed, it
could be caſt off by unſcrewing the rod, which faſtened it
upon the top of the tube.

Behind the ſub-marine veſſel, was a place, above the
rudder, for carrying a large powder magazine, this was
made of two pieces of oak timber, large enough when hollowed out to contain one hundred and fifty pounds of powder, with the apparatus uſed in firing it, and was ſecured
in its place by a ſcrew, turned by the operator.   A ſtrong
piece of rope extended from the magazine to the wood-
ſcrew (A) above mentioned, and was faſtened to both.

S ſ 2   When

When the wood-screw was fixed, and to be caft off from its tube, the magazine was to be caft off likewife by unfcrewing it, leaving it hanging to the wood-fcrew; it was lighter than the water, that it might rife up againft the object, to which the wood-fcrew and itfelf were faftened.

Within the magazine was an apparatus, conftructed to run any propofed length of time, under twelve hours; when it had run out its time, it unpinioned a ftrong lock refembling a gun lock, which gave fire to the powder. This apparatus was fo pinioned, that it could not poffibly move, till, by cafting off the magazine from the veffel, it was fet in motion.

The fkilful operator could fwim fo low on the furface of the water, as to approach very near a fhip, in the night, without fear of being difcovered, and might, if he chofe, approach the ftem or ftern above water, with very little danger. He could fink very quickly, keep at any depth he pleafed, and row a great diftance in any direction he defired, without coming to the furface, and when he rofe to the furface, he could foon obtain a frefh fupply of air, when, if neceffary, he might defcend again, and purfue his courfe.

No. 3.

Experiments made to prove the nature and ufe of a fub-marine veffel.

The firft experiment I made, was with about two ounces of gun powder, which I exploded 4 feet under water, to prove to fome of the firft perfonages in Connecticut, that powder would take fire under water.

The fecond experiment was made with two pounds of powder, inclofed in a wooden bottle, and fixed under a hogfhead, with a two inch oak plank between the hogfhead

and

and the powder; the hogfhead was loaded with ftones as deep as it could fwim; a wooden pipe defcending through the lower head of the hogfhead, and through the plank, into the powder contained in the bottle, was primed with powder. A match put to the priming, exploded the powder, which produced a very great effect, rending the plank into pieces; demolifhing the hogfhead; and cafting the ftones and the ruins of the hogfhead, with a body of water, many feet into the air, to the aftonifhment of the fpectators. This experiment was likewife made for the fatisfaction of the gentlemen above mentioned.

I afterwards made many experiments of a fimilar nature, fome of them with large quantities of powder; they all produced very violent explofions, much more than fufficient for any purpofe I had in view.

In the firft effays with the fub-marine veffel, I took care to prove its ftrength to fuftain the great preffure of the incumbent water, when funk deep, before I trufted any perfon to defcend much below the furface: and I never fuffered any perfon to go under water, without having a ftrong piece of rigging made faft to it, until I found him well acquainted with the operations neceffary for his fafety. After that, I made him defcend and continue at particular depths, without rifing or finking, row by the compafs, approach a veffel, go under her, and fix the *wood-fcrew* mentioned in No. 2, and marked A, into her bottom, &c. until I thought him fufficiently expert to put my defign into execution.

I found, agreeably to my expectations, that it required many trials to make a perfon of common ingenuity, a fkilful operator: the firft I employed, was very ingenious, and made himfelf mafter of the bufinefs, but was taken fick in the campaign of 1776, at New-York, before he had an opportunity to make ufe of his fkill, and never recovered his health fufficiently, afterwards.

Experiments

Experiments made with a sub-marine veffel.

After various attempts to find an operator to my wifh, I fent one who appeared more expert than the reft, from New-York, to a 50 gun fhip lying not far from Governor's Ifland. He went under the fhip, and attempted to fix the wooden fcrew into her bottom, but ftruck, as he fuppofes, a bar of iron, which paffes from the rudder hinge, and is fpiked under the fhip's quarter. Had he moved a few inches, which he might have done, without rowing, I have no doubt but he would have found wood where he might have fixed the fcrew; or if the fhip were fheathed with copper, he might eafily have pierced it: but not being well fkilled in the management of the veffel, in attempting to move to another place, he loft the fhip; after feeking her in vain, for fome time, he rowed fome diftance, and rofe to the furface of the water, but found day light had advanced fo far, that he durft not renew the attempt. He fays that he could eafily have faftened the magazine under the ftem of the fhip, above water, as he rowed up to the ftern, and touched it before he defcended. Had he faftened it there, the explofion of one hundred and fifty pounds of powder, (the quantity contained in the magazine), muft have been fatal to the fhip. In his return from the fhip to New-York, he paffed near Governor's Ifland, and thought he was difcovered by the enemy, on the ifland; being in hafte to avoid the danger he feared, he caft off the magazine, as he imagined it retarded him in the fwell, which was very confiderable. After the magazine had been caft off one hour, the time the internal apparatus was fet to run, it blew up with great violence.

Afterwards, there were two attempts made in Hudfon's river, above the city, but they effected nothing. One of them was by the aforementioned perfon. In going towards

wards the ship, he loft fight of her, and went a great diftance beyond her: when he at length found her,
the tide ran fo ftrong, that as he defcended under water,
for the ship's bottom—it fwept him away. Soon after this,
the enemy went up the river, and purfued the boat which
had the fub-marine veffel on board—and funk it with
their fhot. Though I afterwards recovered the veffel, I found
it impoffible, at that time, to profecute the defign any farther.
I had been in a bad ftate of health, from the beginning
of my undertaking, and was now very unwell; the fitua-
tion of public affairs was fuch, that I defpaired of obtaining the public attention, and the affiftance neceffary.
I was unable to fupport myfelf, and the perfons I muft
have employed, had I proceeded. Befides, I found it ab-
folutely neceffary, that the operators fhould acquire more
fkill in the management of the veffel, before I could expect fuccefs; which would have taken up fome time, and
made no fmall additional expenfe. I therefore gave over
the purfuit for that time, and waited for a more favorable
opportunity, which never arrived.

——————

Other Experiments made with a defign to fire Shipping.

In the year 1777, I made an attempt from a whaleboat, againft the Cerberus frigate, then lying at anchor
between Connecticut river and New London, by drawing
a machine againft her fide, by means of a line. The
machine was loaded with powder, to be exploded by a
gun-lock, which was to be unpinioned by an apparatus,
to be turned by being brought along fide of the frigate.
This machine fell in with a fchooner at anchor, aftern
of the frigate, and concealed from my fight. By fome
means or other, it was fired, and demolifhed the fchooner
and

and three men—and blew the only one left alive, over-board, who was taken up very much hurt.

After this, I fixed several kegs, under water, charged with powder, to explode upon touching any thing, as they floated along with the tide: I set them afloat in the Delaware, above the English shipping at Philadelphia, in December, 1777. I was unacquainted with the river, and obliged to depend upon a gentleman very imperfectly acquainted with that part of it, as I afterwards found. We went as near the shipping as he durst venture; I believe the darkness of the night greatly deceived him, as it did me. We set them adrift, to fall with the ebb, upon the shipping. Had we been within sixty rods, I believe they must have fallen in with them immediately, as I designed; but as I afterwards found, they were set adrift much too far distant, and did not arrive, until after being detained some time by frost, they advanced in the day time, in a dispersed situation, and under great disadvantages. One of them blew up a boat, with several persons in it, who imprudently handled it too freely, and thus gave the British that alarm, which brought on *the battle of the Kegs.*

————————

The above Vessel, Magazine, &c. were projected in the year 1771, but not completed, until the year 1775.

D. BUSHNELL.

The

APPENDIX B

The following is the August 7, 1775, letter from Benjamin Gale to Benjamin Franklin as reprinted in *Naval Documents of the American Revolution* (Clark 1964, 1088–1089). Clark's footnote to Gales's letter explains the use of brackets surrounding some of the text. This letter is also found in *The Papers of Benjamin Franklin* (Wilcox 1982), where the editor has inserted different interpretations of the missing text.

DR. BENJAMIN GALE TO BENJAMIN FRANKLIN[1]

Killingsworth 7 Augt 1775

You[r Congress dou]btless have [had intim]ations of [the Inven]tion of [a new machine] for [the Destru]ction of [Enemy Ships, but I sit] down to Give [you an Account] of that [Machine and] what Exper[iments have] been alr[eady made wit]h it, what I relate y[ou] may Intire[ly rely] upon to be fact—I will not at this time attempt to Give You a Minute Description of the Form, as the Post is now Waiting, thus Much, it doth not Exceed 7 feet in Length, and the Depth not more than 5 1/2 feet, the Person who Navigates it, sits on a Bench in the Center of the Machine—The Person who invented it, is a student of Yale College, and is Graduated this Year—Lives within five Mile of me. I was the second Person who ever was permitted to see it, there being no other Workman but himself & Brother, Excepting what Iron Work is wanted, which was done by His direction, His Plan is to place the Cask Containing the Powder on the Outside of the Machine, and it is so Contrived, as when it strikes the Ship, which he proposes shall be at the Keil it Grapples fast to the Keils—and is Wholly Disengag'd from the Machine, he then Rows off, the Powder is to be fired by a Gun Lock within the Cask which is sprung by a Watch work, which he can so order as to have that take place at any Distance of Time he pleases—The Experiments that has as Yet been Made are as follows—In the Most Private Manner he Convey'd it on Board a Sloop In the Night and Went out into the Sound, He then sunk under Water, where he Continued about 45 Minutes without any Inconveniency as to Breathing, he Can Row it either Backward or forward Under water about 3 Miles an Hour—And Can steer to what Poi[nt o]f Compass he [pl]eases—he can

Rise to the [Surface of] the Water w[here and] when he Pleases to [get a fr]esh supply [of air wh]en that is Exhausted [Inside th]e Machine [is a Barometer] by which he can [tell the dep]th under w[ater and can] admit water if [needed] to Bring [the Machine] into a perfect [Equilibr]ium with [the water] he has also another Pair of Oars by whi[ch he] can Rowe it either up or Down—and a forcing Pump by which he Can free himself from the Water which he Admits to bring the Machine to a Proper Equilibrium with the Water at the Top he has a pair of Glass Eyes by which he sees Objects Under Water—These Parts are all Compleat and these Experiments he has Already Made I might add, he has an Anchor by which he Can remain in Any Place to Wait for Tide Oppy &c and again Weigh it at Pleasure—about 1000 wt of Lead is his Ballast, part of which is his Anchor, which he Carries on the outside at Bottom of the Machine, this story may Appear Romantic, but thus far is Compleated and All these Experiments above related has been Actually Made, He is now at New Haven with Mr Doolittle an Ingenious Mechanic in Clocks &c Making those Parts which Conveys the Powder, and secures the same to the Bottom of the Ship, and the Watchwork which fires it—I every Minute Expect his return, when a full Tryal will be made, and Give me Leave to Say, it is all Constructed with Great simplicity, and upon Principles of Natural Philosophy, and I Conceive is not Equall'd by any thing I ever heard of or Saw, Except Dr Franklins Electrical Experiments—he Builds it on his own Acct, he was Urged to Ask some Assistance from the Government, Upon the Leiut Govrs seeing it they Offered him Assistance, but it was so Inconsiderable a sum, he refused it, and Says he will go through with it at his own Risque—the Only Objections in my Mind from what I have seen of the Machinery of it is that he Cannot see under Water so Deep so perfectly as to fix it right, and wh 100 wt of Powder will force its way through the ship I fear the Water will give way before the Bottom of the Ship, and the force of the Explosion Eluded—the Whole Machine may be Transported in a Cart—I might have added he has made the Experiment of firing Powde[r] Under Water after remaining there 25 Minutes—I have been Long Urging him for permission to Acquaint You with these facts He at Length has Consented with this Condition that I request You would not Mention the Affair Untill he has made the Experiment, when Compleated, if Agreeable I will Acquaint You with the Experiments he makes before he goes with it down to Boston, He is Quite Certain he

Can Effect the thing and his reasoning so Philosophically and Answering every Objection I ever made that In truth I have great relyance upon it—

. . . I ask Ten Thousand Pardons for presuming to Trouble You with this Long Acct which I fear will Appear to You too Romantic to Obtain Beleiff—but have Endeavoured in the Strictest Sense to relate Facts Truly . . .

---

[1]*Franklin Papers*, IV, 61, APS. Parts of this letter are torn or blurred. The words within brackets are supplied largely on the basis of a subsequent letter from Dr. Gale to Silas Deane, which described the same machine.

APPENDIX C

Between August 1775 and February 1776, Benjamin Gale and Silas Deane, Connecticut delegate to the Continental Congress, corresponded on matters related to David Bushnell and his submarine. Gale was a friend and physician of the Deane family. The five letters that have survived were published in *Collections of the Connecticut Historical Society* (Vol. 2, J. Hammond Trumbull, editor) and are transcribed below.

SILAS DEANE TO BENJAMIN GALE

Wethersfield, Aug. 10th, 1775.

Dear Sir,—I wrote you several letters from Philadelphia, and was happy in yours; but you have not acknowledg'd the rec't of mine, informing you that I had by me the letters you wrote me formerly, and would at any time produce them, if necessary for the vindication of your character against the malevolent aspersions of which you complain. I wish to know if you received those letters.

The Congress as you have heard make but a short recess; before the expiration of which, pray favor me with a line, and say what ground is there for the report of a certain new invention for destroying Ships. You are in the neighborhood, and therefore presume you can give me the particulars.

I can give you nothing new more than is in public papers. The Journals of the Congress at large, until their adjournment, will be published soon.

I am, Dear Sir, Your most humble Servt,

S. Deane.

FOUR LETTERS FROM BENJAMIN GALE TO SILAS DEANE

Killingworth, 9th Nov., 1775

Dear Sir,—In your last you requested I would give you an account of the progress of our machine, and whether anything may be expected of it. I now sit down to give you a succinct but imperfect account of its structure, which is so complicated that it is impossible to give a perfect idea of it.

The Body, when standing upright in the position in which it is navigated, has the nearest resemblance to the two upper shells of a Tortoise joined together. In length, it doth not exceed 7 1/2 feet from the stem to the higher part of the rudder: the height not exceeding 6 feet. The person who navigates it enters at the top. It has a brass top or cover, which receives the person's head as he sits on a seat, and is fastened on the inside by screws. In this brass head is fixed eight glasses, viz. two before, two on each side, one behind, and one to look out upwards. In the same brass head are fixed two brass tubes, to admit fresh air when requisite, and a ventilator at the side to free the machine from the air rendered unfit for respiration. On the inside is fixed a Barometer, by which he can tell the depth he is under water; a Compass, by which he knows the course he steers. In the barometer and on the needles of the compass is fixed *fox-fire*, i.e. wood that gives light in the dark. His ballast consists of about 900 wt. of lead which he carries at the bottom and on the outside of the machine, part of which is so fixed as he can let run down to the Bottom, and serves as an anchor, by which he can ride *ad libitum*. He has a sounding lead fixed at the bow, by which he can take the depth of water under him; and to bring the machine into perfect equilibrium with the water, he can admit so much water as is necessary, and has a forcing pump by which he can free the machine at pleasure, and can rise above water, and again immerge, as occasion requires.

In the bow, he has a pair of oars fixed like the two opposite arms of a wind mill, with which he can row forward, and turning them the opposite way, row the machine backward; another pair fixed upon the same model, with which he can row the machine round, either to the right or left; and a third, by which he can row the machine either up or down: all which are turn'd by foot, like a spinning wheel. The rudder by which he steers, he manages by hand, within board. All these shafts which pass through the machine are so curiously fix'd as not to admit any water to incommode the machine. The magazine for the powder is carried on the hinder part of the machine, without-board, and so contrived, that when he comes under the side of the Ship, he rubs down the side until he comes to the keel, and a hook so fix'd as that when it touches the keel it raises a spring which frees the magazine from the machine and fastens it to the side of the Ship; at the same time, it draws a pin, which sets the watch-work agoing which, at a given time, springs the lock and the explosion ensues.

Three magazines are prepared; the first, the explosion takes place in twelve,—the second in eight,—the third in six hours, after being fixed to the ship. He proposes to fix these three before the first explosion takes place. He has made such a trial of the effects of the explosion of gunpowder under water, since Dr. Franklin did me the honor to call upon me, as has exceeded his most sanguine expectations, and is now convinced his magazines will contain three times so much powder as is necessary to destroy the largest ship in the navy.

I now write with the greater freedom, as I conclude by the time this reaches you the machine will be in the camp. Lately he has conducted matters and his designs with the greatest secrecy, both for the personal safety of the navigator and to produce the greater astonishment to those against whom it is designed,—if this projection succeeds, of which I make no doubt, as I well know the man and have seen the machine while in embryo, and every addition made to it fills me with fresh astonishment and surprize. And you may call me a visionary, an enthusiast, or what you please,—I do insist upon it, that I believe the inspiration of the Almighty has given him understanding for this very purpose and design. If he succeeds, a stipend for life, and if he fails, a reasonable compensation for time and expense is his due from the public.

What astonishment it will produce and what advantages may be made by those on the spot, if it succeeds, is more easy for you to conceive than for me to describe.

I congratulate you and my country in the begun success of our Arms to the northward, and the prospects of further success. Make my most respectful compliments to Dr. Franklin, and our Delegates, your associates; and am, most respectfully,
Your sincere friend and
Most humble servt,
Benjn Gale.

Killingworth, Nov. 22d, 1775
Sir.—I have to ask pardon for the wrong information I gave you. At the time of my last writing, I supposed the Machine was gone, but since find one proving the navigation of it in Connecticut River. The forcing pump made by Mr. Doolittle, not being made according to order given, did not answer; which has delayed him. The trials I mentioned to have been made since Dr. Franklin's being here, was the explosion, which prov'd beyond expectation.

I suppose he sets off this day with his new constructed pump, in order to prove the navigation, and if not prevented by ice in the River, will proceed soon. So far as you may have made known the contents of my letter, you may add this supplement.

He is by no means discouraged in the attempt. I had not seen him myself since Dr. Franklin was here, and his movement I had only from common report, but have since seen him myself. But few know the cause of his present delay.

I may not add, but to congratulate you and my country on our happy success to the Northward; and just to acquaint you that a report was industriously spread among us, that Col. Dyer and yourself were both confined in irons, for being Tories. The story gained great credit by your being left out by the Assembly, which was a wicked scheme. One of my neighbors came in one morning, looking much dejected, and told me the story. I told him I had just read in one of our papers you was *committed*; took it up, and read him your being appointed of the Committee to collect and assess the damages. He laughed, and went away much satisfied. I hope the Congress will not suffer you and Col. Dyer to return without some marks of respect. I forgot almost to subscribe,
Your most obt humble servt
B. Gale
P.S. Now we may write post free,—pray let me know something new, and when we may expect to see our little States new modeled. I want to see it come abroad.

We never more shall again be united with Great Britain. They are a devoted Kingdom.

Ut supra, B.G.

Killingworth, 7th Dec, 1775
Dear Sir,—According to your request I wrote you sometime since respecting our machine, supposing it was gone to the eastward. On finding that on proof of the navigation one instrument failed performing what was expected from it I then by letter acquainted you the proceeding was delayed until that could be repaired; which when done, another proof has been made which answers well, and every trial made requisite to the attempt respecting navigation, and everything answers well, but

still he fails on one account. He proposes going in the night, on account of safety. He always depended on fox-wood, which gives light in the dark, to fix on the points of the needle of his compass, and in his barometer, by which he may know what course to steer and the depth he is under water, both which are of absolute necessity for personal safety of the navigator: but he now finds that the frost wholly destroys that quality in that wood, of which he was before ignorant, and for that reason and that alone he is obliged to desist. He was detained near two months for want of money, and before he could obtain it the season was so far advanced he was, in the manner I have now related, frustrated. I write you this with two views, first that you and those to whom you may have communicated what I wrote, may not think I have imposed upon you an idle story, and in the next place to have you enquire of Dr. Franklin wr he knows of any kind of phosphorous which will give light in the dark and not consume the air. He has tried a candle, but that destroys the air so fast he cannot remain under water long enough to effect the thing. This you may rely upon, he has made every requisite experiment in proof of the machine, and it answers expectations; what I mentioned above is only wanting.

As I have not received a line from you since I first wrote, I know not whether you ever have received it, for I have not received one letter by the post for two months past, from any quarter whatever. As our tavern-keeper where the post leaves his letters is one of our Committee, they are all intercepted; and if I send any, I am obliged to deliver them out of town or have them superscribed by some other person. I take it hard that neither Col. Saltonstall nor Mr. Beers, by you nominated Post-Master in New Haven, can order the post-riders to deliver their letters at some other house in Killingworth excepting that house. All my letters on business and other affairs are intercepted by them, and the posts themselves I believe connive with them. The town are affronted, and the whole Committee will to a man be left out next week; but that will not help the matter for six months yet to come, as he will remain tavern-keeper until June next. If therefore I have no interest in your friendship in that matter, I must conclude to sweat it out.

The person, the inventor of this machine, now makes all his affairs a secret even to his best friends, and I have liberty to communicate this much from him only with a view to know if Dr. Franklin knows of any kind of phosphorous that will answer his purpose; otherwise the execution must be omitted until next spring, after the frosts are past. I am therefore to request your strictest silence in that matter.

I am Sir, your humble servant,

B. Gale.

Killingworth, Feby 1st, 1776

Dear Sir,—I have to acknowledge the receipt of yours of ye 13th ult. I make no doubt of your pressure by the important concerns lying before you, many of which I trust if you had leisure you might not communicate; but with regard to the matter of principal concern, if the Philosopher's Lanthorn may be attained, and will give a better light than what is proposed, should be glad you would get what knowledge you can from Dr. Franklin respecting it. Light is of absolute necessity; not to perform the operation,—that can be effected, if he hits the object right, as well in the dark as at noon day,—but to get free from the object when the operation is performed,—for this, light is absolutely necessary,—what point to steer, and to know whether he rises or sinks deeper, for the personal safety of the operator. You will well understand my meaning, if I am not more explicit. I have lately seen the man, and conversed freely with him. He is no enthusiast; a perfect philosopher, and by no means doubtful of succeeding. I wish Col. Dyer and you were to remain where you are, even altho' the other gentlemen were added. I have no objection to that addition, but if I may judge from what I hear, your countrymen are not suited with your recall.

Let me hear from you on this subject as soon as you can. I may ask, and you may refuse to tell me,—Are we well provided with Warlike Stores? Shall we have, or can we have if desired Foreign Aid? What are French troops to do in the West Indies? Is there a channel of communication open with our friends at home?

By the public accounts there is some prospect of a rupture at hand. Can the British Nation suffer such wicked work, and tamely look on?

I am, Dear Sir, Your Most Obt Humble Servt,

Benj. Gale.

APPENDIX D

The following is Ezra Lee's account of the *Turtle,* written in 1815 to Bushnell's good friend, David Humphreys. The transcription was produced from a copy of the manuscript letter on display at the Thomas Lee House (ca. 1660), East Lyme Historical Society, East Lyme, Connecticut. Ezra Lee was a descendant of Thomas Lee. There are seven manuscript notes at the end, possibly added by Humphreys after receiving the letter. The author of the notes also inserted corresponding footnote numbers (1) through (7) within the text of Lee's letter.

Lyme 20th Feby 1815
Dr. Sir.
Judge Griswold & Charles Griswold Esq both informed me that you wished to have an account of a machine invented by David Bushnell of Saybrook at the commencement of our Revolutionary war. In the Summer of 1776, he went to New York with it to try the Asia man of war; his brother being acquainted with the working of the machine, was to try the first experiment with it, but having spent until the middle of August, he gave out in consequence of indisposition. Mr Bushnell then came to General Parsons (of Lyme) to get some one to go and learn the ways & mystery of this new machine and to make a trial of it.

General Parsons sent for me, & two others, who had given in our names to go in a fire ship if wanted, to see if we would undertake the enterprise—we agreed to it, but first returned with the machine down Sound, and on our way practised with it in several harbors—we returned as far back as SayBrook with Mr Bushnell, where Some Little alterations were made in it—in the course of which time, (it being 8 or 10 days) the British had got possession of Long Island & Governor's Island—We went back as far as New Rochelle and had it carted over by land to the North River.

Before I proceed further, I will endeavour to give you some idea of the construction of this machine, turtle or torpedo, as it has since been called.

(1) Its shape was most like a round clam, but longer, and set up on its square side. It was high enough to stand in or sit as you had occasion,

with a (2) composition head hanging on hinges—it had six glasses inserted in the head, and made water tight, each the size of a half Dollar piece, to admit light—in a clear day, a person might see to read in three Fathoms of water—The machine was steered by a rudder having a crooked tiller, which led in by your side, through a water joint—(3) then sitting on the seat, the navigator rows with one hand, & steers with the other—it had two oars, of about 12 inches in length, & 4 or 5 in width, shaped like the arms of a windmill, which led also inside through water joints, in front of the person steering, and were worked by means of a wench (or crank) and with hard labor, the machine might be impelled at the rate of 3 nots an hour for a short time—Seven hundred pounds of lead were fixed on the bottom for ballast, and two hundred weight of it was so contrived, as to let it go in case the pumps choaked, so that you could rise at the surface of the water. It was sunk by letting in water by a spring near the bottom, by placing your foot against which, the water would rush in and when sinking take off your foot & it would cease to come in & you would sink no further, but if you had sunk too far, pump out water until you got the necessary depth—These pumps forced the water out at the bottom, one being on each side of you as you rowed— A pocket compass was fixed in the side, with a piece of light (4) wood on the north side, thus +, and another on the east side thus -, to steer by while under water—Three round doors were cut in the head (each 3 inches diameter) to let in fresh air untill you wished to sink, and then they were shut down & fastened—There was also a glass tube (5) 12 inches long & 1 inch diameter, with a cork in it, with a piece of light wood, fixed to it, and another piece at the bottom of the tube, to tell the depth of descent—one inch rise of the cork in the tube gave about one fathom water.—It had a screw, that pierced through the top of the machine, with a water joint, which was so very sharp that it would enter wood, with very little force, and this was turned with a wench, or crank, and when entered fast in the bottom of the ship, the screw is then left, and the machine is disengaged by unscrewing another one inside that held the other From the screw now fixed on the bottom of the ship, a line—led to & fastened to the mazagine, to prevent its escape either side of the ship. The magazine was directly behind you on the outside, and that was freed from you by unscrewing a screw inside—Inside the magazine was a Clock and [ma]chinery, which immediately sets a going after it is disengaged & a gun lock is fixed to strike fire to the powder at the set time after the Clock should run down—The clock might be set to go longer or shorter—20 or 30 minutes was the usual time, to let the navigator escape—This magazine was shaped like an egg & made of oak dug out in two pieces, bound together with bands of iron, corked & pav'd over with tar so as to be perfectly tight, and the clock was bound so as not to run untill this magazine was unscrewed.

I will now endeavor to give you a short account of my voyage in this machine.—The first night after we got down to New York with it that was favourable, (for the times for a trial must be when it is slack water, & calm, as it is unmanagable in a Swell or a Strong tide) the British Fleet lay a little above Staten Island. We set off from the City—the whale boats towed me as nigh the ships as they dared to go, and then cast me off. I soon found that it was too early in the tide, as it carried me down by the Ships. I however hove about, and rowed for 5 glasses, by the ships' bells, before the tide slacked so that I could get along side of the man of war, which lay above the transports. The Moon was about 2 hours high, and the daylight about one—when I rowed under the stern of the ship. could see the men on deck, and hear them talk—

I then shut down all the doors, sunk down, and came under the bottom of the ship, up with the screw against the bottom but found that it would not enter. (6) I pulled along to try another place, but deviated a little one side, and immediately rose with great velocity and came above the surface 2 or 3 feet between the ship and the daylight—then sunk again like a porpoise. I hove partly about to try again, but on further thought I gave out, knowing that as soon as it was light the ship's boats would be rowing in all directions, and I thought the best generalship was to retreat, as fast as I could as I had 4 miles to go, before passing Governor's Island.—So I jogg'd on as fast as I could, and my compass being then of no use to me, I was obliged to rise up every few minutes to see that I sailed in the right direction, and for this purpose keeping the machine on the surface of the water, and the doors open—I was much afraid of getting aground on the island as the Tide of the flood set on the north point. While on my passage up to the City, my course owing to the above circumstances, was very crooked & zig zag, and the enemy's attention was drawn towards me, from Governor's Island—When I was abreast of the fort on the island 3 or 400 men got upon the parapet to observe me—at length a number came down to the shore, shoved off a 12 oar'd barge with 5 or 6 sitters and pulled for me—I eyed them, and when they had got within 50 or 60 yards of me, I let loose the magazine in hopes that if they should take me, they would likewise pick up the magazine, and then we should all be blown up together, but as kind Providence would have it, they took fright, and returned to the island to my infinite joy. I then weathered the Island and our people seeing me, came off with a whale boat, and towed me in—The Magazine after getting a little past the Island, went off with a tremendous explosion throwing up large bodies of water to an immense height. (7)

Before we had another opportunity to try an experiment our army evacuated New York and we retreated up the North River as far as fort Lee—A Frigate came up and anchored off Bloomingdale. I now made another attempt upon a new plan. My intention was to have gone under the ship's stern, and screwed on the magazine close to the water's edge, but I was discovered by the watch and was obliged to abandon the scheme, then shutting my doors, I dove under her, but my cork in the tube (by which I ascertained my depth) got obstructed, and deceived me, and I descended too deep & did not touch the ship and I then left her—Soon after the Frigate came up the river, drove our Crane galley on shore, and sunk our Sloop, from which we escaped to the shore—
I am &c. E. Lee
For General David Humphreys

(1) This machine was built of oak in the strongest manner possible, corked and tarred, and though its sides were at least six inches thick, the writer of the foregoing, told me that the pressure of the water, against it, at the depth of two fathoms was so great, that it oozed quite through, as mercury will by means of the air pump. Mr Bushnell's machine was no larger than just to admit one person to navigate—the extreme length was not more than 7 feet.—When lying in the water, in its ordinary state without ballast, its upper works did not rise more than 6 or 7 inches out of water.

(2) This composition head means a composition of metals—something like bell metal, and was fixed on the top of the machine, and which afforded the only admission to the inside.

(3) The steering of this machine was done on the same principles with ordinary vessels, but the rowing her through the water, was on a very different plan. These oars were fixed on the end of a shaft like windmill arms projected out forward and turned at right angles with the

course of the machine, and upon the same principles that windmill arms are turned by the wind these oars, when put in motion as the writer describes, draws the machine slowly after it—this moving power is small and every attendant circumstance must cooperate with it, to answer the purpose, calm water & no current—

(4) This light wood is what we sometimes call foxfire, and is the dry wood that shines in the dark, this was necessary as the points of the compass could not readily be seen without.

(5) The glass tube here mentioned, which was a sort of thermometer, to ascertain the depth of water the machine descended, is the only part that is without explanation—the writer of the foregoing could not recollect the principles on which such an effect was produced, nor the mechanical contrivance of it. He only knows that it was so contrived that the cork & light wood rose or fell in the tube by the ascent or descent of the machine—

(6) The reason why the screw would not enter was that the ship's bottom being coppered it would have been difficult under any circumstances to have peirced through it—but on attempting to bore with the auger, the force necessary to be used in pressing against the ships bottom, caused the machine to rebound off. This difficulty defeated the whole— the screw could not enter the bottom, and of course the magazine could not be kept there in the mode desired—

(7) When the explosion took place, General Parsons was vastly pleased, and cried out in his peculiar way—"God's curse 'em. That'll do it for 'em!"—

APPENDIX E

At some point toward the end of his life, Ezra Lee related his adventures with the *Turtle* to Charles Griswold, who in 1815 had encouraged Lee to write the story to David Humphreys (appendix D). In February 1820, Griswold wrote an account that may have been a combination of Lee's letter to Humphreys, plus details gained during a subsequent interview with Lee. Griswold then sent this account to Professor Benjamin Silliman, editor of *The American Journal of Science, and Arts*. The majority of the information in Lee's two accounts is very similar. While we considered Griswold's version to be moderately credible "hearsay" evidence, there were aspects that helped us with our creation of a *Turtle* replica. For that reason, we are including a transcription of this letter as published in Silliman's *Journal*.

PHYSICS, MECHANICS, CHEMISTRY, AND THE ARTS

Art. VIII. *Description of a Machine, invented and constructed* by David Bushnell, *a native of Saybrook, at the commencement of the American revolutionary war, for the purpose of submarine navigation, and for the destruction of ships of war; with an account of the first attempt with it, in August 1776, by* Ezra Lee, *a sergeant in the American army, to destroy some of the British ships then lying at New-York. Communicated by* Charles Griswold, Esq.

To PROFESSOR SILLIMAN.

Lyme, *Conn. Feb.* 21*st.* 1820.
Sir,

It is to be presumed that every person who has paid any attention to the mechanical inventions of this country, or has looked over the history of her revolutionary war, has heard of the machine invented by David Bushnell, for submarine navigation, and the destruction of hostile shipping. I have thought that a correct and full account of that novel and original invention, would not be unacceptable to the public, and particularly to those devoted to the pursuit of science and the arts.

If the idea of submarine warfare had ever occurred to any one, before the epoch of Bushnell's invention, yet it may be safely stated, that no ideas but his own ever came to any practical results. To him, I believe, the whole merit of this invention is unanimously agreed to belong.

But such an account as I have mentioned, must derive an additional value, and an increased interest from the fact, that all the information contained on the following pages, has been received from the only person in existence possessed of that information, and who was the very same that first embarked in this novel and perilous navigation.

Mr. Ezra Lee, first a sergeant and afterwards an ensign in the revolutionary army, a respectable, worthy, and elderly citizen of this town, is the person to whom I have alluded; to him was committed the first essay for destroying a hostile ship by submarine explosion, and upon his statements an implicit reliance may be placed.

Considering Bushnell's machine as the first of its kind, I think it will be pronounced to be remarkably complete throughout in its construction, and that such an invention furnishes evidence of those resources and creative powers, which must rank him as a mechanical genius of the first order.

I shall first attend to a description of this machine, and afterwards to a relation of the enterprise in it by sergeant Lee; confining myself in each case, strictly to the facts with which he has supplied me.

Yours, &c.

CHARLES GRISWOLD.

Bushnell's machine was composed of several pieces of large oak timber, scooped out and fitted together, and its shape my informer compares to that of a round clam. It was bound around thoroughly with iron bands, the seams were corked, and the whole was smeared over with tar, so as to prevent the possibility of the admission of water to the inside.

It was of a capacity to contain one engineer, who might stand or sit, and enjoy sufficient elbow room for its proper management.

The top or head was made of a metallic composition, exactly suited to its body, so as to be water-tight; this opened upon hinges, and formed the entrance to the machine. Six small pieces of thick glass were inserted in this head, for the admission of light: in a clear day and clear seawater, says my informer, he could see to read at the depth of three fathoms. To keep it upright and properly balanced, seven hundred pounds of lead were fastened to its bottom, two hundred pounds of which were so contrived as to be discharged at any moment, to increase the buoyancy of the machine.

But to enable the navigator when under water, to rise or sink at pleasure, there were two forcing pumps, by which water could be pressed out at the bottom; and also a spring, by applying the foot to which, a passage was formed for the admission of water. If the pumps should get deranged, then resort was had to letting off the lead ballast from the bottom.

The navigator steered by a rudder, the tiller of which passed through the back of the machine at a water joint, and in one side was fixed a small pocket compass, with two pieces of shining wood, (sometimes called foxfire,) crossed upon its north point, and a single piece upon the last point. In the night, when no light entered through the head, this compass thus lighted, was all that served to guide the helmsman in his course.

The ingenious inventor also provided a method for determining the depth of water at which the machine might at any time be. This was achieved by means of a glass tube, twelve inches in length, and about four in diameter, which was also attached to the side of the machine; this tube enclosed a piece of cork, that rose with the descent of the machine, and fell with its ascent, and one inch rise of the cork denoted a depth of about one fathom. The principle upon which such a result was produced, and also the mechanical contrivance of this tube, entirely escaped the observation of Mr. Lee, amidst the hurry and constant anxiety attendant upon such a perilous navigation.

But not the least ingenious part of this curious machine, was that by which the horizontal motion was communicated to it. This object was effected by means of two oars or paddles, formed precisely like the arms of a wind-mill, which revolved perpendicularly upon an axeltree that projected in front; this axeltree passed into the machine at a water joint, and was furnished with a crank, by which it was turned: the navigator being seated inside, with one hand laboured at the crank, and with the other steered by the tiller.

The effect of paddles so constructed, and turned in the manner stated, by propelling or rather drawing a body after them under water, will readily occur to any one without explanation.

These paddles were but twelve inches long, and about four wide. Two smaller paddles of the same description, also projected near the head, provided with a crank inside, by which the ascent of the machine could be assisted.

By vigorous turning of the crank, says my informer, the machine could be propelled at the rate of about three miles an hour in still water.

When beyond the reach of danger, or observation of an enemy, the machine was suffered to float with its head just rising from the water's surface, and while in this situation, air was constantly admitted through three small orifices in the head, which were closed when a descent was commenced.

The efficient part of this engine of devastation, its magazine, remains to be spoken of. This was separate and distinct from the machine. It was shaped like an egg, and like the machine itself, was composed of solid pieces of oak scooped out, and in the same manner fitted together, and secured by iron bands, &c. One hundred and thirty pounds of gun powder, a clock, and a gun lock, provided with a good flint that would not miss fire, were the apparatus which it enclosed. This magazine was attached to the back of the machine, a little above the rudder, by means of a screw, one end of which passed quite into the magazine, and there operated as a stop upon the movements of the clock, whilst its other end entered the machine. This screw could be withdrawn from the magazine, by which the latter was immediately detached, and the clock commenced going. The clock was set for running twenty or thirty minutes, at the end of which time, the lock struck, and fired the powder, and in the meantime the adventurer effected his escape.

But the most difficult point of all to be gained, was to fasten this magazine to the bottom of a ship. Here a difficulty arose, which, and which alone, as will appear in the ensuing narrative, defeated the successful operations of this warlike apparatus.

Mr. Bushnell's contrivance was this—A very sharp iron screw was made to pass out from the top of the machine, communicating inside by a water joint; it was provided with a crank at its lower end, by which the engineer was to force it into the ship's bottom: this screw was next to be disengaged from the machine, and left adhering to the ship's bottom. A line leading from this screw to the magazine, kept the latter in its destined position for blowing up the vessel.

I shall now proceed to the account of the first attempt that was made to destroy a ship of war, all the facts of which, as already stated, I received from the bold adventurer himself.

It was in the month of August, 1776, when Admiral Howe lay with a formidable British fleet in New-York bay, a little above the Narrows, and a numerous British force upon Staten Island, commanded by General Howe, threatened annihilation to the troops under Washington, that Mr.

Bushnell requested General Parsons of the American army, to furnish him with two or three men to learn the navigation of his new machine, with a view of destroying the enemy's shipping.

Gen. Parsons immediately sent for Lee, then a sergeant, and two others who had offered their services to go on board a fire ship; and on Bushnell's request being made known to them, they enlisted themselves under him for this novel piece of service. The party went up Long Island Sound with the machine, and made various experiments with it in the different harbors along shore, and after having become pretty thoroughly acquainted with the mode of navigating it, they returned through the Sound; but during their absence, the enemy had got possession of Long-Island and Governor's-Island. They therefore had the machine conveyed by land across from New-Rochelle to the Hudson river, and afterwards arrived with it at New-York.

The British fleet now lay to the north of Staten-Island, with a large number of transports, and were the objects against which this new mode of warfare was destined to act; the first serene night was fixed upon for the execution of this perilous enterprise, and sergeant Lee was to be the engineer. After a lapse of a few days, a favorable night arrived, and at 11 o'clock, a party embarked in two or three whale boats, with Bushnell's machine in tow. They rowed down as near the fleet as they dared, when sergeant Lee entered the machine, was cast off, and the boats returned.

Lee now found the ebb tide rather too strong, and before he was aware, had drifted him down past the men of war; he however immediately *got the machine about*, and by hard labour at the crank for the space of five glasses by the ship's bells, or two and a half hours, he arrived under the stern of one of the ships at about slack water. Day had now dawned, and by the light of the moon he could see the people on board, and heard their conversation. This was the moment for diving: he accordingly closed up overhead, let in water, and descended under the ship's bottom.

He now applied the screw, and did all in his power to make it enter, but owing probably in part to the ship's copper, and want of an adequate pressure, to enable the screw to get a hold upon the bottom, his attempts all failed; at each essay the machine rebounded from the ship's bottom, not having sufficient power to resist the impulse thus given to it.*

He next paddled along to a different part of her bottom, but in this manoeuvre he made a deviation, and instantly arose to the water's surface on the east side of the ship, exposed to the increasing light of the morning, and in imminent hazard of being discovered. He immediately made another descent, with a view of making one more trial, but the fast approach of day, which would expose him to the enemy's boats, and render his escape difficult, if not impossible, deterred him; and he concluded that the best generalship would be to commence an immediate retreat.

He now had before him a distance of more than four miles to traverse, but the tide was favourable. At Governor's-Island great danger awaited him, for his compass having got out of order, he was under the necessity of looking out from the top of the machine very frequently to ascertain his course, and at best made a very irregular zigzag track.

The soldiers at Governor's-Island espied the machine, and curiosity drew several hundreds upon the parapet to watch its motions. At last a party came down to the beach, shoved off a barge, and rowed towards it. At that moment sergeant Lee thought he saw his certain destruction, and as a last act of defence, let go the magazine, expecting that they would seize that likewise, and thus all would be blown to atoms together.

Providence however otherwise directed it: the enemy, after approaching within fifty or sixty yards of the machine, and seeing the magazine detached, began to suspect a *yankee trick*, took alarm and returned to the island.

Approaching the city, he soon made a signal, the boats came to him and brought him safe and sound to the shore. The magazine in the mean time had drifted past Governor's-Island into the East river, where it exploded with tremendous violence, throwing large columns of water and pieces of wood that composed it high into the air. Gen. Putnam, with many other officers, stood on the shore spectators of this explosion.

In a few days the American army evacuated New-York, and the machine was taken up the North river. Another attempt was afterwards made by Lee upon a frigate that lay opposite Bloomingdale: his object now was to fasten the magazine to the stern of the ship, close at the water's edge. But while attempting this, the watch discovered him, raised an alarm, and compelled him to abandon his enterprise. He then endeavoured to get under the frigate's bottom, but in this he failed, having descended too deep. This terminated his experiments.

*It yet remains a problem, whether the difficulty here spoken of will ever be fully obviated. Mr. Fulton's torpedoes were never fairly brought to the test of experiment, though he and his friends entertained perfect confidence that they would not be found defective in any of their operations.

APPENDIX F

While Ambassador to France, Thomas Jefferson took an interest in Bushnell's inventions and made an inquiry of George Washington about the *Turtle* and the events around New York. Washington sent a brief reply, suggesting that Jefferson contact Bushnell's friend Colonel David Humphreys, who had also served as Washington's aide-de-camp. The following are transcripts of Jefferson's letter from *The Papers of Thomas Jefferson* (Julian P. Boyd, ed., 1953, vol. 8, p. 301), and Washington's response as found in the 1966 facsimile edition of *Beginning of Modern Submarine Warfare* by Henry L. Abbot (1881, 4–5).

THOMAS JEFFERSON TO GEORGE WASHINGTON

Paris July 17, 1785
Sir

Permit me to add, what I forgot in my former letter, a request to you to be so kind as to communicate to me what you can recollect of Bushnell's experiments in submarine navigation during the late war, and whether you think his method capable of being used successfully for the destruction of vessels of war. It's not having been actually used for this purpose by us, who were so peculiarly in want of such an agent seems to prove it did not promise success. I am with the highest esteem Sir Your most obedt. & most humble servt., Th: Jefferson

GEORGE WASHINGTON TO THOMAS JEFFERSON

Mount Vernon, 26 Sept., 1785

I am sorry that I cannot give you full information respecting Bushnell's projects for the destruction of ships. No interesting experiments having been made, and my memory being bad, I may in some measure be mistaken in what I am about to relate. Bushnell is a man of great mechanical powers, fertile in inventions and master of execution. He came to me in 1776, recommended by Governor Trumbull and other respectable characters, who were converts to his plan. Although I wanted faith myself, I furnished him with money and other aids to carry his plan into execution. He labored for some time ineffectually; and though the advocates for his schemes continued sanguine, he never did succeed.

One accident or another always intervened. I then thought, and still think, that it was an effort of genius, but that too many things were necessary to be combined to expect much from the issue against an enemy who are always upon guard.

That he had a machine so contrived as to carry him under water at any depth he chose, and for a considerable time and distance, with an appendage charged with powder, which he could fasten to a ship and give fire to it in time sufficient for his returning, and by means thereof destroy it, are facts, I believe, which admit of little doubt. But then, where it was to operate against an enemy, it was no easy matter to get a person hardy enough to encounter the variety of dangers to which he would be exposed; first, from the novelty; secondly, from the difficulty of conducting the machine, and governing it under water, on account of the current; and thirdly, from the consequent uncertainty of hitting the object devoted to destruction, without rising frequently above water for fresh observations, which when near the vessel, would expose the adventurer to discovery and to almost certain death. To these causes I always ascribe the failure of his plan, as he wanted nothing that I could furnish to insure the success of it. This, to the best of my recollection, is a true state of the case; but Humphreys, if I mistake not, being one of his converts, will be able to give you a more perfect account of it than I have done.

GLOSSARY

"When I use a word," Humpty Dumpty said in a rather scornful tone, "it means just what I choose it to mean—nothing more nor less."

"You seem very clever at explaining words, Sir," said Alice. "Would you kindly explain the meaning of the poem 'Jabberwocky.'"

"Let's hear it," said Humpty Dumpty.

This sounded very hopeful, so Alice repeated the first verse:

"Twas brillig, and the slithy toves
Did gyre and gimble in the wabe;
All mimsey were the borogroves,
And the mome raths outgrabe."

Humpty Dumpty interrupted: "there are plenty of hard words there. '*Brillig*' means four o'clock in the afternoon—the time when you begin broiling things for dinner."

—Lewis Carrol, 1877

ANEMOSCOPE. (MARTIN)

Benjamin Martin (1747, 82) described a machine that resembled a windmill whereby the torque generated by wind could be measured. "The best Method, that I know of, to bring the Force of the Wind to a Mathematical Calculation and Certainty, is by the following new contrived ANEMOSCOPE, of which I had the first Hint from my ingenious and generous Friend Dr. *Burton* of *Windsor*." What soon became known as the "anemometer" began as several devices, much simpler than Martin's anemoscope, for determining the strength of the wind. The seventeenth-century interest in scientific experimentation led to many mechanical instruments that began life as some type of ". . . scope" or ". . . meter." Robert Boyle had devised his baroscope and barometer in the 1660s, the hygroscope became the hygrometer, and Bailey's dictionary noted that thermoscope and thermometer were synonyms. (See also Lowthorp (1716, 2–46), and Bud and Warner (1998) various pages.)

AXEL-TREE. The pin which passes through the midst of the wheel, on which the circumvolutions of the wheel are performed. (Johnson)

This was the term used by Charles Griswold (appendix E) to describe Bushnell's propeller shaft: "But not the least ingenious part of this curious machine, was that by which the horizontal motion was communicated to it. The object was effected by means of two oars or paddles, formed precisely like the arms of a wind-mill, which revolved perpendicularly upon an axeltree that projected in front; this axeltree passed into the machine at a water joint, and was furnished with a crank, by which it was turned."

BARREL. (Johnson)

1. A round wooden vessel to be stopped close.

2. A barrel of wine is thirty-one gallons and a half; of ale, thirty-two gallons; of beer, thirty-six gallons; and of beer vinegar, thirty-four gallons.

Bailey's definition is essentially the same, but adds confusion regarding the physical size of an eighteenth-century barrel, by noting specifically that "a BARREL of *Essex* Butter, contains 106 lbs," whereas a barrel of "*Sussex* Butter [contains] 256 lb." See below for definitions of other containers: "hogshead," "kilderkin," and "firkin."

BELL-*Metal*, a Mixture of Tin and Copper (Bailey)

Tin, rather than zinc, is the alloying ingredient used to produce what was then, and still is, referred to as bell metal. The use of tin produces an alloy with lower damping qualities than zinc, hence a better material for bells.

BELLS: (Frazar)

Douglas Frazar in his *Practical Boat-Sailing* (1879) describes the use of the ship's bell to specify the half-hour time intervals throughout each four-hour watch. This system would have been familiar to Ezra Lee. By counting the rings of the *Eagle*'s bell, Lee would be able to estimate the time duration of his operation.

> Time at sea is divided differently than on shore; and the day commences at twelve o'clock noon. The reason of this is, that at that time usually at sea, the navigator determines and ascertains the position of the ship, hence the true time; and the clock is corrected from the difference in longitude from noon of the preceding day.

The time of twelve o'clock is denoted by striking the vessel's bell eight times in a particular manner thus: by sets of twos, one, two, rapidly following each other, then a pause of three or four seconds, and then the next set of twos, thus: one, two—one, two—one, two—one, two; whilst seven bells would be struck thus: one, two—one, two—one, two—one; and three bells: one, two—one: four bells: one, two—one, two.

This system of eight strokes of the bell does for the whole twenty-four hours, each stroke denoting one half-hour.: hence eight bells cover a space of four hours, which is termed a watch, and if each watch was four hours long, of course there would be six such watches in the twenty-four hours; and the crew, divided as they always are into starboard and port watches, would, during the whole voyage, have just the same hours on deck. That is to say, the starboard watch would come on deck at twelve o'clock noon every day of the voyage, and stay till four P.M.

This would not be fair to the other watch; and to avoid this repetition, and to divide the time differently each day, the hours from four to eight in the afternoon are divided up into what are called dog-watches of two hours each, which breaks up the daily regularity, and changes the hours; so that the starboard watch who happen to be on deck from twelve to four P.M. one day are the next day below during the same hours, and the port watch on deck; and thus the same watches come round every forty-eight hours.

BLOCKHEAD. A stupid fellow; a dolt; a man without parts. (Johnson) Literacy was an important social responsibility within the eighteenth-century family. While this colorful term carries the same implications today as it did in Bushnell's day, we have added new versions such as knucklehead, bonehead, meathead, and cementhead.

BRASS. A factitious Metal of Copper mix'd with *Lapis Calaminaris*. (Bailey) Bailey defines "Factitious" as "any Thing made by Art in Opposition to the Product of Nature," i.e., creating a metallic composition that is not naturally occurring. *Lapis Caliminaris* (see below) was known to be the ore that when alloyed with copper, produced brass.

CALAMINARIS *Lapis*, the Calamine Stone, which being mixed with Copper, turns it into yellow Brass. (Bailey)

CALAMINE, or *Lapis Calaminaris*. A kind of fossile, bituminous earth, which, being mixed with copper, changes it into brass. (Johnson)
The early mining industry supplied this ore to brass foundries, but had no idea what its composition was. The only fact of interest was that when calamine was added to molten copper, brass was produced. It was not until the early nineteenth century, when James Smithson found that calamine was a term used interchangeably for two minerals, zinc carbonate and zinc silicate. Calamine lotion, a mixture of zinc oxide and iron oxide, has no relationship with calamine ore.
Note: Johnson defines the adjective "FOSSIL[E]" as "that which is dug out of the earth," and when used as a noun, is defined as "many bodies, because we discover them by digging into the bowels of the earth, are called *fossils*."

CAMPANA URINATORIA:
Martin Triewald, a member of the Royal Society, noted in *Philosophical Transactions* (1747, 634) that "Experience has shown me, that no Invention built upon any other Principles than those of the *Campana Urinatoria*, can be of Use in any considerable Depths; or that the Diver, in any other Invention whatever, can be a single Moment safe." Benjamin Martin (1747, 53) also provided descriptions whereby, "the Reader may have a just Idea of the *Campana Urinatoria* or Diving-Bell, according to the latest Improvements by *Dr. Halley* and *Mr. Triewald* of *Stockholm*." [*See also* "URINATOR."]

CENTER (of Gravity) is a Point which, if a Body were suspended, all its parts would be in *Aequilibrio*. (Bailey).
To the eighteenth-century scientist studying windmill technology (or designing a propeller blade), an object's center of gravity is equivalent to the modern term "centroid."

CLOCK. (JOHNSON)
1. The instrument which tells the hour.
2. It is an usual expression to say, *What is it of the clock?* for *What hour is it?* Or, *ten o'clock for the tenth hour.*
It is interesting to note that by the mid-eighteenth century, the contraction "o'clock" for "of the clock" was becoming sufficiently accepted to be included in Johnson's dictionary.

CONJUGATE *Diameter*, is the shortest Axis or Diameter in an Ellipsis or Oval. (Bailey)

When Bushnell used this term when describing his submarine to Thomas Jefferson, his assumption was that mathematical terms such as this would be familiar to the educated readers of his day.

COPPER. One of the six primitive metals. *Copper* is the most ductile and malleable metal, after gold and silver. Of *copper* and lapis calaminaris is formed brass; of *copper* and tin bell metal; *copper* and brass, what the French call bronze, used for figures and statues. (Johnson)

DEAD LIGHTS. A kind of window shutter for the windows in the stern of a ship, used in very bad weather only. (Bailey)

We have used this term to define what Bushnell incorporated into his hatch for daytime illumination and that he termed "glass windows with covers to secure them," and the "three round doors . . . when open they admitted fresh air; their shutters were ground perfectly tight." The term "Bulls Eye" would also apply, particularly for the "windows" at the top of the hatch (and used by Captain Joy in his 1776 description of a submarine, see chapter 4). An 1881 edition of *A Naval Encyclopaedia* (no author, Philadelphia, L. R. Hammersly & Co.) defines Bulls Eye as "a small thick, circular piece of glass inserted in the decks, port lids, etc., for the admission of light."

DEADRISE. The vertical distance from the point of intersection of the top of the keel to the turn of the bilge. (Rogers)

DEAD RISING. The line along the bottom of the interior of a vessel where the *floor-timbers* join the lower futtocks. (Ansted)

DEAD-RISING in shipbuilding, is that part of the ship that lies aft between the keel and her floor-timbers; generally it is applied to those parts of the bottom, throughout the ship's length, where the sweep or curve at the head of the floor-timbers terminates, or intersects to join the keel. (Smyth)

This term reflects the flatness of the bottom of a ship's hull, and varies along its length. When Ezra Lee attempted to position and buoy himself up against the *Eagle*'s hull, he had to be in a flat region, i.e., with the minimal dead-rising amidships. In 1776, the Connecticut Council of Safety specified "that the dead rising for the row-gallies ordered to be built . . . be about 7 inches." This included row-galley *Crane* that was in the Hudson River with the sloop that supported Bushnell's *Turtle*, giving it a very flat bottom contour, which facilitated its maneuverability when operating with oars in shallow water.

DOODLE. A trifler; an idler. (Johnson)
When the British military referred to a New Englander as a "Yankee Doodle," the intent was to characterize the colonial farmer/soldiers as not worthy of serious concern. A similar word "Noodle," which Johnson defined as "a fool; a simpleton" was also used during the eighteenth century.

ELLIPSIS [in *Geometry*] is an Oval Figure, produced from the Section of a Cone, by a Plane cutting both Sides of a Cone (but not parallel to the Base) and which may be described upon a Plane, by a Line made by a loose Cord carried round upon two Centers or Pins. (Bailey)
Bushnell used this term to describe the shape of his submarine hull. He had to balance the need to build a strong hull yet create a shape that was maneuverable and would minimize the effort expended by the operator.

EMERY. A Sort of Stone used in burnishing Metals &c. (Bailey)

EMERY. Emery is an iron ore. It is prepared by grinding in mills. It is useful in cleaning and polishing steel. (Johnson)
In the absence of o-rings used in modern designs to create a watertight seal, Bushnell had to rely on well-fitting surfaces tightly secured. Emery was used to polish the surfaces in contact, such as where the "three round doors" were secured into the brass hatch. Because the pilot relied on these doors to provide fresh air, they would have been opened and closed many times. The mating surfaces had to be precise, or significant leaking would have occurred when submerged. Bushnell may also have greased these surfaces, using some form of animal fat to fill any imperfections.

An ENGINE. Any mechanick instrument composed of Wheels, Screws, &c. in order to raise, cast, or sustain any Weight, &c. (Bailey)

ENGINEER. A Person skilled in Fortification, Building, Attacking, Defending Castles, Forts &c. also in making Fire-works. (Bailey)
In the eighteenth century, "engineer" was essentially a military term. In August 1779, David Bushnell was commissioned captain-lieutenant in

the newly established Corps of Sappers and Miners, a unit assigned to Washington's Corps of Engineers. In 1820 when Charles Griswold related Lee's account of the attack, he noted that the *Turtle* "was of a capacity to contain one engineer." The various contemporary writers also used the terms "operator" and "navigator" to describe Lee's function.

ENTHUSIAST one who fancies himself inspired with the Divine Spirit, and so to have a true Sight and Knowledge of Things. (Bailey)

ENTHUSIAST (Johnson)
1. One who vainly imagines a private revelation; one who has a vain confidence in his intercourse with God.
2. One of a hot imagination, or violent passions.
3. Elevated in fancy; exalted in ideas.
This term was used by Benjamin Gale after describing Bushnell's submarine concept to Silas Deane. When Gale ended his letter: "you may call me a visionary, an enthusiast, or what you please,—I do insist upon it, that I believe the inspiration of the Almighty has given him understanding for this very purpose and design," he was reassuring Deane that the ideas being related were not simply the writings of a "whimsical, fanciful person" or a "fanatick" (see VISIONARIES), but that Bushnell may very well have had divine inspiration.

ERGONOMICS, an applied science concerned with the characteristics of people that need to be considered in designing and arranging things that they use, in order that people and things will interact most effectively and safely—called also *biotechnology, human engineering*. (Gove)
There was no such "applied science" in Bushnell's time. He did, however, accommodate the mechanical systems in the *Turtle* to suit his brother who was to conduct the anticipated combat missions. To expect success, Bushnell's instincts drove him to design his submarine whereby, "people and things will interact most effectively."

EXPERIMENT. Trial of any thing; something done in order to discover an uncertain or unknown effect. (Johnson)
The term was used to describe any activity where the outcome was uncertain, not just in relation to eighteenth-century scientific investigation (i.e., experimental philosophy). Bushnell defines as an "experiment" every phase of his military operations with the *Turtle* and later his floating mines.

FIRKIN. (. . . it being the fourth Part of a Barrel) a Measure containing eight Gallons of Ale, and nine of Beer. (Bailey)

FIRKIN-*Man*, one who buys Small Beer of the Brewer, and sells it again to his Customers. (Bailey), i.e. the beer distributor.

FOXFIRE. The phosphorescent light given forth by decayed or foxed timber. (Whitney)
There are few contemporary references to the term "foxfire." The word, undoubtedly a colloquial term, appears in a memoir by Joseph Dodderidge (1972, 283) chapter 36 titled, "Attack on Rice's Fort," first published in 1824, describing an attack on a small fort in the western frontier during September 1782. "If they had seen any thing like fire, between that and the fort, it must have been fox fire." Dodderidge was referring to claims by several individuals that they had seen fire when seeking refuge in the fort from Indian attacks.

GLASSES: (Middlebrook)
A ship's crew was required to stand "watch" (see Bailey's definition below) at four-hour intervals, twenty-four hours a day. Louis F. Middlebrook, in his *History of Maritime Connecticut during the American Revolution* describes timekeeping onboard ship: "The keeping of time was a somewhat crude ordeal. The half-hour sand-glasses were commonly used, and necessarily required constant and careful vigilance, for turning, as well as the counts and their records. Eight glasses made a watch." The ubiquitous "hour glass" was also available in smaller sizes to indicate half-hour intervals when used onboard ship. Every half hour when the sand ran out, the "sand-glass" would be inverted and the ship's bell rung according to a specific sequence (see Frazar's definition for BELLS above). The number of times the bell was rung told the crew the amount of time that had transpired during that particular watch.

To HEAVE *at the capstan* [*Sea term*] is to turn it about. (Bailey)

To HEAVE *and Set* [*Sea Phrase*] is said of a Ship, when, being at Anchor, she rises and falls by the Force of the Waves. (Bailey)

To HEAVE *at the Top-sails*, is to put them abroad. (Bailey)

HEAVE ABOUT. To go upon the other tack suddenly (Smyth)

Hove—Heave in the past tense, i.e., 'we hove to during the squall.' (Ansted)

Samuel Johnson may not have been a seafaring man, preferring lexicography to oceanography. Unlike Bailey, there are no references in Johnson's dictionary to the many maritime uses of the word "heave" other than the following that may have been describing an uncomfortable experience out on the ocean:

To Heave: to keck; to feel a tendency to vomit. (Johnson)

Hogshead. A Vessel containing of liquids 63 Gallons. (Bailey)

During the eighteenth century, a hogshead was the standard container used for the molasses and rum trade between New England and the West Indies. The wholesale price of the best West Indies rum at the beginning of the Revolution was six shillings per gallon, making a hogshead valued at nearly £19, while locally produced New England rum would have been £11. By 1778, inflation had increased the cost of this essential part of the colonial diet threefold.

Infernal, belonging to Hell, hellish, low, nethermost. (Bailey}

It is not surprising that in a religious eighteenth-century society, the technologies designed for submarine warfare would have been described as infernal [hellish] machines as did British Commodore Symons, after Bushnell attempted to sink the HMS *Cerberus* with a floating mine in 1777.

Jill. A Quarter of a Pint (Bailey) [Bailey also defines the word "Gill" as a quarter of a pint.]

A soldier was allowed a ration of one jill of rum after a day of "fatigue." Four ounces of rum would have relaxed anyone after a tough day on the battlefield. A militia company, typically about sixty men, might consume 240 ounces or nearly two gallons per day (one hogshead per month) during a military campaign—a tall order for their quartermaster.

Kilderkin, a liquid Measure, containing eighteen Gallons, or two Firkins. (Bailey)

A kilderkin approximates the volume that Bushnell would have used to contain the two cubic feet (150 pounds) of compacted gunpowder.

Machine. Any complicated piece of workmanship, An engine. (Bailey)

All of the contemporary writers referred to Bushnell's submarine as a "machine," or more specifically a "machine for the destruction of ships of war."

MAGAZINE a Storehouse for Arms and Ammunition of War. Also now used for the Name of several periodical miscellaneous Pamphlets. (Bailey)

MAGAZINE (Johnson)
1. A storehouse, commonly an arsenal or armory, or repository of provisions.
2. Of late this word has signified a miscellaneous pamphlet, from a periodical miscellany named the *Gentleman's Magazine*, by Edward Cave.
From the French "magazin" or storehouse, this word has been associated with the storage of gunpowder, and as such it is not surprising that Bushnell and his contemporaries used the term to describe the weapon carried on the back of his *Turtle*. It is also understandable why "magazine" was selected to describe the popular eighteenth-century monthly periodicals that were self-described storehouses of information, e.g., *The Universal Museum and Complete Magazine of Knowledge and Pleasure*.

MECHANICAL *Philosophy*, is that which explains the Phenomena or Appearances of Nature by mechanical Principles. (Bailey)

MECHANICK. A Handicraftsman. (Bailey)

MECHANICK *Principals*, are commonly reckoned six, the Balance, the Lever, the Pulley, the Screw, the Wedge and the Wheel. (Bailey)
Certainly, Bushnell was familiar with all of these six "mechanick principals." Had he any questions as to how to apply these principals to his vessel, he could have consulted his team of "mechanicks," including Isaac Doolittle, Phineas Pratt, and his brother Ezra.

A MINE [*in military affairs*] a hole dug to be filled with Barrels of Gunpowder, in order to blow it up. (Bailey)

MINE *Ships*, Ships filled with Gun-powder, inclosed in strong Vaults of Brick or Stone, to be fired in the midst of an Enemy's Fleet. (Bailey)
By the time Bushnell was designing his "magazine," there was certainly precedent for the word "mine" to describe this type of underwater

weapon. In 1626, one hundred fifty years before Bushnell, Cornelius
Drebbel was contracted by King Charles I to construct what was referred
to as a "water-petard" or "water myne." In 1585 Frederico Gianabelli
demonstrated the use of the "mine ship," detonated with a clockwork
mechanism, against the Spanish during their siege of Antwerp. See Alex
Roland (1978) for a thorough discussion of these "infernal machines" of
Gianabelli and Drebbel.

MOCK TURTLE SOUP: (Henderson)
> Scald a calf's head with the skin on, and take off the horny part,
> which must be cut into pieces about two inches square. Let these
> be well washed and cleaned, then dry them in a cloth, and put
> them into a stew-pan, with four quarts of water made as follows:
> Take six or seven pounds of beef, a calf's foot, a shank of ham, an
> onion, two carrots, a turnip, a head of celery, some cloves and
> whole pepper, a bunch of sweet herbs, a little lemon-peel, and a
> few truffles. Put these into eight quarts of water, and let it stew
> gently till the broth is reduced one half; then strain it off, and put
> it into the stew-pan, with the horny parts of the calf's head. Add
> some knotted marjoram, a little savory, thyme, and parsley, all
> chopped small together, with some cloves and mace pounded, a
> little chyan pepper, some green onions, a shallot cut fine, a few
> chopped mushrooms, and half a pint of Madeira wine. Stew all
> these together gently till the soup is reduced to two quarts; then
> heat a little broth, mix some flour smooth in it, with the yolks of
> two eggs, and keep it stirring over a gentle fire till it is near boil-
> ing. Add this to the soup, keeping it stirring as you pour it in, and
> let them all stew together for another hour. When you take it off
> the fire, squeeze in the juice of half a lemon, and half an orange,
> and throw in some broiled forcemeat balls. Pour the whole into
> your tureen, and serve it up hot to table.—This is a rich soup, and
> to most palates deliciously gratifying.

A contemporary recipe left out the calf's head but increased the quantity of feet, and added veal knuckles and oysters. The popularity of this
"deliciously gratifying" soup became manifest in a character from Lewis
Carroll's *Alice in Wonderland* known as "Mock Turtle," a creature composed of the soup's various animal ingredients.

NATURAL *Philosophy*, that Science which contemplates the Powers of Nature, the Properties of natural Bodies, and their mutual Action one upon another. (Bailey)

This was the phrase that until well into the nineteenth century referred to general scientific study. Benjamin Gale, when discoursing on the legitimacy of Bushnell's submarine, asks Franklin: "Give me Leave to Say, it is all Constructed with Great simplicity, and upon Principles of Natural Philosophy . . ." For an eighteenth-century scientist, this would have been a great compliment to bestow on Bushnell and his submarine. To add more insight, this glossary also includes definitions of "Naturalist," "Philosopher," "Philosophy," "Physical," "Physick," and "Physicotheology."

NATURALIST. A student in physicks. (Johnson)

OIL. Any fat, greasy, unctuous, thin matter. (Johnson)

To OIL. To smear or lubricate with oil. (Johnson)

Bushnell, concerned about keeping the penetrations through the hull watertight, ensured that "the joints were also kept full of oil, to prevent rust and leaking."

PETARD. An engine of metal, almost in the shape of a hat, about seven inches deep, and about five inches over at the mouth: when charged with fine powder well beaten, it is covered with a madrier or plank, bound down fast with ropes, running through handles, which are round the rim near the mouth of it: this *petard* is applied to gates or barriers of such places as are designed to be surprised, to blow them up. (Johnson)

According to Bailey, the word "petard" was derived from the French: "*Pet,* a Fart, a farting Engine."

PHILOSOPHER. A man deep in knowledge, either moral or natural. (Johnson)

PHILOSOPHY. The Knowledge of Things Natural and Moral, grounded upon Reason and Experience. (Bailey)

PHILOSOPHY. (Johnson)
1. Knowledge natural or moral.
2. Hypothesis or system upon which natural effects are explained.
3. Reasoning; argumentation.

4. The course of sciences read in the schools.
Benjamin Gale finishes his letter to Franklin with "his reasoning so Philosophically and Answering every Objection I ever made that in truth I have great relyance upon it."

PHOSPHOROUS, a Preparation which is kept in Water, and being taken out and exposed to Air, shines in the Dark, and actually takes Fire of itself. (Bailey)
Bushnell refers to the material he used to illuminate his compass and depth gauge as "phosphorous." Because phosphorous ignites in air, it is most likely that he was referring to the bioluminescent properties of "foxfire," which his mentor Benjamin Gale noted in his letters was used in the *Turtle*.

PHYSICAL. Natural, belonging to Natural Philosophy, or the Art of Physick. (Bailey)

PHYSICK. Is in general the Science of all material Beings, or whatever concerns the System of this visible World; though in a more limited and improper Sense, is applied to the Science of Medicine. (Bailey)

PHYSICOTHEOLOGY. Divinity enforced or illustrated by natural philosophy. (Johnson)
Yale recognized the interest during the seventeenth and eighteenth centuries to demonstrate the relationship between science and theology. Students would have had the opportunity to study both, and may have read William Derham's *Physicotheology* at the college library.

ROMANCE a feigned Story, a mere Fiction. (Bailey)

To ROMANCE to tell a magnificent Lie. (Bailey)

ROMANCER a Teller of Lies or false Stories. (Bailey)

ROMANCE (Johnson)
1. A tale of wild adventures in war and love.
2. A lie; a fiction.

ROMANTICK (Johnson)
1. Resembling the tales of romances; wild.
2. Improbable; false.

3. Fanciful; full of wild scenery.

4. A rude, awkward, boisterous, untaught girl.

5. Rough rude play.

Benjamin Gale ends his lengthy description of the *Turtle* to Benjamin Franklin with: "I ask Ten Thousand Pardons for presuming to Trouble You with this Long Acct which I fear will Appear to You too Romantic to Obtain Beleiff—but have Endeavoured in the Strictest Sense to relate Facts Truly." He hopes Franklin will not think of the submarine concept as being wild, improbable, false, fanciful, and mere fiction, or simply a tale of wild adventure.

Sanguine: (Bailey and Johnson)In relationship to an individual's personality, the word "sanguine" in the eighteenth century was defined as being "brisk, forward, vigorous" (Bailey), and "ardent, confident" (Johnson), as opposed to modern definitions of "cheerful" and "hopeful." When Washington tells Jefferson that "he [Bushnell] labored some time ineffectually; and, though the advocates for his scheme continued sanguine, he never did succeed," he is implying that the advocacy was vigorous and ardent, well beyond simply cheerful and hopeful.

To Sap, to undermine , or dig into. (Bailey)

A Sap, (*Fortification*) a digging at the Foot of a Work to undermine it; a deep Trench cut in order to make a Passage into a Covert Way, &c. (Bailey)

Bushnell received a commission in the Corps of Sappers and Miners in August 1779. It was probably preferable to refer to them as sappers rather than ditch diggers.

Scheme. (Johnson)

1. A plan; a combination of various things into one view, design, or purpose.

2. A project; a contrivance; a design.

3. A representation of the aspects of the celestial bodies; any lineal or mathematical diagram.

George Washington referred to Bushnell's inventions as "schemes" in his letter to Thomas Jefferson. The inference here is simply that Bushnell had a design that was of particular interest, without any implications of it being impractical or devious.

STATESMAN: (Johnson)
1. A politician; one versed in the arts of government.
2. One employed in public affairs.

STATESWOMAN: (Johnson)
A woman who meddles in public affairs.

SUBMARINE lying or acting under the Sea. (Bailey)

SUBMARINE. Lying or acting under the sea. (Johnson)
Both of these eighteenth-century dictionary definitions are included here, in spite of the fact that they are identical. This emphasizes the fact that the word "submarine" was only considered an adjective, as in Bushnell's reference to his "sub-marine vessel." In Johnson's folio edition, he quotes the Reverend John Wilkins (see Lynch, 2004). "This contrivance [referring to Cornelius Drebbel's "Ark for submarine navigation"] may seem difficult, because these submarine navigators will want winds and tides for motion, and the sight of the heavens for direction." The use of the word "submarine" exclusively as an adjective continued into the early twentieth century, as indicated in the following definition from *The Century Dictionary* published by William Whitney in 1903:

SUBMARINE. 1. Situated or living under or in the sea, either at the bottom or below the surface; below the surface of the sea: as, *submarine* plants; a *submarine* telegraph. 2. Occurring or carried on or below the surface: as, *submarine* explorations; designed for use under the sea: as, SUBMARINE BOAT, a boat which is so fitted that it can be propelled when entirely submerged, and carries a sufficient amount of compressed air to admit of remaining below the surface for several hours. The chief object sought is the carrying and operating of torpedoes. (Whitney)
When *The Century Dictionary Supplement* was published by Benjamin Smith in 1909, the term was finally established as a standalone noun, but could still be used as an adjective when it was necessary to be more specific:

SUBMARINE. A vessel designed to navigate either entirely under water or on the surface, with a crew contained within it and with its own motive power; specifically a submarine torpedo-boat. (Smith)
An illustration of a "Submarine Torpedo-boat of the Holland Type" followed the text. John Holland sold his first submarine to the U.S. Navy in 1900.

To SUBMERGE to drown, dip, or plunge under the sea (Bailey)

To SUBMERGE. To drown; to put under water. (Johnson)

TORPEDO. A fish that hath the nature to make the handes of them that touche it to be astonyed [numbed], though he doe it with a long pole. (Cooper)

TORPEDO. A fish which while alive, if touched even with a long stick, benumbs the hand that so touches it, but when dead is eaten safely. (Johnson)

Thomas Cooper's 1578 Latin-English dictionary shows the Latin definition essentially the same as found in Johnson (and other eighteenth-century dictionaries). An alternate name "cramp-fish" was also common at that time. Conrad Gesner illustrated the "Torpedo" fish in 1560 in his *Icones Animalium Aquatilium*, and the name has continued to the present as the taxonomic genus "Torpedo" associated with elasmobranch rays in the family "Torpedinae." It wouldn't be until the beginning of the nineteenth century when "torpedo" became associated with an underwater weapon:

TORPEDO, a machine so called, invented by Mr. Fulton of the United States, and destined to blow up the largest ships. It is an apparatus of which the principal piece is a copper box, inclosing a certain quantity of gunpowder, and prepared with an interior spring which sets fire to the powder, at the same time that the whole is inclosed in a covering of cork, or some other light wood, to make the torpedo float under the surface of the water. It is placed under the keel of the vessel to be destroyed, by means of a harpoon directed against the side of the ship. (Rees)

Robert Fulton had first used the term "torpedo" about 1807 in letters regarding his proposed undersea weapon. In 1810, a United States Senate committee reported their findings regarding Fulton's recently published pamphlet *Torpedo War and Submarine Explosions*. The committee was moderately impressed, commenting that while they "deem it premature to offer any opinion upon the expediency of incorporating it into the naval or military preparations," they were willing to encourage Fulton to carry on experiments. The committee recommended "that a sum ought to be appropriated by government, and experiments actually made under the direction of the secretary of the navy, to enable him [to determine if Congress should] employ the torpedo or submarine explosions for the better defence of the ports and harbors of the United States." (U.S. Senate 1810, 4)

TORTOISE. An animal covered with a hard shell; there are tortoises both of land and water. (Johnson)

TORTOISE. Testudo (Littleton)
Littleton's English–Latin dictionary (1684) indicates the English word "tortoise" from the Latin "testudo" and as such the derivation of the modern biological Order "Testudinata" that includes the marine turtles. Following Conrad Gesner (1560), Linneaus described one of the Green Turtle species as *Testudo mydas* (later changed to *Chelonia mydas*). Thomas Cooper (1578) defined "Testudo" as "a Fishe called a Torteise", also: "The bellie of a lute; A lute which somewhat resembleth a Torteise his shell."

TRAIN BANDS. The militia; that part of a community trained to martial exercise. (Johnson)
The use of citizen soldiers in a trained militia was a tradition in England carried to the colonies. These militia units, or trained bands, were essential to the survival of the settlements during the seventeenth and eighteenth centuries. An early law (see Connecticut *Acts and Laws* (1715, 155–156)) stipulated that "there shall be a Guard appointed [from the Train Band] in every Town and Plantation in this Colony, to attend on the Sabbath, and other days of Public Worship, compleat in their Arms, with Powder and Bullets." At the onset of the Revolutionary War, Connecticut had laws on the books (see Connecticut *Acts and Laws* (1769, 155–156; 1777, 441–442)) specifying a military establishment modeled on English standing armies, yet still manned by militia units. For example, Connecticut's 2nd Brigade consisted of the 2nd, 7th, 10th, and 23rd Regiments. In 1776, Colonel Charles Webb's 7th Regiment included the 3rd Company commanded by Capt. Nathan Hale and included among his sixty-eight enlisted men, Sergeant Ezra Bushnell.

TURTLE: (Johnson)
1. A species of dove.
2. It is used among sailors and gluttons for a tortoise.
It appears that a "turtle" was considered of some culinary importance (see

Mock Turtle Soup). Johnson and Bailey both indicate the etymology of this word as derived from the Latin "turtur." Cooper (1578) confirms this with his definition for "Turtur" as "A birde called a turtill."

Urinator. A diver. (Johnson)

Urination, a diving or swimming under Water. (Bailey)

Vrinator, one that diveth or swimmeth under the water. (Cooper) Derived from the word "Vrinare, To dive or swimme under the water" (Cooper, 1578). The word "Urinor, -aris," is found in Littleton (1684): "to duck under water, and to spring up again; to dive", not to be confused with "Urina, -ae, water of a man or beast."

Ventilator, an Instrument to supply close Places with Air. (Bailey)

Ventilator. An instrument contrived by Dr. *Hale* to supply close places with fresh air. (Johnson)
Bushnell provided air to the interior of the *Turtle* using a hand-operated ventilation system. Our assumption is that it was similar to a centrifugal bellows described in the *Philosophical Transactions*, (Martyn 1747, 270), and the "Hessian rotary sucker and forcer" that Denis Papin used in his submarine (ca. 1695) (see chapter 3).

Visionaries. whimsical, fanciful Persons, Fanaticks. (Bailey)
See also *Enthusiast.*

A Watch [among *Sailors*] is the Space of four Hours; during which time one half of the Ship's Company watch in their Turns, and are relieved by the other half for four Hours more. (Bailey)

Water-Gage, an Instrument to measure the Depth and Quantity of any Water. (Bailey)
Bushnell, and his predecessor Denis Papin, understood the need to monitor the depth that their submarines had descended. Synonyms such as "sea-gage" and "water barometer" were also used in the eighteenth century.

Yankee, origin uncertain . . . [possibly] a var. of *Yenkees* or *Yengees* or *Yaunghees*, a name said to have been given by the Massachusetts Indians to the English colonists, being, it is supposed, an Indian corruption of the E. word *English* or . . . the F. *Anglais* . . . The word is said to have been adopted by the Dutch on the Hudson, who applied it to the people of New England (in 'contempt,' but prob. not more in contempt than any other designation of them). (Whitney)

While spoken by outsiders intending negative connotations, New Englanders considered the term a compliment.

"And what does '*outgrabe*' mean?"

"Well, '*outgribing*' is something between bellowing and whistling" . . . "Who's been repeating all that hard stuff to you?"

"I read it in a book," said Alice.

NOTES

Introduction

1. See Joseph Plumb Martin, *Private Yankee Doodle, Being a Narrative of Some of the Adventures, Dangers and Sufferings of a Revolutionary Soldier* (Boston & Toronto: Little, Brown, 1962); Frederick Wagner, ed., *Submarine Fighter of the American Revolution* (New York: Dodd, Mead, 1963) for details about Bushnell's service in the Corps of Sappers and Miners.

2. George Eleazer Bushnell, comp., *Bushnell Family Geneology—Ancestry and Posterity of Francis Bushnell (1580—1640)* (Nashville, Tenn.: n.p., 1945), p. 189—91.

3. In 1788, David Humphreys sent the manuscript for Putnam's biography to the Society of Cincinnati, a Connecticut organization of veteran Revolutionary War officers, where both Humphreys and Bushnell were members. See Col. David Humphreys, *An Essay on the Life of the Honorable Major-General Israel Putnam* (Middletown, CT: Moses H. Woodward, 1794), p. iv.

4. From Henry L. Abbot, *Beginning of Modern Submarine Warfare under Captain-Lieutenant David Bushnell, Sappers and miners, Army of the Revolution* (Battalion Press 1881; facsimile reprint, ed. Frank Anderson, Hamden, CT: Archon Books, 1966), p.

5. (See appendix F for the correspondence between Washington and Jefferson.)

5. Wagner, *Submarine Fighter*, p. 116. Wagner quotes Fulton in a letter to Lord Grenville, that Bushnell "certainly did not compose his machines so as to make them of any use, nor did he organize anything like a system . . ."

6. Jack Lynch, ed., *Samuel Johnson's Dictionary, Selections from the 1755 Work that Defined the English Language,* abridged ed. (Delray Beach, FL: Levenger Press, 2004). Original editions of Johnson's folio are rare and difficult to access.

Chapter One

1. Conrad Gesner, *Icones Animalium Aquatilium* (Christopher Frosch, 1560), pp. 182–84; Gesner, *Historiae Animalium Liber Primus De Quadrupedibus Viviparis* (Frankfort: Henry Laurent, 1620), [bound with] *Historiae Animalium Liber II Qui Est De Quadrupedibus Oviparis* (Frankfort: Henry Laurent, 1617), pp. 107–18.

2. Humphreys, *Essay on the Life*, pp. 111–16.

3. David Brewster, ed., *The Edinburgh Encyclopaedia, Conducted by David Brewster, L.L.D. F.R.S., with the Assistance of Gentlemen Eminent in Science and Literature; the First American Edition* (Philadelphia: Joseph and Edward Parker, 1832), vol. 7, pp. 621—24, and plate CCXXXI [231].

4. The Naval Torpedo Station had established a press to train seaman printers and publish the numerous pamphlets, including Barber's, that were used for the torpedo school. The press operated until 1903, when the staff transferred to Washington.

5. Barber's association with Holland is in Richard K. Morris, *John P. Holland, 1841–1914, Inventor of the Modern Submarine* (Annapolis: United States Naval Institute, 1966), pp. 72–73.

6. Maurice Delpeuch was an experienced submariner, having taken command of the *Korrigan* in 1902. His knowledge of the history of submarine warfare, as recorded in his book *La Navigation Sous-Marine á travers les Siécles* (Paris: Félix Juven, [1902?]) was highly recognized by his contemporaries. He died January 19, 1905.

7. Alan H. Burgoyne, *Submarine Navigation Past and Present* (New York: E. P. Dutton, 1903), vol. 1, pp. 12—13.

8. *Regulations Governing the Uniform of Commissioned Officers, Warrant Officers, and Enlisted Men of the Navy of the United States* (Washington, DC: Government Printing Office, 1886), p. 20. The description reads: "*For Petty Officers First Class* (Plates I and X); Dark navy-blue cloth, double breasted sack pattern, rolling collar, front and back of skirt to descend to top of inseam of trousers, lined with dark blue flannel or black Italian cloth. A pocket in the left breast and one in each front near bottom. Five gilt buttons, seven-tenths of an inch in diameter, on each breast, equally spaced. Coat to be worn with the four lower buttons buttoned." The *Regulations* also included an illustration of the jacket alone, buttoned left over right as in Barber's drawing.

9. Burgoyne, *Submarine Navigation*, vol. 1, p. 12.

10. F. Forest and H. Noalhat, *La Bateaux Sous-Marine Historique* (Paris: Dunlop, 1900), pp. 13—14.

11. Delpeuch, *La Navigation*, pp. 56—57.

12. G.-L. Pesce, *La Navigation Sou-Marine* (Paris: Vuibert & Nony, 1906), p. 157.

13. Farnham Bishop, *The Story of the Submarine* (New York: The Century, 1916), pp. 16—17. Bishop made the assumption that the *Turtle* propeller was made of wood, comparing them to the wooden propellers (tractors) used on early airplanes.

CHAPTER TWO

1. *The Gentleman's Magazine and Historical Chronicle* (London: Edward Cave) 19 (1749): 249, 312, 411–13. The twelve monthly issues of *Gentleman's Magazine*, paginated consecutively, were typically bound into one volume. Subsequent articles related to the one referenced here appeared in the July issue (p. 312) and September issue (pp. 411—13).

2. John Wilkins (1614—1672) was a scientist, educator, and member of the clergy of the Church of England during a turbulent period. He had a close association with British aristocracy and was a tutor to many of their children. He began publishing various scientific discourses in 1638, speculating that the moon may be able to support life, and that there would come a time when humans would devise a method of

traveling there. Wilkins's relationship with the royalists came to an abrupt end with the beheading of King Charles I in 1649 (a year after his *Mathematical Magic* was first published). He immediately moved his sentiments to the new government and consummated his relation with Cromwell and the parliamentarians by marrying Oliver's sister. Shortly after the Restoration in 1660, he returned to being a friend of the Crown, and was one of the founders and first secretary of the Royal Society of London.

3. John Wilkins, *Mathematical Magick: or the Wonders that may be Performed by Mechanical Geometry,* 4th ed. (London: Ric. Baldwin, 1691), pp. 178—90.

4. Ben Jonson, *The Staple of News,* act 3, scene 2.

5. Benjamin Jonson, *The Works of Ben. Johnson* [sic], 6 vols. (London: For J. Walthoe et al., 1714), vol. 4, p. 194.

6. Francis Bacon, *Silva Sylvarum, or a Natural Philosophy in Ten Centuries,* [bound with] *New Atlantis,* 9th ed. (London: Printed by J. R. for William Lee, 1670), p. 27.

7. Thomas Sprat, *The History of the Royal-Society of London for the Improving of Natural Knowledge,* 2nd ed. (London: for Rob. Scot, et al.), p. 33.

8. Ibid., p. 321.

9. Ibid., p. 35.

10. Ibid., pp. 378—79.

11. Ibid., p. 393.

12. Ibid., p. 248—49.

13. Ibid., pp. 219—20.

14. Excerpt from an eighteenth-century edition of the *New England Primer,* reprinted in Paul Leicester Ford, The *New England Primer, A History of its Origin and Development* (New York: Dodd, Mead, 1897), p. 47.

15. Connecticut, *Acts and Laws of His Majesties Colony of Connecticut in New-England* (New-London: Timothy Green, 1715), p. 16 (see also p. 110).

16. Connecticut, *Acts and Laws of His Majesty's English Colony of Connecticut in New-England in America, 1750* (New-London: Timothy Green, 1769), p. 212 (hereafter cite as *Conn. Acts and Laws 1750*).

17. Samuel Webster, *A Sermon Preached Before the Honorable Council, and the Honorable House of Representatives of the Colony of Massachusetts-Bay in New-England, at Boston, May 28, 1777 Being the Anniversary for the Elections of the Honorable Council, By Samuel Webster, A. M. Pastor of a Church in Salisbury* (Boston: Edes & Gill, 1777), p. 29.

18. James Dana, *A Sermon, Preached before the General Assembly of the State of Connecticut on the Day of the Anniversary Election, May 13, 1779, by James Dana, D. D. Pastor of the First Church of Wallingford* (Hartford: Hudson and Goodwin, 1779), p. 39.

19. Marion Hepburn Grant, *The Infernal Machines of Saybrook's David Bushnell* (Old Saybrook, CT: Bicentennial Committee of Old Saybrook, 1976), pp. 7—8.

20. Charles J. Hoadley, *The Public Records of the Colony of Connecticut, From May, 1717, to October, 1725* (Hartford: Case, Lockwood & Brainard, 1872), vol. 6, pp. 91—97.

21. Edward Atwater, *History of the City of New Haven to the Present Time* (New York: W. W. Munsell, 1887), pp. 164—67.

22. Louis Leonard Tucker, *Connecticut's Seminary of Sedition: Yale College* (Chester, Connecticut: Pequot Press, 1974), pp. 26, 52-53.

23. Ibid., p. 29.

Chapter Three

1. Charles J. Hoadley, *The Public Records of the State of Connecticut, From October, 1776, to February, 1778* (Hartford: Case, Lockwood & Brainard, 1894), vol. 1, p. 178.

2. Rollin G. Osterweis, *Three Centuries of New Haven, 1638—1938* (New Haven: Yale Univ. Press, 1953), pp. 130-31. See also Henry P. Johnston, *Yale and Her Honor-Role in the American Revolution, 1775-1783* (New York: G. P. Putnam's Sons, 1888), p. 13.

3. Henry Howe, *Memoirs of the Most Eminent American Mechanics* (New York: J.C. Derby, 1856), p. 137.

4. George Dudley Seymour, *Documentary Life of Nathan Hale* (New Haven: privately printed for the author, 1941), pp. 151—53.

5. *The Annual Register or a View of the History, Politics and Literature for the Year 1774,* 2nd ed. (London: for J. Dodsley, 1778), pp. 245–48.

6. Alex Roland, *Underwater Warfare in the Age of Sail* (Bloomington & London: Indiana Univ. Press, 1978), p. 24.

Chapter Four

1. Ralph Waldo Emerson, 1837.

2. The Fitch and Stiles diary quotes are reprinted from Tucker, *Connecticut's Seminary of Sedition*, pp. 46–48.

3. Marion Hepburn Grant, *Infernal Machines*, pp. 9—10.

4. *The Universal Magazine of Knowledge and Pleasure* 56 (1775): 276-77.

5. Ibid., pp. 331-32.

6. Stanza from the state song of Connecticut.

7. Ruth Mack Wilson, *Connecticut's Music in the Revolutionary Era* (Hartford: The American Revolution Bicentennial Commission of Connecticut, 1979), pp. 92-94.

8. *The Town and Country Magazine or Universal Repository of Knowledge, Instruction, and Entertainment* 4 (1772): 242–43.

9. Thomas Paine, *Collected Writings* (New York: Literary Classics of the United States, 1995), p. 6.

10. Ibid., p. 29.

11. W. E. Woodward, *Tom Paine: America's Godfather 1737—1809* (New York: E. P. Dutton, 1945; reprint ed., Westport, Conn.: Greenwood Press, 1972), p. 66.

12. Ibid., p. 91.

13. Samuel West, *A Sermon Preached Before the Honorable Council, and the Honorable House of Representatives of the Colony of Massachusetts-Bay in New-England, May 29, 1776 Being the Anniversary for the Elections of the Honorable Council for the Colony by Samuel West, A. M. Pastor of a Church in Dartmouth* (Boston: John Gill, 1776). The quotes are from pages 14, 54, 59, and 66. The "tyranny" theme occurred throughout the sermon.

14. Webster, *Sermon Preached*, p. 26.

15. Penrose R. Hoopes, *Connecticut Clockmakers of the Eighteenth Century* (Rutland, Vermont: Charles E. Tuttle, 1975), p. 71.

16. Ibid., p. 71.

17. Charles J. Hoadley, *The Public Records of the Colony of Connecticut, From May, 1775, to June, 1776* (Hartford: Case, Lockwood & Brainard, 1890), vol. 15, pp. 440, 488, 497, 507, and 508.

18. Rev. F. W. Chapman, *The Pratt Family: or the Descendants of Lieut. William Pratt, One of the First Settlers of Hartford and Say-Brook* (Hartford: Case, Lockwood, 1864), p. 352.

19. Hoopes, *Connecticut Clockmakers*, p. 105.

20. Phineas Pratt Jr., typescript letter to the editor, *Boston Journal of Chemistry*, June 1870.

21. Phineas Pratt Jr., typescript document, ca. 1870, "Such of the History of Deacon Phineas Pratt Son of Mr. Aziriah and Mrs. Agnes Pratt of Saybrook, Conn."

22. The transcription of the Latin and its translation are from Abbot, *Modern Submarine Warfare*, p. 171 (p. 9 of the facsimile edition). The use of italics for the Latin is from the current authors.

23. Gardner W. Allen, *A Naval History of the American Revolution* (Boston and New York: Houghton Mifflin, 1913), p. 153.

24. Louis F. Middlebrook, *History of Maritime Connecticut during the American Revolution 1775—1783* (Salem, Massachusetts: The Essex Institute, 1925), p. 160.

25. From the Pennsylvania Archives quoted in Lincoln Diamant, *DIVE! The Story of David Bushnell and His Remarkable 1776 Submarine (and Torpedo)* (Fleischmanns, New York: Purple Mountain Press, 2003), p. 22.

CHAPTER FIVE

1. Hoadley, *Public Records of the Colony*, vol. 15, p.39.

2. Ibid., pp. 233—34.

3. Ibid, p. 235.

4. Connecticut, *Acts and Laws, Made and passed by the General Court or Assembly of the State of Connecticut, in New-England, in America; holden at Hartford, by special Order of the Governor, on the 19th Day of November, A. Dom. 1776* (New-London: Timothy Green, 1776), pp. 437-38.

5. Connecticut, *Acts and Laws, Made and passed by the General Court or Assembly of the Governor and Company of the State of Connecticut, in America; holden at Hartford, (by Adjournment) on the twelfth Day of February, Anno Domini, 1778* (New-London: Timothy Green, 1778), pp. 485–87 (hereafter cited as *Conn. Acts and Laws 1778*).

6. Hoadley, *Public Records of the Colony,* vol. 15, pp. 296—97.

7. Ibid., pp. 15—16.

8. See M. L. Brown, *Firearms in Colonial America* (Washington: Smithsonian Institution Press, 1980), pp. 299—305, for a discussion of gunpowder in the colonies.

9. Sprat, "The History of the Making of Salt-Peter by Mr. Henshaw"; and "The History of Making Gun-Powder," in *History of the Royal-Society*, pp. 260–83.

10. Hoadley, *Public Records of the Colony,* vol. 15, pp. 213, 459.

11. Ibid., p. 459.

12. The information on copper, brass, and bronze in the colonies comes from J. Leander Bishop, *History of American Manufactures* (Philadelphia: Edward Young, 1868; facsimile edition by Johnson Reprint Corp., 1967); William G. Lathrop, *The Brass Industry in Connecticut* (New Haven: Price, Lee and Adams, 1909); and primarily James A. Mulholland, *A History of Metals in Colonial America* (University, Alabama: The Univ. of Alabama Press, 1981).

13. Richard H. Phelps, *Newgate of Connecticut: Its Origin and Early History* (Hartford: American Publishing, 1890), p. 53.

14. Bishop, *History of American Manufactures*, p. 572.

15. Adam Ward Rome, *Connecticut's Cannon, The Salisbury Furnace in the American Revolution* (Hartford: The American Revolution Bicentennial Commission of Connecticut, 1977), pp. 24, 30—31.

16. From Mulholland, *History of Metals*, p. 130; and William James Morgan, ed., *Naval Documents of the American Revolution* (Washington, D.C.: U.S. Government Printing Office, 1970), vol. 5, pp. 710-12.

17. *Conn. Acts and Laws 1778*, p. 486.

18. Brown, *Firearms in Colonial America*, p. 243.

19. See Louis I. Kuslan, *Connecticut Science, Technology, and Medicine in the Era of the American Revolution* (Hartford: The American Revolution Bicentennial Commission of Connecticut, 1978), pp. 32 and 58, for these references to Eliot and Gale.

20. The various glassmaking processes in the eighteenth century are described and illustrated in *The Universal Magazine of Knowledge and Pleasure*, 1747, pp. 149-153 and 284-285.

21. Kenneth M. Wilson, *New England Glass and Glassmaking* (New York: Thomas Y. Crowell Company, 1972), pp. 41—51.

22. The information about the excavations at Germantown and the linen smoother is from Richmond Morcom, "The First Germantown Glassworks, 1746-1755," *Antique Bottles and Glass Collectors Magazine*, http://www.glswrk-auction.com.

23. *Universal Magazine* 1 (1747): 119—20.

24. N. Bailey, *An Universal Etymological English Dictionary* (London: for J. Buckland et al., 1773) notes that the word "line" as a unit of measure is from the French "ligne" meaning the "twelfth part of an inch."

25. Robert Bud and Deborah Jean Warner, eds., *Instruments of Science, An Historical Encyclopedia* (New York & London: Garland Publishing, 1998) p. 219.

26. Arlene Palmer, *The Wistars and their Glass* (Millville, New Jersey: Wheaton Historical Assoc., 1989), p. 20.

27. Page Talbot, ed., *Benjamin Franklin In Search of a Better World* (New Haven: Yale Univ. Press, 2004), pp. 182-85, includes an illustration of this tube sent by Collinson to Franklin in 1747.

CHAPTER SIX

1. Conrad Gesner, *Icones Animalium Aquatilium,* pp. 175–81.

2. Tom Gidwitz, "*Turtle* Dives Again," *Archaeology,* May/June 2005, pp. 36—41. Additional information about the Handshouse Studio *Turtle* replica can be found at their Web site, http://www.handshouse.org.

CHAPTER SEVEN

1. Wilkins, *Mathematical Magick,* p. 182.

2. William Emerson, *The Principals of Mechanics Explaining and Demonstrating the General Laws of Motion,* 3rd ed. (London: for G. Robinson, 1773), p. 200.

3. Robert Hooke, *Posthumous Works* (London: Richard Waller, 1705), pp. 561–62.

4. From Robert Macfarlane, "First Steam Screw Boats," in *History of Propellers and Steam Navigation* (New York: George P. Putnam, 1851); and Francis B. Stevens, "The First Steam Screw Boats to Navigate the Waters of Any Country," *Stevens Indicator* 10, no. 2 (April 1893).

5. All quotes regarding the Anemoscope are from B. Martin, *Philosophia Britannica,* (Reading, England: C. Micklewright, 1747), vol. 2, pp. 82—87.

6. Stevens, "The First Steam Screw Boats," p. 115.

7. H. W. Dickinson, *Robert Fulton Engineer and Artist His Life and Works* (London: John Lane, 1913), p. 295.

8. Emerson, *The Principals of Mechanics,* p 212.

9. The blades for this propeller were cast by Sharon Hertzler at the Mystic River Foundry, Mystic, Connecticut. Randy Hale of Hale Propellers in Old Saybrook, Connecticut, modified a bronze propeller hub and mounted the two triangular blades.

CHAPTER EIGHT

1. James P. Walsh, *Connecticut Industry and the American Revolution* (Hartford: The American Revolution Bicentennial Commission of Connecticut, 1978), p. 54.

2. Wilkins, *Mathematical Magick, p.* 182.

3. The ballast for the *Turtle* was originally used on the USS *Nautilus* (SSN 571) during a scientific mission in the 1970s when author Roy Manstan was on board for the Mobile Acoustic Communication System (MACS) testing. Some of this ballast is visible in Figure 12.6.

4. Emerson, *Principals of Mechanics*, p. 195. The term "clack" used by Emerson as a synonym for "valve" is not found in contemporary dictionaries. The word is likely a vernacular term derived from the metallic sound produced when the brass valves open and close.

CHAPTER NINE

1. Hoopes, *Connecticut Clockmakers*, p. 71.

2. *The Universal Magazine of Knowledge and Pleasure* 13 (1753): 129.

3. Wilkins, *Mathematical Magick*, pp. 183-85.

4. Martyn, *Philosophical Transactions (From the Year 1732, to the Year 1744) Abridged* 8 (1747): 270–74.

CHAPTER TEN

1. Lunar predictions for September 1776 were obtained from the U.S. Naval Observatory Web site, http://aa.usno.navy.mil.

2. John Lowthorp, ed., *The Philosophical Transactions and Collections to the End of the Year 1700 Abridged* 2:2–16 (London: for Robert Knaplock, et. al, 1716).

3. Ibid., p. 201.

4. Ibid., p. 202.

5. Martin, *Philosophia Britannica*, pp. 23—25.

6. Lowthorp, *Philosophical Transactions* 2: 201-2.

7. Brewster, *Edinburgh Encyclopaedia*, vol. 7, p. 629.

8. Bacon, *Silva Sylvarum*, pp. 76—77.

9. Robert Boyle, "Observations about the Resemblances and Differences between Coal and shining Wood," in *The Philosophical Transactions and Collections to the End of the Year 1700 Abridged* 3: 646–49, 5th ed. (London 1749).

10. The information regarding the biology of foxfire and *Armillaria mellea* comes from Elio Schaechter's, *"In the Company of Mushrooms,"* and Dr. Kim D. Coder's, "FOXFIRE: Bioluminescence in the Forest," University of Georgia School of Forest Resources publication FOR99-21, as provided in their Web site, http://www.forestry.uga.edu/efr.

CHAPTER ELEVEN

1. Robert Fulton, *Torpedo War, and Submarine Explosions*, facsimile reproduction [New York: William Elliot, 1810; ed. Herman Henkle (Chicago, The Swallow Press, 1971)], pp. 5-6.

2. From the United States Navy Bureau of Ordnance, "Instructions for Care and Preparation of Ammunition," 1874, p. 3, the proportions were approximately 75% potassium nitrate, 15% charcoal, and 10% sulfur.

3. Middlebrook, *History of Maritime Connecticut*, pp. 201—3.

4. Roland, *Underwater Warfare*, pp. 23—26.

5. Richard Elton, *The Compleat Body of the Art Military* (London: for W.L., 1668), p. 254. See also Sprat, *History of the Royal-Society*, pp. 281—83, for a discussion of the separation of artillery powder from the finer grained musket and pistol powder.

6. *Gentleman's Magazine* 17 (1747): 581.

7. Lowthorp, *Philosophical Transactions* 2: 250.

8. We found in Lieutenant Commander J. S. Barnes's *Submarine Warfare, Offensive and Defensive* (New York: Van Nostrand, 1869), pp. 199—205, several references to the size of post Civil War mines or "torpedoes" and the quantity of artillery grade gunpowder used. Based on estimates of the interior volumes of Barnes's torpedoes, a range of densities was calculated to be from fifty-seven to sixty-one pounds per cubic foot. We also found references to the density of artillery powder in the United States Navy Bureau of Ordnance "Instructions for Care and Preparation of Ammunition," 1874, p. 40. Based on the description of copper containers used for storing two-hundred pounds of gunpowder onboard ships, the density of artillery powder would be approximately fifty-five to sixty pounds per cubic foot.

9. Fulton, *Torpedo War*, p. 7.

10. For information about Confederate torpedoes, see Jack Bell, *Civil War Heavy Explosive Ordnance* (Denton, Texas: Univ. of North Texas Press, 2003); and Michael P. Kochan and John C. Wideman, *Torpedoes, Another Look at the Infernal Machines of the Civil War* (2004). An excellent contemporary account is in Barnes, *Submarine Warfare*.

11. According to Kochan and Wideman, *Torpedoes*, a clockwork mine was carried onto the Union supply depot at City Point, Virginia, in July 1864 by Confederate spy, John Maxwell. The explosion killed more than fifty-five. One of these timing mechanisms is on display at the City Point National Military Park. The device was described and illustrated by Lieutenant G. A. Converse in a pamphlet titled "Notes on Torpedo Fuses," published at the U.S. Torpedo Station, Newport, Rhode Island, in 1875.

12. Moore, "The First Submarine Ever Built in America," pp. 22-23, 117-18.

CHAPTER TWELVE

1. Kenneth C. Barnaby, *Basic Naval Architecture* (London: Hutchinson, 1967), pp. 292—98.

CHAPTER THIRTEEN

1. Stevens, "The First Steam Screw Boats," p. 115.

2. Thomas C. Gillmer and Bruce Johnson, *Introduction to Naval Architecture* (Annapolis, Maryland: Naval Institute Press, 1982), p. 233.

3. David Gerr, *Propeller Handbook* (Camden, Maine: International Marine, 2001), p. 60.

4. After five minutes with propeller (2), the test director suspended the test after noticing that the pilot was rapidly tiring. The pilot was provided with a fifteen-minute rest period while propeller (3) was installed, and the subsequent static thrust measurements were completed as scheduled with the pilot reporting no ill effects.

CHAPTER FOURTEEN

1. H. W. Dickinson, *Robert Fulton Engineer and Artist His Life and Works* (London: John Lane, 1913), pp. 296-308.

2. John P. Comstock, ed., *Principles of Naval Architecture* (New York: The Society of Naval Architects and Marine Engineers, 1967), p. 288.

CHAPTER FIFTEEN

1. As reprinted in Kuslan, *Connecticut Science*, p. 17.

2. The dispute was between John Ericsson and John B. Emerson, early propeller designers. In 1847, the case was initially decided by the Circuit Court in favor of Emerson, but on appeal, the Supreme Court reversed the decision in favor of Ericsson.

3. Stevens, "The First Steam Screw Boats," pp. 101—30.

CHAPTER SIXTEEN

1. The underwater cables and connectors for the communications buoy were donated to the *Turtle* project by Mark Warren, Electronic Sales of New England.

2. Morgan, *Naval Documents*, p. 736.

3. See Roland, *Underwater Warfare*, p. 81, also p. 198n18, noting his source as communication from N. E. Upham, Department of Ships, National Maritime Museum, London. Howard I. Chapelle, *The History of the American Sailing Navy, The Ships and Their Development* (New York: Bonanza Books, 1949), p. 21, notes that copper sheathing was introduced into the British navy after 1761.

CHAPTER SEVENTEEN

1. Henry P. Johnston, *Record of Service of Connecticut Men in the Military and Naval Service During the War of the Revolution 1775—1783* (Hartford: Case Lockwood & Brainard, 1889), pp. 79—80.

2. Ibid., pp. 99—100. Colonel Parsons was promoted to Brigadier General on August 9, 1776 probably just days before Bushnell approached him about the need for volunteers.

3. Authors note: Because the *Turtle* was held in a stationary position during the endurance run, there was no requirement to control rudder position. As the pilot indicated, he used the crank and treadle to establish his operating rhythm. In a sim-

ilar fashion, the test pilot for the transit runs also combined the hand crank and foot treadle, yet was able to maintain steerage with his free hand on the tiller.

4. The analysis of the *Turtle* test data drew primarily from Per-Olaf Åstrand and Kaare Rodahl, "Physical Performance," "Evaluation of Physical Performance on the Basis of Tests," and "Applied Work Physiology" in *Textbook of Work Physiology, Physiological Bases of Exercise* (New York: McGraw-Hill, 1986).

5. Åstrand and Rodahl, *Textbook of Work Physiology,* pp. 296—297; see also p. 369 for heart rate efficiency data related to training.

6. Ibid., pp. 299—300.

7. Ibid., p 502.

8. Ibid., p. 308.

9. Ibid., p. 491.

10. Ibid., p 302.

11. Ibid., p 308.

12. Ibid., p 307.

Chapter Eighteen

1. Personal communication with Paul Johnston, Maritime Historian, 2008.

2. The *Connecticut Courant and Hartford Weekly Intelligencer* began publication in 1764 and continues today as the *Hartford Courant.*

3. *Conn. Acts and Laws 1750,* p. 159.

4. Johnston, *Record of Service,* pp. 99–100, 145–46.

5. Henry P. Johnston (1889, 229) notes that in March 1778, George Washington ordered an increase in his Life Guards specifying they be "men of five feet eight to five feet ten; age from twenty to thirty, and men of established character for sobriety and fidelity. They must be American born."

6. See Per-Olaf Åstrand and Kaare Rodahl, "The Muscle and its Contraction," in *Textbook of Work Physiology.*

7. See Per-Olaf Åstrand and Kaare Rodahl, "Neuromuscular Function," in *Textbook of Work Physiology,* p 115–17.

8. Ibid., p. 122.

Chapter Nineteen

1. Everett Tomlinson, *David Bushnell and his American Turtle* (New York, Akron & Chicago: The Werner, 1899). Book is not paginated, but the quote is found on the first page of part 2.

2. Tomlinson, *David Bushnell and his American Turtle,* whose name does not appear on the book's title page, is attributed as the author by Frank Anderson in the bibliography for his 1966 facsimile reprint of Abbot's 1881 book, *Beginning of Modern Submarine Warfare.* Tomlinson was an author who wrote several books about the Revolutionary War for young readers, including *Washington's Young Aides* (1897), *A Jersey Boy in the Revolution* (1899), and *In the Hands of the Redcoats* (1900). The copy

of Tomlinson's *Bushnell* book used in the *Turtle* Project is inscribed on the front fly-leaf: *A Merry Christmas to Clay From Jen*. Sure sounds like a writer of "once upon a time" books.

3. Wagner, *Submarine Fighter*, pp. 29 and 126n.

4. From J. W. Adams, "Map of the Mouth of the Connecticut River and Saybrook Harbor," engr. Capt. W. H. Smith UST (Washington: Hood, 1838). (Courtesy of Connecticut River Museum, Essex, Connecticut.)

5. The copy of the chart studied is from Lawrence C. Wroth, *Abel Buell of Connecticut, Silversmith Typefounder & Engraver* (Middletown: Wesleyan Univ. Press, 1958), p. 57. According to Wroth, it is thought that Buell engraved Parker's chart in 1774, having submitted his bill in October of that year. The only known copy is in the collections of the Connecticut Historical Society. This is the same Abner Parker that Phineas Pratt had apprenticed with in learning the ship carpentry business.

6. "Mouth of the Connecticut River," F. R. Hassler & A. D. Bache, Superintendents of the Survey of the Coast of the United States, 1853. (Courtesy of Connecticut River Museum, Essex, Connecticut.)

7. From "Saybrook During the Revolutionary War, Transcripts from the Journal of Samuel Tully Farmer and Town Clerk of Saybrook, July 1775 to May 1777" (Old Saybrook Historical Society, June 2007). The author was able to view the original diary in the historical society archives, and is grateful for their assistance. There are several subsequent Tully diaries, all of which have been transcribed.

8. Johnston, *Yale and Her Honor Role*, pp. 317—18, and Seymour, *Documentary Life of Nathan Hale*, p. 138.

9. George G. Sill, *Geneology of the Descendants of John Sill, Who Settled in Cambridge, Mass., in 1637* (Albany: Munsell & Rowland, 1859), p. 63. George G. Sill "had gathered in the course of years . . . well authenticated facts from individuals in the Sill connection, with whom I have been acquainted within the last fifty years." During this time Richard Sill's son, Elisha, and grandson, Charles, lived at the Ayer's Point farm and likely provided the Bushnell story to George Sill. We give this genealogy enough credibility to speculate that after graduating from Yale, Bushnell split his time between the family farm with his brother, Ezra, and the Sill house where he "perfected" his submarine in the Connecticut River.

10. Gilman C. Gates, *Saybrook at the mouth of the Connecticut: the first one hundred years* (Orange, Conn., New Haven, Conn.: Press of the Wilson H. Lee Co., 1935), pp. 238—39.

11. The history of the Sill house is from Grant, *Infernal Machines*, p. 10.

12. Pratt's original home is long gone, but was located about four miles from Bushnell's home along what is now Connecticut Route 156, near the exit ramp from Route 9.

13. Note that the original tavern sign is in the Connecticut River Museum, Essex, Connecticut.

14. The map studied is in *Lyme Connecticut: From Founding to Independence*, 1976, by Bruce P. Stark, p. 42. According to Stark, the original is in the Stiles Papers, Beinecke Library, Yale University.

CHAPTER TWENTY

1. Charles S. Hall, *Life and Letters of General Samuel Holden Parsons* (Binghamton, N.Y.: Otseningo, 1905), p. 60.

2. Abbot, *Modern Submarine Warfare*, p. 4.

3. Chapelle, *History of American Sailing Ships*, pp. 19-28.

4. Information about Continental galleys, including the illustration of the *Washington*, is from Chapelle, *History of American Sailing Ships*, pp. 71-75. See also Chapelle, *The History of the American Sailing Navy, The Ships and Their Development* (New York: Bonanza Books, 1949), pp. 106 and 557.

5. Hoadley, *Public Records of the Colony*, vol. 15, p. 222.

6. Ibid., p.227.

7. Ibid, p. 230.

8. Ibid., pp. 250, 454, 468.

9. Ibid., p. 407.

10. Ibid., pp. 456 and 458.

11. Ibid., p.473.

12. Ibid., p. 481.

13. Major General Heath later wrote about the naval engagements in the Hudson River noting that, in addition to the galleys, the British had "sunk a sloop, which had on board the machine, invented by, and under the direction of, a Mr. Bushnell, intended to blow up the British ships. This machine was worked under water. It conveyed a magazine of powder, which was to be fixed under the keel of a ship, then freed from the machine, and left with clock work going, which was to produce fire when the machine had got out of the way. Mr. Bushnell had great confidence of its success, and had made several experiments which seemed to give him confidence; but its fate was truly a contrast to its design [i.e. it sank]." William James Morgan, ed., *Naval Documents of the American Revolution* (Washington, D.C.: U.S. Government Printing Office, 1972), vol. 6, p. 1181–85.

CHAPTER TWENTY-ONE

1. Humphreys, *Essay of the Life*, p. 116.

2. Johnston, *Record of Service*, p. 80.

3. David McCullough, *1776* (New York: Simon & Schuster, 2005) p. 31.

4. Abbot, *Modern Submarine Warfare*, p. 5.

5. Johnston, *Record of Service*, p. 121-22.

6. From *The New-York Commercial Advertiser*, November 15, 1821, as reprinted on the Web site, http://www.archive.org.

7. Chapman, *Pratt Family*, p. 352.

8. Ibid., p. 353.

9. Pratt, typescript letter to the editor, *Boston Journal of Chemistry*, June 1870. This document is almost certainly an early transcript of a handwritten letter. Typewriters were not readily available until Remington Arms Co. began producing them in quantity in 1874.

10. Martin, *Private Yankee Doodle*, p. 283.

11. From copy of Ezra Lee manuscript, "Petition for a Gov't Stipend, District of Connecticut, 26th Day of June 1820," available at the Submarine Force Museum and Library, Groton, Connecticut.

Chapter Twenty-Two

1. Wagner, *Submarine Fighter*, p. 52.

2. McCullough, *1776*, p. 148.

3. See, e.g., Wagner, *Submarine Fighter*; Frank Anderson's commentary in the 1966 facsimile edition of Abbot, *Modern Submarine Warfare*; Roland, *Underwater Warfare*; and Grant, *Infernal Machines*.

4. Morgan, *Naval Documents*, vol. 6, pp. 724—25 and 736—37.

Chapter Twenty-Three

1. The illustration of HMS *Agamemnon* was provided by Paul Dobson from a series of views on his Web site, http://www.prdobson.com. The *Eagle* and *Agamemnon* were nearly identical in size, both with a length of about 160 ft., beam of 44 ft., and a depth of hold of 19 ft. Both held crews of five hundred officers and men, and carried sixty-four guns with twenty-six 24-pounders on the gun deck. The *Agamemnon* entered the Revolutionary War after Yorktown, being involved in action against the French fleet under Comte de Grasse off the island of Dominica in 1782.

2. From the Ships Plans Archive at the Mystic Seaport Museum for an unnamed eighty-gun warship built in 1765 (archive number 14.18), only slightly larger than HMS *Eagle*.

3. From notes, presumably by Humphreys, appended to Lee's letter (see appendix D).

4. Babcock's letter is quoted in Rodman Gilder, *The Battery* (Boston: Houghton Mifflin, 1936), p. 92—93.

Epilogue

1. Hoadley, *Public Records of the State*, vol. 1, p. 212.

2. Ibid., p. 212.

3. Chapman, *Pratt Family*, p. 352.

4. Symons's complete report to Rear Admiral Parker, including a detailed description of the "infernal" device is in Abbot, *Modern Submarine Warfare*, pp. 30—31. Wagner,

Submarine Fighter, pp. 79—82, describes Bushnell's floating mine and the attack on *Cerberus*.

5. In 1998, the Naval Undersea Warfare Center and the Naval Historical Center signed a Memorandum of Understanding (MOU) enabling NUWC to assist the Rhode Island Marine Archaeology Project (RIMAP) with investigations into historic naval shipwreck sites within Rhode Island waters. The NUWC dive team then joined with RIMAP in the survey of the remains of *Cerberus*, locating and documenting several cannons and remnants of the hull. The divers also participated in the search for Captain James Cook's ship *Endeavor*, which was among several vessels scuttled off Newport to restrict the French fleet from advancing within cannon range of the town.

6. David Osgood, *A Sermon Preached on the Day of Annual and National Thanksgiving December 11, 1783 by David Osgood, A. M., Pastor of the Church in Medford* (Boston: T. and S. Fleet, 1784), p. 12.

7. Charles J. Hoadley, *The Public Records of the State of Connecticut, From May, 1778, to April, 1780* (Hartford: Case, Lockwood & Brainard, 1895), vol. 2, pp. 289—90.

8. Abbot, *Modern Submarine Warfare*, p. 6, quoting from the Journals of Congress.

9. Johnston, *Record of Service*, p. 298.

10. References to Martin's *Private Yankee Doodle* are from a 1962 reprint that provides the diary in its entirety.

11. Martin, *Private Yankee Doodle*, p. 263.

12. William B. Wilcox, ed., *The Papers of Benjamin Franklin* (New Haven: Yale Univ. Press, 1982), vol. 22, p. 185. Franklin, concerned about America's vulnerability to British naval power, was interested in any ideas that could be used to even the playing field, including Silas Deane's mechanical genius, David Bushnell.

BIBLIOGRAPHY

Abbot, Henry L. *Beginning of Modern Submarine Warfare under Captain-Lieutenant David Bushnell, Sappers and Miners, Army of the Revolution.* Battalion Press, 1881. A facsimile reprint by Frank Anderson. Hamden, CT: Archon Books, 1966. Citations are to the Archon edition.

Allen, Gardner W. *A Naval History of the American Revolution.* 2 vols. Boston and New York: Houghton Mifflin, 1913.

Ansted, A. *A Dictionary of Sea Terms.* Glasgow: Brown, Son & Ferguson, 1933.

Åstrand, Per-Olof and Kaare Rodahl. *Textbook of Work Physiology: Physiological Bases of Exercise.* New York: McGraw-Hill, 1986.

Atwater, Edward. *History of the City of New Haven to the Present Time.* New York: W. W. Munsell, 1887.

Bacon, Francis. *Silva Sylvarum, or a Natural Philosophy in Ten Centuries,* [bound with] *New Atlantis.* 9th ed. London: Printed by J. R. for William Lee, 1670. Paginated separately.

Bailey, N. *An Universal Etymological English Dictionary.* London: for J. Buckland et al., 1773.

Barber, Lieutenant Francis M. *Lecture on Submarine Boats and their Application to Torpedo Operations.* Newport, Rhode Island: U.S. Torpedo Station, 1875.

Barnaby, Kenneth C. *Basic Naval Architecture.* London: Hutchinson, 1967.

Barnes, J.S., Lieut. Commander, USN. *Submarine Warfare, Offensive and Defensive.* New York: Van Nostrand, 1869.

Bell, Jack. *Civil War Heavy Explosive Ordnance.* Denton, Texas: Univ. of North Texas Press, 2003.

Bishop, Farnham. *The Story of the Submarine.* New York: The Century, 1916.

Bishop, J. Leander. *A History of American Manufactures from 1608 to 1860.* 3 vols. Philadelphia: Edward Young, 1868. Facsimile edition by Johnson Reprint Corp., 1967.

Brewster, David, ed. *The Edinburgh Encyclopaedia, Conducted by David Brewster, L.L.D. F.R.S., with the Assistance of Gentlemen Eminent in Science and Literature; the First American Edition.* 18 vols. Philadelphia: Joseph and Edward Parker, 1832.

Brown, M. L. *Firearms in Colonial America.* Washington: Smithsonian Institution Press, 1980.

Bud, Robert and Deborah Jean Warner, eds. *Instruments of Science, An Historical Encyclopedia.* New York & London: Garland Publishing, 1998.

Burgoyne, Alan H. *Submarine Navigation Past and Present.* 2 vols. New York: E. P. Dutton, 1903.

Bushnell, David. "General Principals and Construction of a Sub-marine Vessel." *Transactions of the American Philosophical Society* 4. Philadelphia, 1799.

Bushnell, George Eleazer, comp. *Bushnell Family Geneology—Ancestry and Posterity of Francis Bushnell (1580—1640) of Horsham, England and Guilford, Connecticut.* Nashville, Tennessee: n.p., 1945.

Carroll, Lewis. *Through the Looking Glass, and what Alice Found There.* Illustrated by John Tenniel. London and New York: MacMillan, 1877.

Catalog of Books in the Library of Yale-College New-Haven. New London: T. & S. Green, 1791.

Chapelle, Howard I. *The History of American Sailing Ships.* New York: Bonanza Books, 1935.

———. *The History of the American Sailing Navy, The Ships and Their Development.* New York: Bonanza Books, 1949.

———. *The National Watercraft Collection.* Washington, D.C.: Smithsonian Institution Press, 1976.

Chapman, Rev. F. W. *The Pratt Family: or the Descendants of Lieut. William Pratt, One of the First Settlers of Hartford and Say-Brook.* Hartford: Case, Lockwood, 1864.

Clap, Thomas. *A Catalog of the Library of Yale-College in New Haven.* New London: T. Green, 1743.

Clark, William B., ed. *Naval Documents of the American Revolution.* Vol. 1. Washington, D.C.: U.S. Government Printing Office, 1964.

Collins, A.Frederick and Virgil D. Collins. *The Boys' Book of Submarines.* New York: Frederick A. Stokes, 1917.

Comstock, John P., ed. *Principles of Naval Architecture.* New York: The Society of Naval Architects and Marine Engineers, 1967.

Connecticut. *Acts and Laws of His Majesties Colony of Connecticut in New-England.* New-London: Timothy Green, 1715.

———. *Acts and Laws of His Majesty's English Colony of Connecticut in New-England in America, 1750. Revision of 1769.* New-London: Timothy Green, 1769.

———. *Acts and Laws, Made and passed by the General Court or Assembly of the State of Connecticut, in New-England, in America; holden at Hartford, by special Order of the Governor, on the 19th Day of November, A. Dom. 1776.* New-London: Timothy Green, 1776. [Acts passed at an individual session of the General Assembly.]

———. *Acts and Laws, Made and passed by the General Court or Assembly of the State of Connecticut, in New-England, in America; holden at Middletown, by Adjournment, on the 18ᵗʰ Day of December, Anno Domini, 1776.* New-London: Timothy Green, 1777. [Acts passed at an individual session of the General Assembly.]

———. *Acts and Laws, Made and passed by the General Court or Assembly of the Governor and Company of the State of Connecticut, in America; holden at Hartford, (by Adjournment) on the twelfth Day of February, Anno Domini, 1778.* New-London: Timothy Green, 1778. [Acts passed at an individual session of the General Assembly.]

Cooper, Thomas. *Thesaurus Linguae Romanae et Britannicae.* London: 1578.

Delpeuch, Maurice. *La Navigation Sous-Marine á travers les Siécles.* Paris: Félix Juven, [1902?]. [Contemporary author Herbert C. Fyfe (1907) attributes the publishing date as 1902.]

Department of the Navy, United States Naval History Division. *The American Revolution, 1775—1783; An Atlas of 18ᵗʰ Century Maps and Charts: Theaters of Operations.* Washington, D.C.: U. S. Government Printing Office, 1972.

Diamant, Lincoln. *Chaining the Hudson, The Fight for the River in the American Revolution.* New York: Citadel Press, 1994.

———. *DIVE! The Story of David Bushnell and His Remarkable 1776 Submarine (and Torpedo).* Fleischmanns, New York: Purple Mountain Press, 2003.

Dickinson, H. W. *Robert Fulton Engineer and Artist His Life and Works.* London: John Lane, 1913.

Dodderidge, Joseph. *Notes on the Settlement and Indian Wars of the Western Parts of Virginia and Pennsylvania from 1763 to 1783 Inclusive with a Memoir of the Author by his Daughter.* Reprint. Edited by Alfred Williams. New York: Lennox Hill Pub. And Dist. Co., 1972.

Dodson, John E., ed. *Dodson's Coast and Harbor Charts Calais to Sandy Hook.* Stonington, CT: Dodson's Nautical Charts, 1961.

Elton, Richard. *The Compleat Body of the Art Military.* London: for W.L., 1668.

Emerson, William. *The Principals of Mechanics Explaining and Demonstrating the General Laws of Motion.* 3rd ed. London: for G. Robinson, 1773.

Field, Cyril, Lieutenant Colonel. *The Story of the Submarine from the Earliest Ages to the Present Day.* London: Sampson Low, Marston, 1908.

Ford, Paul Leicester. *The New England Primer, A History of its Origin and Development.* New York: Dodd, Mead, 1897.

Forest, F. and H. Noalhat. *Les Bateaux Sous-Marins Historique.* Edited by Ch. Dunod. Paris: n.p., 1900.

Frazar, Douglas. *Practical Boat-Sailing*. New York: Charles T. Dillingham, 1879.

Fulton, Robert. *Torpedo War, and Submarine Explosions*. New York: William Elliot, 1810. Facsimile reproduction. Edited by Herman Henkle. Chicago: The Swallow Press, 1971.

Fyfe, Herbert C. *Submarine Warfare Past and Present*. 2nd ed. New York: E.P. Dutton, [1902] 1907.

Gaget, M. *La Navigation Sous-Marine Généralités et Historique*. Paris: Librairie Polytechnique, 1901.

Gates, Gilman C. *Saybrook at the mouth of the Connecticut: the first one hundred years*. Orange, Conn., New Haven, Conn.: Press of the Wilson H. Lee Co., 1935.

Gerr, David. *Propeller Handbook*. Camden, Maine: International Marine (A Division of The McGraw-Hill Companies), 2001.

Gesner, Conrad. *Historiae Animalium Liber Primus De Quadrupedibus Viviparis*. Frankfort: Henry Laurent, 1620, [bound with] *Historiae Animalium Liber II Qui Est De Quadrupedibus Oviparis*. Frankfort: Henry Laurent, 1617.

———. *Icones Animalium Aquatilium*. Christopher Frosch, 1560.

Gilder, Rodman. *The Battery*. Boston: Houghton Mifflin, 1936.

Gillmer, Thomas C. and Bruce Johnson. *Introduction to Naval Architecture*. Annapolis, Maryland: Naval Institute Press, 1982.

Gove, Philip Babcock, ed. *Webster's Third New International Dictionary of the English Language Unabridged*. Springfield, MA: Merriam-Webster, 1993.

Grant, Marion Hepburn. *The Infernal Machines of Saybrook's David Bushnell*. Old Saybrook, CT: Bicentennial Committee of Old Saybrook, 1976.

Hall, Charles S. *Life and Letters of General Samuel Holden Parsons*. Binghamton, N.Y.: Otseningo, 1905.

Harris, Brayton. *The Navy Times Book of Submarines*. Edited by Walter Boyne. New York: Berkely Books, 2001.

Henderson, W. A. *The Housekeeper's Instructor, or, Universal Family Cook*. London: J. Stratford, n.d., ca. 1804. (Mock Turtle Soup recipe on pages 22–23.)

Hennebert, Lt.-Colonel. *Les Torpilles*. Paris: Librairie Hachette Et Cie., 1888.

Hoadley, Charles J. *The Public Records of the Colony of Connecticut, From May, 1717, to October, 1725*. Vol. 6. Hartford: Case, Lockwood & Brainard, 1872.

———. *The Public Records of the Colony of Connecticut, From May, 1775, to June, 1776*. Vol. 15. Hartford: Case, Lockwood & Brainard, 1890.

————. *The Public Records of the State of Connecticut, From October, 1776, to February, 1778*. Vol. 1. Hartford: Case, Lockwood & Brainard, 1894.

————. *The Public Records of the State of Connecticut, From May, 1778, to April, 1780*. Vol. 2. Hartford: Case, Lockwood & Brainard, 1895.

Hooke, Robert. *Posthumous Works*. London: Richard Waller, 1705. [Facsimile ed. by Johnson Reprint Corp., 1969.]

Hoopes, Penrose R. *Connecticut Clockmakers of the Eighteenth Century*. Rutland, Vermont: Charles E. Tuttle, 1975.

Hovgaard, G. W. *Submarine Boats*. London: E. and F.N. Spon, 1887.

Howe, Henry. *Memoirs of the Most Eminent American Mechanics*. New York: J.C. Derby, 1856.

Humphreys, Col. David. *An Essay on the Life of the Honorable Major-General Israel Putnam*. Middletown, CT: Moses H. Woodward, [1788] 1794.

Johnson, Samuel. *A Dictionary of the English Language*. London: for A. Millar, et al., 1766.

Johnston, Henry P. *Record of Service of Connecticut Men in the Military and Naval Service During the War of the Revolution 1775—1783*. Hartford: Case Lockwood & Brainard, 1889.

————. *Yale and Her Honor-Role in the American Revolution, 1775—1783*. New York: G. P. Putnam's Sons, 1888.

Jonson, Benjamin. *The Works of Ben. Johnson* [sic]. 6 vols. London: for J. Walthoe et al., 1716. [A six volume set was at Yale; Vol. 4 contained "The Staple of News" that referenced Cornelius Drebbel's submarine as quoted in this book.]

Kochan, Michael P. and John C. Wideman. *Torpedoes, Another Look at the Infernal Machines of the Civil War*. Np.:n.p., 2004.

Kuslan, Louis I. *Connecticut Science, Technology, and Medicine in the Era of the American Revolution*. Hartford: The American Revolution Bicentennial Commission of Connecticut, 1978.

Lathrop, William G. *The Brass Industry in Connecticut*. New Haven: Price, Lee and Adams, 1909.

Laubeuf, M. *Sous-Marines et Submersibles*. Paris: Librairie Delegrave, 1918.

Lefkowitz, Arthur S. *Bushnell's Submarine: The Best Kept Secret of the American Revolution*. New York: Scholastic, 2006.

Littleton, Adam. *Linguae Latinae*. London: T. Basset, et al., 1684.

Lowthorp, John, ed. *Philosophical Transactions and Collections to the End of the Year 1700 Abridged . . . 2*. London: for Robert Knaplock, et al, 1716.

————, ed. *Philosophical Transactions and Collections to the End of the Year 1700 Abridged . . . 3*. 5th ed. London: for W. Innys, et al, 1749.

Lynch, Jack, ed. *Samuel Johnson's Dictionary, Selections from the 1755 Work that Defined the English Language.* Delray Beach, FL: Levenger Press, 2004.

Macfarlane, R. *History of Propellers and Steam Navigation.* New York: George P. Putnam, 1851.

Marshall, Douglas W. and Howard H. Peckham. *Campaigns of the American Revolution: An Atlas of Manuscript Maps.* Ann Arbor, MI: Univ. of Michigan Press, 1976.

Martin, B. *Philosophia Britannica.* Vol. 2. Reading, England: C. Micklewright, 1747.

Martin, Joseph Plumb. *Private Yankee Doodle, Being a Narrative of Some of the Adventures, Dangers and Sufferings of a Revolutionary Soldier.* Edited by George F. Scheer. Boston & Toronto: Little, Brown, 1962.

Martyn, John, ed. *Philosophical Transactions (From the Year 1732, to the Year 1744) Abridged.* Vols. 8—9. London: for W. Innys, et al, 1747.

McCullough, David. *1776.* New York: Simon & Schuster, 2005.

Middlebrook, Louis F. *History of Maritime Connecticut during the American Revolution 1775—1783.* 2 vols. Salem, Massachusetts: The Essex Institute, 1925.

Monturiol, Narciso. *Ensayo Sobre El Arte De Navegar Por Debajo Del Agua.* Barcelona: n.p., 1891.

Morgan, William James, ed. *Naval Documents of the American Revolution.* Vol. 5. Washington, D.C.: U.S. Government Printing Office, 1970.

———, ed. *Naval Documents of the American Revolution.* Vol. 6. Washington, D.C.: U.S. Government Printing Office, 1972.

Morris, Richard K. *John P. Holland, 1841—1914, Inventor of the Modern Submarine.* Annapolis: United States Naval Institute, 1966.

Moxon, Joseph, *Mechanick Exercises of the Doctrine of Handy-Works.* London: Midwinter & Leigh, 1703. [Facsimile reprint by Early American Industries Association, Scardsdale, NY, 1979.]

Mulholland, James A. *A History of Metals in Colonial America.* University, Alabama: The Univ. of Alabama Press, 1981.

Officer of the Army. *The History of the Civil War in AMERICA.* London: T. Payne and Son, 1780. (Authorship uncertain, but possibly attributed to William Cornwallis Hall.)

Osterweis, Rollin G. *Three Centuries of New Haven, 1638—1938.* New Haven: Yale Univ. Press, 1953.

Paine, Thomas. *Collected Writings.* New York: Literary Classics of the United States, 1995. (The Library of America Series, selections and notes by Eric Foner.)

Palmer, Arlene. *The Wistars and their Glass.* Millville, New Jersey: Wheaton Historical Assoc., 1989.

Parsons, Wm. Barclay. *Robert Fulton and the Submarine*. New York: Columbia Univ. Press, 1922.

Pesce, G.-L. *La Navigation Sou-Marine*. Paris: Vuibert & Nony, 1906.

Phelps, Richard H. *Newgate of Connecticut: Its Origin and Early History*. Hartford: American Publishing, 1890.

Rees, Abraham. *The Cyclopedia; or Universal Dictionary of Arts Science and Literature in Forty-One Volumes*. Philadelphia: Samuel F. Bradford and Murray, Fairman, n.d., ca. 1820.

Regulations Governing the Uniform of Commissioned Officers, Warrant Officers, and Enlisted Men of the Navy of the United States. Washington, D.C.: Government Printing Office, 1886.

Rogers, John G. *Origins of Sea Terms*. Boston: Nimrod Press, 1985.

Roland, Alex. *Underwater Warfare in the Age of Sail*. Bloomington & London: Indiana Univ. Press, 1978.

Rome, Adam Ward. *Connecticut's Cannon, The Salisbury Furnace in the American Revolution*. Hartford: The American Revolution Bicentennial Commission of Connecticut, 1977.

Schaechter, Elio. *In the Company of Mushrooms*. Cambridge, Massachusetts: Harvard Univ. Press, 1997.

Seymour, George Dudley. *Documentary Life of Nathan Hale*. New Haven: privately printed for the author, 1941.

Sill, George G. *Geneology of the Descendants of John Sill, Who Settled in Cambridge, Mass., in 1637*. Albany: Munsell & Rowland, 1859.

Sleeman, C. L. *Torpedoes and Torpedo Warfare: Containing a Complete and Concise Account of the Rise and Progress of Submarine Warfare*. Portsmouth [UK]: Griffin, 1880.

Smith, Benjamin E., ed. *The Century Dictionary Supplement*. New York: The Century, 1909.

Smyth, Admiral W. H. *Sailors Word-Book, An Alphabetical Digest of Nautical Terms*. London: Blackie and Son, 1867.

Sprat, Thomas. *The History of the Royal-Society of London for the Improving of Natural Knowledge*. 2nd ed. London: for Rob. Scot, et al, 1702.

Sutcliffe, Alice Crary. *Robert Fulton and the "Cleremont."* New York: The Century, 1909.

Talbot, Page, ed. *Benjamin Franklin In Search of a Better World*. New Haven: Yale Univ. Press, 2004.

Terry, Marian Dickinson, ed. *Old Inns of Connecticut*. Hartford: The Prospect Press, 1937.

Thatcher, James, M.D. *A Military Journal During The American Revolutionary War, From 1775 to 1783, Describing Interesting Events and Transactions of this Period, with Numerous facts and Anecdotes, from the Original Manuscript*. Boston: Richardson and Lord, 1823.

Tomlinson, Everett. *David Bushnell and his American Turtle*. New York, Akron & Chicago: The Werner, 1899.

Trumbull, J. Hammond, ed. *Collections of the Connecticut Historical Society*. Hartford: Connecticut Historical Society, 1870.

Tucker, Louis Leonard. *Connecticut's Seminary of Sedition: Yale College*. Chester, Connecticut: Pequot Press, 1974.

Tully, Samuel. *Saybrook During the Revolutionary War, Transcripts from the Journal of Samuel Tully Farmer and Town Clerk of Saybrook, July 1775 to May 1777*. Old Saybrook, Connecticut: Old Saybrook Historical Society, June 2007.

United States Navy Bureau of Ordnance. *Instructions for Care and Preparation of Ammunition*. 1874.

Van Erman, Eduard. *The United States in Old Maps and Prints*. Wilmington, Delaware: Atomium Books, 1990.

Verne, Jules. *Twenty Thousand Leagues Under the Sea*. London: Sampson Low, Marston, Searle, & Rivington, 1876.

Wagner, Frederick. *Submarine Fighter of the American Revolution*. New York: Dodd, Mead & Company, 1963.

Walsh, James P. *Connecticut Industry and the American Revolution*. Hartford: The American Revolution Bicentennial Commission of Connecticut, 1978.

Whitney, William Dwight, ed. *The Century Dictionary, an Encyclopedic Lexicon of the English Language*. New York: The Century, 1903.

Wilcox, William B., ed. *The Papers of Benjamin Franklin*. New Haven: Yale Univ. Press, 1982.

Wilkins, John. *Mathematical Magick: or the Wonders that may be Performed by Mechanical Geometry*. 4th ed . London: Ric. Baldwin, 1691.

———. *The Mathematical and Philosophical Works of the Right Reverend John Wilkins, Late Lord Bishop of Chester*. London: for J. Nicholson et al., 1708.

Wilson, Kenneth M. *New England Glass and Glassmaking*. New York: Thomas Y. Crowell Company, 1972.

Wilson, Ruth Mack. *Connecticut's Music in the Revolutionary Era*. Hartford: The American Revolution Bicentennial Commission of Connecticut, 1979.

Woodward, W. E. *Tom Paine: America's Godfather 1737—1809*. New York, E.P. Dutton & Co., 1945. A reprint of the first edition. Westport, Connecticut: Greenwood Press, 1972.

Wroth, Lawrence C. *Abel Buell of Connecticut, Silversmith Typefounder & Engraver*. Middletown: Wesleyan Univ. Press, 1958.

PERIODICALS, PAMPHLETS, AND MANUSCRIPTS

The American Journal of Science and Arts 2, no. 2 (November 1820): 94—100. Edited by Benjamin Silliman.

Annual Register or a View of the History, Politics and Literature for the Year 1774. 2nd ed. London: for J. Dodsley, 1778.

The Boston Journal of Chemistry 4, nos. 9–10, (March 1 and April 1, 1870).

The Connecticut Courant, and Hartford Weekly Intelligencer, no. 641 (May 5, 1777).

Connecticut Gazette; and the Universal Intelligencer 14, no. 708 (June 6, 1777); no. 712 (July 4, 1777).

Dana, James, D. D. *A Sermon, Preached before the General Assembly of the State of Connecticut on the Day of the Anniversary Election, May 13, 1779, by James Dana, D. D. Pastor of the First Church of Wallingford.* Hartford: Hudson and Goodwin, 1779.

The Gentleman's Magazine and Historical Chronicle 17. London: Edward Cave, 1747.

The Gentleman's Magazine and Historical Chronicle 19. London: Edward Cave, 1749.

Lee, Ezra. Copy of manuscript letter to David Humphreys, February 20, 1815. On display at the Thomas Lee House, East Lyme (CT) Historical Society; included here as appendix D.

———. Copy of manuscript, "Petition for a Gov't Stipend, District of Connecticut, 26th Day of June 1820," Submarine Force Museum, Groton, Connecticut.

Liebknecht, William and Damon Tvaryanas. *Archaeological Investigations at The Wistarburgh Glassworks Site, Alloway Township, Salem County, New Jersey.* Prepared for the Wheaton Museum of American Glass, Millville, New Jersey: Hunter Research, 1999.

Liebknecht, William, et al. *2001 Archaeological Investigations at The Wistarburgh Glassworks Site [28SA134], Alloway Township, Salem County, New Jersey.* Prepared for the Wheaton Museum of American Glass, Millville, New Jersey: Hunter Research, 2004.

Moore, H. Edward. "The First Submarine Ever Built in America," *Mechanics and Handicraft*, Autumn 1934.

Morris, Richard K. "An Effort of Genius, A Tale for the American Bicentennial." *Trinity Reporter* 6, no. 6 (May 1976).

Osgood, David, A. M. *A Sermon Preached on the Day of Annual and National Thanksgiving December 11, 1783 by David Osgood, A. M., Pastor of the Church in Medford.* Boston: T. and S. Fleet, 1784.

Pratt, Phineas Jr. Typescript letter to the editor, *Boston Journal of Chemistry*. Sent from Deep River, CT, June 1870.

————. Typescript document, "Such of the History of Deacon Phineas Pratt Son of Mr. Aziriah and Mrs. Agnes Pratt of Saybrook, Conn." Undated (believed ca. 1870).

Robbins, Peggy. "Bushnell's Remarkable Turtle." *Military History*, October 1989.

Stevens, Francis B. "The First Steam Screw Boats to Navigate the Waters of Any Country." *Stevens Indicator* 10, no. 2 *(*April 1893).

Thomson, David Whittet. "David Bushnell and the First American Submarine." *United States Naval Institute Proceedings* 68, no. 468. Annapolis: Naval Institute Press, 1942.

The Town and Country Magazine or Universal Repository of Knowledge, Instruction, and Entertainment 4:242–243. London: for A. Hamilton, Jr., 1772.

Tully, Samuel. Manuscript diary written during the Revolutionary War, Old Saybrook [CT] Historical Society. (The original was viewed when researching this book, although a transcription has been published and is available through the historical society.)

United States Senate. *Report of the Committee, to whom was Referred by the Senate, A Resolution to Inquire into the Expediency of Employing the Torpedo or Submarine Explosions, February 26th, 1810.* Washington City: Roger C. Weightman, 1810.

The Universal Magazine of Knowledge and Pleasure 1. London: For John Hinton, 1747.

The Universal Magazine of Knowledge and Pleasure 13. London: For John Hinton, 1753.

The Universal Magazine of Knowledge and Pleasure 56. London: For John Hinton, 1775.

Webster, Samuel. A. M. *A Sermon Preached Before the Honorable Council, and the Honorable House of Representatives of the Colony of Massachusetts-Bay in New-England, at Boston, May 28, 1777 Being the Anniversary for the Elections of the Honorable Council, By Samuel Webster, A. M. Pastor of a Church in Salisbury.* Boston: Edes & Gill, 1777.

West, Samuel, A. M. *A Sermon Preached Before the Honorable Council, and the Honorable House of Representatives of the Colony of Massachusetts-Bay in New-England, May 29, 1776 Being the Anniversary for the Elections of the Honorable Council for the Colony by Samuel West, A. M. Pastor of a Church in Dartmouth.* Boston: John Gill, 1776.

INDEX

Abbot, Henry L., xvi, 15, 215
Adams, John, 56
Agamemnon, 259
Alligator, 40–41
American Philosophical Society, 24
American Turtle. see *Turtle*; *Turtle*
  replicas
anchor, 103–104
Anderson, Frank, 215
Anemoscope, 89–90, 89*fig*, 311
Archimedian screw, 86, 87–88, 152.
  see also propellers
Arnold, Benedict, 33
Asia, 57, 248
Åstrand, Per-Olof, 200–203, 211–212
attachment mechanism, 136–138,
  139*fig*, 140*fig*, 187–191, 189*fig*,
  260–262
Atwater, Edward, 28
Atwater, Jeremiah, 53
axel-tree, 312
Ayer's Point, 42*map*, 218–219,
  220–223, 220*fig*

Babcock, Oliver, 266
Bacon, Francis, 21–23, 125–126
Bailey, Nathan, xviii
Baldwin, Abraham, 34
ballast
  attachment mechanism and, 261
  buoyancy and, 100, 101
  emergency, 103–104
  keel, 102–103
  lead, 63–64, 104–105
  Papin and, 40
  stability and, 102
  water, 105–108, 106*fig*, 107*fig*,
    186–187, 190–191, 259
Barber, Francis M.
  Bishop and, 14
  Burgoyne and, 10*fig*
  Field and, 13
  Fyfe and, 10
  hatch design and, 110

illustration by, 8*fig*
interpretation of *Turtle* by, xvi, 7,
  9, 88, 103
Barlow, Joel, 34
Barnaby, Kenneth C., 154
Barnes, J.S., xvi, 15, 133
barometer, 121–122, 311
Battery
  Lee's return to, 262–263
  map of, 247*map*
  tides and, 249, 251–253, 264
  transportation to, 226, 246, 249
Battle of Long Island, 255
bayonets, source of steel, 66, 67*fig*
Beatrice, Bonnie, 80
Beatrice, Ken, 80, 81, 82*fig*, 216, 217,
  221
Bedlow, Isaac, 251
Bedlow's Island, 247*map*, 248,
  251–252, 258–259, 264
Beers Tavern, xvii, 32*fig*, 33–34
bell metal, 312, 315
bells/time keeping, 312–313, 318
bilge, 105–108, 106*fig*, 107*fig*. *see also*
  ballast, water
Bishop, Farnham, 14
Bishop, J. Leander, 67
Borelli, Giovanni, 17–18, 18*fig*, 86
Bourne, William, 17
Boyle, Robert
  barometer and, 67, 121–122, 311
  depth gauge and, 124*fig*
  on foxfire, 126
  Papin and, 37
  on water depth and pressure,
    118*fig*
Boyle's Law, 122
brass, 64–65, 313–314, 315
bronze, 64–65, 315
Brooklyn Heights, 243
Brooks, David, 230
Brown, Laura, 82, 83*fig*
Brown, Rick, 82, 83*fig*
Buckingham, David, 28

Buell, Abel, 342n5
buoyancy
    attachment mechanism and,
        190–191, 261–262
    ballast and, 73, 100, 102, 106
    control of, 257
    design issues and, 43
    mines and, 134, 136
    neutral, 40–41, 184–187, 186*fig*
    salinity and, 101–102, 258–259
Burgoyne, Alan, 9, 10*fig*, 12–13, 14
Bushnell, David
    career after *Turtle*, 269–275
    Council of Safety and, 59–60
    creation of *Turtle* and, xiii–xiv, 20,
        43–46, 55–56
    description of, xv
    description of *Turtle* by, 279–288
    Doolittle and, 52–53
    education and, 26–27, 28, 30–31
    Hale and, xvi, 34, 268
    Jefferson and, xiv–xv, 277–278
    in Long Island Sound, 217-218
    Papin and, 37–41
    as pilot, 233–234
    postcard regarding, 271*fig*
    in prisoner exchange, 272-273
    Pratt and, 54–55
    propeller innovation and, 174
    Stiles and, 276–277
    terminology of, 3
    testing of *Turtle* and, 217
    woodcut of, xiii*fig*
Bushnell, Dency, 27
Bushnell, Ezra
    education and, 26–27
    ergonomics and, 206–208
    Hale and, 194, 234, 327
    illness of, 225
    Lee and, 34
    as pilot, 44, 162, 234–235
    Pratt and, 54–55
    replacement of, 193
    training of, 211, 221*fig*, 222–223
Bushnell, Lydia, 27
Bushnell, Sarah, 27
Byers, James, 65

calaminaris/calamine, 64, 313–314.
    see also Lapis Calaminaris
Calves Island, 220*fig*, 221–222, 223
Campana Urinatoria. *see* diving bells
Carrol, Lewis, 311, 321
centrifugal bellows, 116, 117*fig*
Cerberus, 270, 319, 345n5
Chapelle, Howard I., 175, 227, 228
Chapman, F.W., 53–54, 238–239
Charles, Landgrave of Hesse-Cassel,
    37–38, 132
Charles I, 39, 131–132, 132*fig*, 321,
    333n2
Charles II, 22
Chatham, 248
check valves, 77, 78, 104, 105, 106,
    116. *see also* ventilation
Clap, Thomas, 28–29
clockwork timing device, 133–136,
    135*fig*, 265, 321
cockpit design, 148–150, 149*fig*,
    205–206
Collins, Frederick, 14, 15*fig*
Collins, Virgil, 14, 15*fig*
Collinson, Peter, 71
compass, 119–120, 148, 323. *see also*
    navigation
Comstock, John P., 164
Concord, Battle at, 45, 47–48, 50
conjugate diameter, 315
Connecticut Council of Safety,
    59–60, 63, 99, 228–230, 269,
    315–316
Connecticut Historical Society, 134
Connecticut Rangers, 236
Connecticut River Museum, 75*fig*,
    82*fig*, 145
Cooper, Thomas, xviii, 326, 327
copper, 64–65, 315
copper sheathing, 188–190, 188*fig*,
    189*fig*, 190*fig*, 257
Cornwallis, Charles, Earl xiv
Corps of Engineers, xiv, 273–274,
    317
Corps of Sappers and Miners, xiv,
    273, 317, 324
Council of Safety. *see* Connecticut
    Council of Safety

Crane, 226, 227–230, 229*fig*, 231, 266, 316
Cromwell, Oliver, 333n2
Cummerow, Charles, 87
currents, 249–250, 251, 252–253

Daily, Eileen, xix, 75*fig*, 145
Dana, James, 26
Daudenart, L.-G., 9, 11, 12*fig*, 13
Day, J., 36
De Barres, J.F.W., 250
deadlights
    cockpit design and, 206
    definition of, 315
    endurance testing and, 195–196
    navigation and, 100, 181
    in replica, 115
    submerging and, 105
    ventilation and, 156, 255
deadrise, 315–316
Deane, Silas, 56, 130, 275, 292–297
Declaration of Independence, 51
Delpeuch, Maurice, 9, 11–13
depth gauge
    accurateness of, 254
    cockpit design and, 150
    description of, 120–123
    diving bells and, 125*fig*
    glass for, 68, 69–72
    illumination of, 77, 119, 125, 253, 323
    Papin and, 40
    in replica, 78, 124*fig*
    submersion and, 105, 187
    water-gage and, 328
Derham, William, 30, 323
Des Cartes, René, 67
Desaguliers, J.T., 116
Devotion, John, 27
Diamant, Lincoln, xvi
Dickenson, H.W., xvii
displacement, 100–101
diving bells
    depth gauge and, 123, 125*fig*
    diving engine and, 113
    Edinburgh Encyclopaedia on, 4, 7
    entry into, 38
    illustration of, 25*fig*

origins of, 20, 23
    support vessels and, 72–73
    Triewald on, 314
diving engines/machines, 35–37, 73–74, 73*fig*, 113–114, 115*fig*
Doane, Mary Brockway, 15, 16*fig*
Dodderidge, Joseph, 318
Doolittle, Isaac
    attachment mechanism and, 138
    brass crown and, 66
    clockwork timing device and, 136
    compass and, 120
    gunpowder and, 63, 131
    hatch and, 110, 112
    location of, 29*fig*
    pumps and, 57, 107
    role of, 44, 52–53
Dorothea, 128*fig*, 130
Drebbel, Cornelius
    Borelli and, 18
    early submarine and, 17, 18*fig*
    Edinburgh Encyclopaedia on, 4
    literary references to, 21–22
    oars and, 86
    Papin and, 40, 41
    underwater mines and, 39, 131–132, 132*fig*, 321
    Wilkins and, 19
drive mechanism, 177. *see also* propulsion
Duncan, Henry, 248
Durham, Victor G., 14

Eagle
    Asia and, 57
    attack on, xiii–xiv, 129, 187–188, 190–191, 214, 247*map*, 248–249, 250–255, 257–262, 263–265
    bells, timing of watch cycle of, 151
    hull shape of, 315
    illustration of, 256*fig*
    submersion and, 120
    weather conditions and, 248
Edinburgh Encyclopaedia, 4–7, 88, 123
education, colonial, 25–27
effort, steady state vs. exhaustion, 201–203. *see also* pilot endurance testing

Elihu Society, 34
Eliot, Jared, 24, 27, 66--67
Emerson, John B., 340n2(ch.15)
Emerson, William, 84*fig*, 86, 91, 91*fig*, 95*fig*, 107–108, 107*fig*
emery, 110, 114, 316. *see also* watertight seals
Endeavor, 345n5
endurance. *see* pilot endurance testing
Engineering and Diving Support Unit (EDSU), 81, 195
 diving support from, 81, 82*fig*, 156*fig*, 168*fig*, 186*fig*
 Turtle pilots from, 81, 195
ergonomics, 317. *see also* pilot ergonomics
Ericsson, John, 88, 340n2(ch.15)
Essex Boat Works, 145

Ferrett, 224*fig*, 227
Field, Cyril, 13, 256*fig*
1st Connecticut Line, 194, 207, 207*fig*
Fitch, Ebenezer, 45–46
flintlock firing mechanism, 141*fig*
flywheel, 94–95, 95*fig*, 160–162, 176–177
forcing pump, 52, 57, 105, 107, 112, 290, 293, 294, 305
Forest, F., 11
Fort George, 245*fig*, 266
Fort Lee, 266
Fort Trumbull, 269
foxfire, 119–120, 124–127, 318, 323
Franklin, Benjamin
 Gale and, 56, 289–291
 glass and, 70–71
 interest in Bushnell by, 275
 inventions of, 174
 Royal Society and, 24
 support of, 130
Frazar, Douglas, 312–313
Frese, Fred
 flywheel and, 177
 photograph of, 111*fig*
 pilots and, 80–81
 replicas and, xix, 2, 4
 Turtle Project and, 78

Fulton, Robert
 clockwork mechanism and, 133–134, 135
 Dorothea and, 128*fig*, 130
 propellers and, 88, 90
 on resistance, 163
 sails and, 86
 torpedoes and, xvi, 134*fig*, 326
funding, 59–60
Fyfe, Herbert C., 10

Gage, Thomas, 47–48
Gaget, Maurice, 9, 10
Gale, Benjamin
 on attachment mechanism, 138
 on ballast, 63, 103
 Bushnell and, xv, 27, 41, 317
 on cockpit, 206
 correspondence of, 56, 57–58, 289–291, 292–297
 on Doolittle, 52
 on foxfire, 124, 127
 on gunpowder, 141
 on hatch, 109
 on hull, 97, 98
 on mines, 130
 on propeller, 88, 94, 95*fig*
 on pumps, 107
 on speed, 169
 steel and, 66--67
 on submerging, 254
 on support vessels, 225
 terminology of, 3
 on testing, 217
 on transportation, 220–221
 trust in, 44
 on *Turtle*, 24, 38, 144, 244, 322, 324
 on ventilation, 114, 116
Galileo, 125
Gates, Gilman, 218
Gerr, David, 155
Gesner, Conrad, 326, 327
Gianabelli, Frederico, 321
Gillmer, Thomas C., 155
glass, 67–72, 71*fig*
Global Positioning System (GPS), 120, 148

Goodwin, Jona., 230
Goose Island, 220*fig*, 221–222,
  221*fig*, 223
Governor's Island
  British possession of, 244, 266
  Bushnell and, 233
  currents and, 251, 265
  Lee and, 263
  weather and, 248
Grant, Marion Hepburn, xvi, 46, 215
Grasso, Ella, 1*fig*
Great Island, 216–217, 220*fig*
Greene, Nathanael, 245, 266
Greenhalgh, Bruce, 139–140
Greenhalgh, Janna, 139–140
Griswold, Charles
  attachment mechanism and, 261
  description by, xvi, 243, 251,
    303–308, 317
  on failed attack, 264
  on hull, 98–99
  on propeller, 88, 312
gunpowder
  Bushnell and, 61–63, 141. *see also*
    mines/magazines
  Doolittle and, 53
  map horn and, 245*fig*
  mines and, 58, 131–133
  prank and, 274
  production of, 131
  storage of, 320
  torpedoes and, 326
  underwater detonation of, 44, 62,
    129–130
  use of, by *Turtle*, 55

Hale, Nathan
  Bushnell and, xiv, 34, 268
  E. Bushnell and, 194, 234, 327
  hanging of, 268
  Knowleton's Rangers and, 236
  Linonian Society and, 34, 218
Halley, Edmund, 23, 25*fig*, 121
Handshouse Studio, 82, 83*fig*, 213*fig*,
  221
  Turtle replica, 82, 83*fig*, 213*fig*, 221
Harris, Brayton, xvi, 18
Hart, David

endurance testing, 195–199
  feedback from, 200
  as test pilot, 85, 175
  testing by, 150
hatch, 38, 109–110, 111*fig*, 112–113.
  see also watertight seals
hatch windows, 68–69, 113*fig*
Hauksbee, Dr. Francis, 67, 70
Hayden, Uriah, 99, 219, 220*fig*
Hayes, *see* Toogood and Hayes
Heath, William, 226–228, 244, 246,
  264, 343n13(ch.20)
Hennebert, Eugene, xvii, 9
Hertzler, Sharon, 111, 112
Hill, William J., 136
Hoadley, Charles J., 230
Holland, John P., xvii, 7, 9, 14, 325
Hooke, Robert, 34, 37, 67, 87
Hopkinson, Francis, 271
horological torpedoes, 135–136. *see*
  also mines/magazines; torpedoes
Hovgaard, G.W., 9
Howe, Henry, 34
Howe, Richard, Lord, xiii, 245*fig*,
  246, 249*fig*
Hughes, Jack, 81
hull
  arch bridge design and, 101*fig*
  construction of, 54–55
  design, 38
  dimensions of, 98
  photograph of, 80*fig*
  riveted, 14, 15*fig*
  shape of *Eagle*'s, 315
  shape of *Turtle*'s, 79*fig*, 97–98,
    183–184, 316
  testing of, 163–167
  thickness of, 98–100
hull resistance testing
  data from, 166*t*
  image from, 167*fig*
  overview of, 163–164
  procedures for, 164–165, 165*fig*
  results of, 165–167, 178
Humphreys, David
  account of, xv–xvi
  on Bushnell, 234
  date of attack and, 246

Griswold and, 303
hull and, 98
Lee's account and, 298, 301–302
terminology and, 4
Washington and, 276, 309
Yale and, 34
Hunley, 135
Huntington, Jedediah, 207
Huygens, Christian, 37, 67, 132
hydraspis, 35, 36*fig*

illumination, 124–127
Independence, 231

James I, 17, 21
Jefferson, Thomas
    on Bushnell, 275
    text of Bushnell's letter to,
        276–278
    Washington and, xv, 309–310, 324
Johnson, Bruce, 155
Johnson, Samuel, xviii, 3
Jonson, Ben, 21
Joy, Daniel, 58

keel, 96*fig*, 102–103
Kennedy, Archibald, 251
Kennedy's Island, 247*map*, 251
Kips Bay, 266
Knowleton, Thomas, 236
Knox, Henry, 65–66

L.A. Dunton, 146, 147*fig*
Lapis Calaminaris, 313–314, 315
Lathrop, William G., 67
Laubeuf, Maxime, 14
lead. *see* ballast
Leary, Joseph, xix, 2, 4
Lee, Ezra
    attachment mechanism and,
        190–191
    attack on *Eagle* by, 129, 214, 246,
        248–249, 250–255, 257–262,
        263–265
    attempted attacks by, 226, 266
    on compass, 120
    Crane and, 227–228, 229*fig*, 231
    on emergency ballast, 103

endurance and, 201, 203–204
ergonomics and, 206–208
gravestone of, 237*fig*
Griswold and, 303
on hatch, 109
hull and, 98–99
Humphreys and, xvi, 4
on illumination, 125
New York evacuation and, 266
pension claim by, xvii, 242
as pilot, 34, 119–120, 155, 181,
    195, 225, 233, 236–237
portrait of, 232*fig*
promotion of, 207, 207*fig*
on propellers, 12, 14, 88, 94, 174,
    175
on pumps, 105
retreat of, 187, 262–263
service of, 235*fig*
Sill and, 219
on support vessel, 226–227
training of, 193–194, 211,
    222–223, 243
on ventilation, 114–115, 116
on windows, 68
written account by, 298–302
Lefkowitz, Arthur S., xvi, 215, 271
Lester, Jonathan, 228–229
Lethbridge, John, 36, 73–74, 73*fig*,
    113, 115*fig*
Lewis, John, 55
Lexington, Battle at, 45–46, 47–48,
    50
Liebknecht, William, 69
Linneaus, 327
Linonian Society, 34, 218
Littleton, Adam, xviii, 327, 328
Long Island, British possession of,
    243–244
Long Island Sound, 217–218, 220*fig*,
    221, 244
Lord Cove, 220*fig*, 221–222, 221*fig*
Lynn, Elmer, 14–15, 16*fig*
Lyttleton, William, 87

magazines. *see* mines/magazines
maneuvering tests
    backing down, 179–180. *see also*

navigation
overview of, 179
S-turn, 182–184
Manstan, Oscar "Ty," 15, 16*fig*
Manstan, Roy
flywheel and, 177
Naval Undersea Warfare Center
(NUWC) and, xix
photograph of, 111*fig*
pilots and, 80–81
recreation of Lee mission by,
257–258
as test pilot, 145, 150, 175, 200,
209, 211
Marolda, Vic, 81
Martin, Benjamin, 31, 89–90, 89*fig*,
92, 311, 314
Martin, Joseph Plumb, 240, 242, 274
Massachusetts College of Art, 82,
83*fig*
McCullough, David, 235, 246
measurement terms, 133, 312, 318,
319
Mersenne, Marin, 17
Middlebrook, Louis F., 318
Mileski, Paul
cockpit design and, 209
on endurance, 200
exhaustion and, 203
feedback from, 174–175
on hand crank, 177
maneuvering tests and, 179–180
photograph of, 158*fig*
on propeller tests, 159–162, 196
on speed tests, 172–174
as test pilot, 81, 178
testing by, 150
mines/magazines. *see also* clockwork
timing device; gunpowder; petards;
torpedoes
attachment mechanism and,
136–138, 187–191
definitions of, 320
detonation of, 263, 265
development of, 129–130
floating, 269–271
gunpowder and, 131–133
photograph of, 141*fig*

replicas and, 138–141, 139*fig*,
140*fig*
use of, 58
Montressor, John, 250, 252
Monturiol, Narciso, 9
Moore, H. Edward, 140–141
Moxon, Joseph, 67
Mulholland, James A., 67
Mystic River Foundry, 110, 111*fig*
Mystic Seaport Museum, 79*fig*,
145–146, 147*fig*, 152*fig*, 158*fig*, 275

National Museum of American
History, 206
National Oceanographic and
Atmospheric Administration
(NOAA), 249, 250, 264
Nautilus (Fulton)
descriptions of, xvii
propellers and, 88
sails and, 86
Turtle and, xvi
Nautilus (Verne), xi–xii
Naval Diving & Salvage Training
Center (NDSTC), 81
Naval Undersea Warfare Center
(NUWC), xix, 81, 110, 111*fig*, 112,
195
diving support from, 81, 195, 345n5
education partnership with, xix,
110, 111*fig*, 112
navigation, 19, 119. *see also* compass;
maneuvering tests; rudder
Navy divers, 80–81
New York Harbor Observing and
Prediction System (NYHOPS), 258
Newton, Isaac, 30
Noalhat, H., 11
Nollet, Abbé, 70, 71–72

oars, 10–12, 11*fig*, 18, 18*fig*, 40–41
Old Saybrook High School, xx, 76.
78, 80*fig*, 111*fig*, 275
Oliver Cromwell, 99, 219, 269
Ommanney, Cornthwaite, 230–231
operational testing, 145–154, 148*fig*
Osgood, David, 272
Osgood, Samuel, 56–57

Oyster Banks, 252, 259

Paine, Thomas, 50
Palmer, Arlene, 71
Papin, Denis, 18, 37–41, 86, 116, 121, 132, 328
Parker, Abner, 54, 216
Parker, Peter, 269, 270
Parsons, Samuel
  Bushnell's capture and, 273
  Lee and, 194, 204, 219, 233
  with Putnam, 244
  support vessel and, 226, 227
  transportation and, 246
  transportation for *Turtle* and, 264
  volunteers selected by, 193
Parsons, Wm. Barclay, xvii
Peel Museum, 140
Penn, William, 65
Percy, Lord Hugh, 48
Pesce, G.-L., 9, 11, 13
petards, 39, 131–132, 132*fig*, 321, 322. *see also* mines/magazines
Phillips, Lodner, xii
Phipps, Charles, 248
Phoenix, 230, 231
phosphorous, 120, 121, 126, 323
pilot endurance testing
  analysis of, 200–201
  data from, 198, 199*fig*
  effort and, 201–203
  ergonomics and, 210–212
  overview of, 193–195
  pilot comments on, 198, 200
  procedures for, 195–197
  results of, 197–198
  test pilot for, 195
pilot ergonomics
  analysis of, 210–212
  E. Bushnell and, 206–208
  Lee and, 206–208
  overview of, 205–206
  test pilots and, 208–210
pilots
  age of, 178
  air supply and, 192*fig*
  attachment mechanism and, 137–138

cockpit design and, 150, 208–210
communication with, 158, 186, 186*fig*, 197
feedback from, 76, 95, 144
importance of, 235–236
monitoring of, 196–197
propeller and, 158–162
selection of, 80–81
speed and, 169
testing of, 147
uniform of, 9–10, 10*fig*
ventilation and, 114
water ballast and, 106–107
weight of, 104
Pitcairn, John, 48
Platt, Daniel, 54
Plocher, Jacob, 5*fig*, 7
Poverty Island, 215–217, 221
Poverty Island Beach, 216–217
Poverty Point, 220*fig*
Pratt, Abbie, 15, 16*fig*
Pratt, Abel, 54
Pratt, Aziriah, 55
Pratt, Phineas
  availability of, 222
  background of, 53–55, 342n5
  building of *Turtle* and, 15, 219
  descendents of, 16*fig*
  gravestone of, 237*fig*
  modifications and, 243
  as pilot, 162, 223, 226, 233, 237–240
  Worthington and, 270
Pratt, Phineas, Jr, 54, 238, 239–240, 241*fig*
pressure, 39–40, 118*fig*, 121–123
prison ships, 267–268, 272
propeller tests
  changes during, 156*fig*
  data from, 159*t*, 160*fig*, 161*fig*
  overview of, 155–156
  pilot comments on, 159–162
  procedures for, 156–158
  results of, 158–159
  thrust, 157*fig*. *see also* propellers
  torque and, 158
propellers. *see also* propeller tests; windmills

Bishop on, 14
blades of, 88–91, 92–93, 93*fig*, 94*t*
Bushnell and, 40, 85–88, 174–175
Daudenart's sketch of, 11–12,
    12*fig*
design and, 10–13, 151–154,
    340n2(ch.15)
driving mechanism of, 93–95
endurance and, 210–212
helical, 8*fig*, 10, 13, 87*fig*
illustration of, 12*fig*, 14*fig*
innovation of, 77
mounting location of, 180–181
pitch, 151–152, 153*fig*, 171–172,
    173*t*
replicas and, 15, 153*fig*
rotation monitoring and, 147–148
Sabino's, 152*fig*
slip, 152–154, 171–172, 173*t*
speed and, 169–170, 171–172,
    173*t*
Stevens's, 176*fig*
torque and, 85, 90–91, 161–162,
    177. *see also* Anemoscope
torque and thrust of, 91
variations of, 144, 150–151
vertical, 184–185, 185*fig*, 186–187
propulsion, 10–13, 19, 40, 86–88,
    178. *see also* propellers; steam
    propulsion
pumps, 105–108, 106*fig*, 107*fig*, 112
Putnam, Israel, xv, 4, 244, 263, 272

Ratzer, Bernard, 247*map*, 250
Raynor, John, 248
Repulse, 244
respiration, 19, 38, 41, 192*fig*, 196. *see
    also* ventilation
Richards, Samuel, 188
Rodahl, Kaare, 200–203, 211–212
Roebuck, 230, 248–249
Rogers, Zabdiel, 33
Roland, Alex, xvi, 18, 271, 321
Rose, 230, 231, 244
Rowe, Thomas, 36, 114
Royal Society of London
    background of, 22–24
    Bushnell and, 30–31

Papin and, 37
pressure and, 121–123
publications and, 86–87
Wilkins and, 19
rudder, 5–6, 180, 182–183, 182*fig*,
    200. *see also* maneuvering tests; navi-
    gation; steering mechanism
rum, 33, 34, 60, 61, 65, 66, 72, 227,
    319
Rumsey, James, 86

Sabino, 93*fig*, 152*fig*
safety
    air supply and, 192*fig*
    Bushnell and, 103
    pilots and, 80–81
    secondary air supply and, 196,
        197*fig*
    swimmers, 197
sailing chariot, 84*fig*, 86
salinity, 100–102, 122–123, 258–259
Salisbury foundry, 33, 65
Saltonstall, Gurdon, 28
saltpeter, 131. *see also* gunpowder
scuba tank, 116–117, 117*fig*, 196,
    197*fig*
7th Connecticut Regiment, 54, 194,
    234, 270, 327
Shark, 231
Shaw, N., Jr., 230
shining wood. *see* foxfire
Shorter, Edward, 87
Shuldham, Molyneux, 57
Siege of Boston, 234
Sill, Charles, 342n9
Sill, David F., 235*fig*
Sill, Elisha, 342n9
Sill, George, 342n9
Sill, Richard, 34, 218–219, 235*fig*
Sill House, 218–220, 219*fig*, 220*fig*,
    243
Silliman, Benjamin, xvi, 264, 303
Sill's Point, 216
Sleeman, C.L., 9
Smith, Benjamin J., 17, 325
Smith, F.P., 87, 87*fig*
Smith, Francis, 47–48
smoke jacks, 90–91, 91*fig*

soldiers
    ads regarding, 207*fig*
    heights of, 206–208, 209*fig*, 210
    as prisoners, 267–268, 272
    treatment of, 240, 242
speed, 156, 169, 180–181. *see also*
    transit speed tests
Spencer, Joseph, 244
spinning wheel, 94–95, 95*fig*
Sprat, Thomas, 22–24, 62
Stanton, Theophilus, 229
static thrust tests, 155–156, 159*t*
steam propulsion, 175, 176*fig. see also*
    propulsion
steel, 66-67
steering mechanism. *see* maneuvering
    tests; navigation; rudder
Stevens, Francis B., 175
Stevens, John, 88, 90, 155, 175,
    176*fig*
Stevens Institute of Technology, 258
Stiles, Ezra
    Ayer's Point map drawn by, 222
    on Battle of Lexington, 46
    Bushnell's letter to, 214, 276–277
    New Haven map drawn by, 27,
        29*fig*
    on propeller, 174
    Yale College and, 44
Strombolo, 248
Strong, Nehemiah, 30–31
submarine
    technology, 13–14, 17–18, 20
    terminology, 17, 163–164, 325
    warfare, 7, 9, 15–16, 43
submersion testing, 184–187, 186*fig.*
    see also ballast; buoyancy
Sutcliffe, Alice Crary, xvii
Symons, John, 269, 270, 319

Taffrail Ship Log, 87
Tallmadge, Benjamin, 34
Tartar, 230–231
10th Continental Regiment, 193,
    194, 207, 219
test pilots. *see* pilots
Thatcher, James, xvi, 4, 271
Thomson, David W., xvi, 15

tides, 249–250, 251, 252–253,
    262–263
timekeeping, 151, 312–313, 314, 318,
    328
timing device. *see* clockwork timing
    device
Tinker, Jahiel, 229, 230
Tomlinson, Everett, xvii, 215–216
Toogood and Hayes, 86
torpedoes. *see also* mines/magazines
    definitions of, 326–327
    development of, 135
    illustration of, 134*fig*, 135*fig*
    use of term, 238–239
Torricelli, Evangelista, 67
Tory Tavern, 32*fig*, 34
tow tests. *see* hull resistance tests
transit speed tests
    overview of, 169–170
    photograph of, 168*fig*, 172*fig*
    pilot and, 171–172
    pilot comments on, 172–174
    procedures for, 170
    results of, 170–172, 171*t*
treadle
    cockpit design and, 148–149
    description of, 94–95, 176–177
    endurance testing and, 202
    ergonomics and, 195, 209–211
    illustrations of, 95*fig*, 149*fig*
    Lee and, 206
    pilot comments on, 200
    static thrust tests and, 158,
        160–162
Triewald, Martin, 314
Trumbull, J. Hammond, xv
Trumbull, John, 49, 269, 272–273
Trumbull, Joseph, 243
Tryal, 230
Tryon, William, 57
Tulley, Samuel, 218
Tully, Elias, 27
Turtle
    Bushnell on training for, 217
    costs of, 59–61, 63, 64
    description of, 4–6, 55–56
    engraving of, 5*fig*
    illustration of, 8*fig*, 10*fig*, 11*fig*,

12*fig*, 13*fig*, 14*fig*, 241*fig*
launching of, 42*map*, 219*fig*,
   220–223, 220*fig*, 224*fig*
materials for, 63–72
naming of, 3–4
Papin's design and, 38–41, 39*fig*
sinking of, 266–267,
   343n13(ch.20)
size of, 260*fig*
source material on, xiv–xvii
submersion and, 260*fig*
support vessel for, 225–227
transportation of, 45, 72–74,
   225–226, 246
Turtle replicas
   comparison of, 82
   early models, 14–16, 16*fig*
   goals for, 77–78
   launching of, 143*fig*, 221
   logo for, 79, 81*fig*
   overview of testing for, xix–xx,
     144. *see also individual tests*
   photograph of, 1*fig*, 75*fig*, 79*fig*,
     82*fig*, 83*fig*, 95*fig*, 96*fig*, 146*fig*,
     158*fig*, 213*fig*

urinator, 20, 114, 327-328

Vandeput, George, 248
velocity, measurement of, 87
ventilation, 38, 40, 114–117, 328. *see
  also* respiration
Verne, Jules, xi–xii
Villeroi, Brutus de, 40–41

Wagner, Frederick, xvi, 215–216, 271
Waller, Richard, 87
Walsh, James, 99
Washington, 228, 229*fig*
Washington, George
   Beers Tavern and, 33
   British invasion and, 243–246
   Bushnell and, xiv, xv, 275, 324
   Connecticut Council of Safety
     and, 230
   Corps of Engineers and, 273–274
   Jefferson and, 276, 309–310
   Lee's gravestone and, 237*fig*

New York evacuation and, 266
Newgate Prison and, 65
on pilot, 235–236
Pratt, Jr. on, 239
10th Continental Regiment and,
   194
on *Turtle*, 214, 226
water ballast. *see under* ballast
water depth, 118*fig*, 121–123
watertight seals, 104, 111, 112–114,
   115*fig*, 192*fig*, 316
Watkins, Francis, 67
weather conditions, 248–249
Webb, Charles, 234, 327
Webster, Samuel, 26
West, Samuel, 51
West Point museum, 135*fig*, 136
Whiting, 230, 231
Whitney, William Dwight, 17, 325
Wilkins, John
   on buoyancy, 100
   Drebbel and, 21
   Edinburgh Encyclopaedia and, 4
   on oars, 86
   on submarine technology, 18–20
   terminology and, 325
   on ventilation, 114
Williams, William, 243
windmills, 88–90, 89*fig*, 91, 245*fig*,
   314
windows, 68–69
Winslow, Job, 228–229
Wistar, Caspar, 71
Woodcroft, Bennet, 87
Worthington, William, 54, 269–270

Yale College, 27–31, 33–34
"Yankee Doodle," 48–49, 316,
   328–329
Young, Samuel, 267

ACKNOWLEDGEMENTS

The *Turtle* Project would never have moved past the idea stage were it not for the Old Saybrook High School administration and its encouragement of "out-of-the-box" thinking by the faculty and students. Particular thanks go to Dr. Salvatore Pascarella, Joseph Onofrio, Scott Schoonmaker, Tara Winch, Albert Mortali, Dana Maccio, and all the students who worked on the *Turtle* Project during their high school career. Special thanks to French teacher Valerie Koif for her translation of early French interpretations of the *Turtle* story. Our appreciation also to Jesse Frese for her help with the manuscript.

In addition, the establishment of an Education Partnership Agreement between the Old Saybrook school system and the Naval Undersea Warfare Center (NUWC) in Newport, Rhode Island, provided access to the specialized knowledge and capabilities of its staff. Thanks to Scott Boyd and the NUWC Industrial Services Enterprise department and to the members of the Engineering and Diving Support Unit (EDSU). Particular appreciation goes to Paul Mileski and David Hart, who followed in Ezra Lee's footsteps by serving as *Turtle* test pilots during two long days of operational testing.

Thanks also to the Toyota TAPESTRY Grants for Science Teachers Program for the initial funding that kicked the project into gear, and to Leighton Lee III for his support to keep the gears turning. The *Turtle* replica now resides in Lee's Museum of Early Engineering Technology in Westbrook, Connecticut. We also want to express our gratitude for the generosity and hard work by Sharon Hertzler and the Mystic River Foundry for converting one hundred pounds of raw material into the *Turtle*'s brass hatch, in much the same way that Isaac Doolittle had done in his bell foundry for David Bushnell in 1775. Thanks to Jerry Roberts, Amy Trout, and Brenda Milkofsky of the Connecticut River Museum, where the christening of our *Turtle* and its launch and initial sea trials were held during the fall of 2007. Our appreciation is extended to New England Dock and Dredge, especially Ned Libby and Danny Rettan, and to William Winterer and the Essex Yacht Club. We thank Doug Teeson, Dana Hewson, and the staff of the Mystic Seaport Museum for their generosity in providing an opportunity to conduct the testing of our *Turtle* replica, opening an entirely new chapter in the understanding of this vessel's contribution to maritime history.

Thanks to David Fothergill and Command Master Diver Rick Donlon of the Naval Submarine Medical Research Lab, Groton, Connecticut, for their valuable insights and suggestions. Our appreciation is extended to Les Jensen of the West Point Museum, to the National Museum of American History, the Old Saybrook Historical Society, the New Haven Museum and Historical Society, the Connecticut Historical Society Museum and Library, the Thomas Lee House and Museum, and the late Don Malcarne for sharing his extensive knowledge of Essex (Potapaug) history. While the resources of many libraries were of great value, particular thanks go to Linda Alexander and the staff of the Phoebe Griffin Noyes Library in Old Lyme, Connecticut; to Mary Barravecchia, NUWC Newport librarian; and to Wendy Gulley, archivist at the Naval Submarine Force Museum and Library in Groton, Connecticut. Also to Barbara Stewart whose research led to the discovery of the woodcut of David Bushnell.

And to Bruce H. Franklin of Westholme Publishing and copy editor Laura Pfost for their encouragement and their interest in providing readers with books and authors who love bringing history to life.

ABOUT THE AUTHORS

During the American Bicentennial, author Fred Frese constructed a working replica of the *Turtle*, christened by then Connecticut Governor Ella Grasso and launched along the waterfront of the Connecticut River in 1977. At about this time, author Roy Manstan, mechanical engineer and navy-trained diver, was at sea onboard USS *Nautilus*. Little did they know that three decades later, their mutual interest in submarines would match them in a project to design and construct yet another *Turtle* replica.

ROY R. MANSTAN received his BS in Mechanical Engineering from Lafayette College, his MS in Mechanical Engineering from the University of Connecticut, and his MA in Zoology from Connecticut College. He began his career in 1967 at the Navy Underwater Sound Laboratory that soon merged with the Navy Underwater Weapons Research and Engineering Station to form the Naval Underwater Systems Center, now known as the Naval Undersea Warfare Center (NUWC) from which he retired in 2006. He became a qualified Navy diver in 1974 and was designated Command Diving Officer in 1984. He led the NUWC dive team on operations worldwide, including the Arctic, Africa, Italy, Scotland, the Azores, the Persian Gulf, Singapore, Cuba, and many other locations. He wrote, directed, and participated in "Human Subjects Testing" protocols related to detection and deterrent simulations of underwater attacks by divers. He was responsible for educational outreach opportunities, including providing the American School for the Deaf with remote underwater video access to various NUWC projects via the Internet.

FREDERIC J. FRESE received his Connecticut Certification in Vocational Education from Central Connecticut State University, and began his teaching career in 1972 at Hillhouse High School in New Haven, Connecticut. He is now a Technical Education teacher at Old Saybrook [CT] High School. His career has also included a variety of experiences with automotive engineering and boat building, fabricating numerous high performance automobiles that have been raced throughout the United States. He has designed, built, and raced high-speed boats, and has held a world record for "Class F Runabouts." In addition to the Bicentennial replica of the *Turtle* now on display at the Connecticut River Museum in Essex, Connecticut, he also produced the full-size cutaway *Turtle* model at the Submarine Force Museum in Groton, Connecticut. He has been the driving force behind the reinvention of the high school technical arts program, and in particular the *Turtle* Project that is the subject of this book. This latest *Turtle* replica was put on display during the biennial International Submarine Races (ISR-10) held at the David Taylor Model Basin in June 2009.